ALSO BY HENRY PETROSKI

Small Things Considered: Why There Is No Perfect Design

Paperboy: Confessions of a Future Engineer

The Book on the Bookshelf

Remaking the World:
Adventures in Engineering

Invention by Design:
How Engineers Get From Thought to Thing

Engineers of Dreams:
Great Bridge Builders and the Spanning of America

Design Paradigms:
Case Histories of Error and Judgment in Engineering

The Evolution of Useful Things

To Engineer Is Human:
The Role of Failure in Successful Design

Beyond Engineering:
Essays and Other Attempts to Figure Without Equations

THE PENCIL

THE PENCIL

A History of Design and Circumstance

by Henry Petroski

ALFRED A. KNOPF
New York
2004

All rights reserved under International and Pan-
American Copyright Conventions. Published in the
United States by Alfred A. Knopf, Inc., New York,
and simultaneously in Canada by Random House of
Canada Limited, Toronto. Distributed by Random
House, Inc., New York.

Portions of this book were originally published in
Across the Board and in *American Heritage of
Invention & Technology.*

Grateful acknowledgment is made to Koh-I-Noor
Rapidograph, Inc., for permission to reprint
excerpts from "How the Pencil Is Made" from *The
Pencil: Its History, Manufacture, and Use* by The
Koh-I-Noor Pencil Company. Reprinted courtesy of
Koh-I-Noor Rapidograph, Inc.

Correspondence between Ralph Waldo Emerson and
Caroline Sturgis quoted by permission of the Ralph
Waldo Emerson Memorial Association and of the
Houghton Library.

Library of Congress Cataloging-in-Publication Data

Petroski, Henry.
    The pencil: a history of design and circumstance/
by Henry Petroski.—1st ed.
        p.   cm.
    Bibliography: p.
    Includes index.
    ISBN 0-679-73415-5
    1. Pencils—History.   I. Title.
TS1268.P47   1989
674'.88—dc20   89-45362
                                              CIP

Manufactured in the United States of America
Published January 26, 1990
Reprinted Two Times
First Paperback Edition Published December 2, 1992
Reprinted Ten Times
Twelfth Printing, January 2004

To Karen

Contents

Preface

All made objects owe their very existence to some kind of engineering, which is essential for civilization. Even the commonest and oldest of artifacts are no less the products of primitive engineering than the artifacts of high technology are the products of modern scientific engineering. But while the practice of engineering has certainly evolved since ancient times, it has also maintained a family resemblance to its ancestors. Although engineers today tend to be more formally mathematical and scientific than their counterparts just a century ago, there are still essential elements of engineering that all ages have in common. A modern engineer and an ancient, even if called an architect or master builder or master craftsman, would find plenty to talk about, and each would be able to learn something from the other.

This timelessness derives from a constant underlying quality inherent in all engineering, a quality that is independent of formal education. The existence of this commonsense aspect of it explains why and how so much ancient and even not so ancient engineering was done by individuals who worried about neither what they themselves nor what they were doing was called. Indeed, such seemingly unlikely persons as the political philosopher Thomas Paine and the philosophical writer Henry David Thoreau effectively acted as if they were engineers and made real contributions to the technology of their times. For this same reason, I believe that anyone today is capable of comprehending the essence of, if not of contributing to, even the latest high technology. Behind all the jargon, mathematics, science, and professionalism of engineering lies a method as accessible and as pervasive as the air we breathe. Certainly business executives with no formal engineering training daily assume this to be the case in making decisions with major technological implications. But that is not to say that professional engineers are dispensable, for it is one thing to understand their method and another to be able to apply it to the details of an increasingly complex and international technological environment and then condense the results in an executive summary.

Because all engineering, past and present, has a common feature to its fabric, the method of engineers and of engineering is embodied in everything ever made and thus is accessible through any single artifact. I believe that a person who is attracted to bridges, for example, can learn more about the method of all engineering—including such seemingly diverse branches as chemical, electrical, mechanical, and nuclear engineering—from a focused study of bridges alone than from a diffuse and cursory survey of all the past and latest wonders of the made world. Yet a focused study need not be overly technical. It need only place the artifact in a proper social, cultural, political, and technological context in order to allow the essence of engineering to be distilled by the receptive mind. For it is attention to all aspects of the long evolutionary process by which such a thing as a rotting log across a stream becomes a corrosion-free suspension bridge across a strait that we discover the essence of engineering and its role in civilization. Just as there is no artifact that is without engineering, so there is no engineering that is free of the rest of society.

In this book I have chosen to approach engineering through the history and symbolism of the common pencil. This ubiquitous and deceptively simple object is something we can all hold in our hands, experiment with, and wonder about. The pencil, like engineering itself, is so familiar as to be a virtually invisible part of our general culture and experience, and it is so common as to be taken up and given away with barely a thought. Although the pencil has been indispensable, or perhaps because of that, its function is beyond comment and directions for its use are unwritten. We all know from childhood what a pencil is and is for, but where did the pencil come from and how is it made? Are today's pencils the same as they were two hundred years ago? Are our pencils as good as we can make them? Are American pencils better than Russian or Japanese pencils?

To reflect on the pencil is to reflect on engineering; a study of the pencil is a study of engineering. And the inescapable conclusion after such reflection and study is that the history of engineering in a political, social, and cultural context, rather than being just a collection of interesting old stories about pencils or bridges or machines, is very relevant to and instructive for engineering and commerce today. The important roles that international conflict, trade, and competition play in the history of the pencil provide lessons for such modern international industries as petroleum, automobiles, steel, and nuclear power. This is so because the engineering and the marketing of the pencil are as inextricably intertwined as they are for any artifact of civilization.

A book is also an artifact, of course, and its author incurs many debts throughout the course of its production. My acknowledgments to works, institutions, and people appear at the end of this book, but some support and encouragement have been too invaluable not to be repeated here. I was able to concentrate on this project through the support of a sabbatical from Duke University and fellowships from the National Endowment for the Humanities and the National Humanities Center. Of the many librarians who have helped me, Eric Smith, of Duke's Vesić Engineering Library, is without peer. My brother, William Petroski, was a constant source of unique information and artifacts. But it was the immeasurable patience and encouragement of my son, Stephen, my daughter, Karen, and most of all my wife, Catherine Petroski, that in the end made this book possible.

Research Triangle Park and
Durham, North Carolina, 1988

THE PENCIL

1/ What We Forget

enry David Thoreau seemed to think of everything when he made a list of essential supplies for a twelve-day excursion into the Maine woods. He included pins, needles, and thread among the items to be carried in an India-rubber knapsack, and he even gave the dimensions of an ample tent: "six by seven feet, and four feet high in the middle, will do." He wanted to be doubly sure to be able to start a fire and to wash up, and so he listed: "matches (some also in a small vial in the waist-coat pocket); soap, two pieces." He specified the number of old newspapers (three or four, presumably to be used for cleaning chores), the length of strong cord (twenty feet), the size of his blanket (seven feet long), and the amount of "soft hardbread" (twenty-eight pounds!). He even noted something to leave behind: "A gun is not worth the carriage, unless you go as a huntsman."

Thoreau actually was a huntsman of sorts, but the insects and botanical specimens that he hunted could be taken without a gun and could be brought back in the knapsack. Thoreau also went into the woods as an observer. He observed the big and the little, and he advised like-minded observers to carry a small spyglass for birds and a pocket microscope for smaller objects. And to capture the true dimensions of those objects that might be too big to be brought back, Thoreau advised carrying a tape measure. The inveterate measurer, note taker, and list maker also reminded other travelers to take paper and stamps, to mail letters back to civilization.

But there is one object that Thoreau neglected to mention,

one that he most certainly carried himself. For without this object Thoreau could not have sketched either the fleeting fauna he would not shoot or the larger flora he could not uproot. Without it he could not label his blotting paper pressing leaves or his insect boxes holding beetles; without it he could not record the measurements he made; without it he could not write home on the paper he brought; without it he could not make his list. Without a pencil Thoreau would have been lost in the Maine woods.

According to his friend Ralph Waldo Emerson, Thoreau seems always to have carried, "in his pocket, his diary and pencil." So why did Thoreau—who had worked with his father to produce the very best lead pencils manufactured in America in the 1840s—neglect to list even one among the essential things to take on an excursion? Perhaps the very object with which he may have been drafting his list was too close to him, too familiar a part of his own everyday outfit, too integral a part of his livelihood, too common a thing for him to think to mention.

Henry Thoreau seems not to be alone in forgetting about the pencil. A shop in London specializes in old carpenter's tools. There are tools everywhere, from floor to ceiling and spilling out of baskets on the sidewalk outside. The shop seems to have an example of every kind of saw used in recent centuries; there are shelves of braces and bins of chisels and piles of levels and rows of planes—everything for the carpenter, or so it seems. What the shop does not have, however, are old carpenter's pencils, items that once got equal billing in Thoreau & Company advertisements with drawing pencils for artists and engineers. The implement that was necessary to draw sketches of the carpentry job, to figure the quantities of materials needed, to mark the length of wood to be cut, to indicate the locations of holes to be drilled, to highlight the edges of wood to be planed, is nowhere to be seen. When asked where he keeps the pencils, the shopkeeper replies that he does not think there are any about. Pencils, he admits, are often found in the toolboxes acquired by the shop, but they are thrown out with the sawdust.

In an American antique shop that deals in, among other things, old scientific and engineering instruments, there is a grand display of polished brass microscopes, telescopes, levels, balances, and scales; there are the precision instruments of physicians, navigators, surveyors, draftsmen, and engineers. The shop also has a collection of old jewelry and silverware and, behind the saltcellars, some old mechanical pencils, which appear to be there for their metal and mystery and not their utility. There are a clever Victorian combination pen and pencil in a single slender, if ornate, gold case; an unassuming little tube of brass less than two inches long that telescopes out to become a mechanical pencil of twice that length; a compact silver pencil case containing points in three colors—black, red, and blue—that can be slid into writing position; and a heavy silver pencil case that hides the half-inch stub of a still-sharpened yellow pencil of high quality. The shopkeeper will proudly show how all these work, but when asked if she has any plain wood-cased drawing pencils that the original owners of the drafting instruments must certainly have used, she will confess that she would not even know what distinguished a nineteenth-century pencil from any other kind.

Not only shops that purport to trade in the past but also museums that ostensibly preserve and display the past can seem to forget or merely ignore the indispensable role of simple objects like the pencil. Recently the Smithsonian Institution's National Museum of American History produced "After the Revolution: Everyday Life in America, 1780–1800," and one group of exhibits in the show consisted of separate worktables on which were displayed the tools of many crafts of the period: cabinetmaker and chairmaker, carpenter and joiner, shipwright, cooper, wheelwright, and others. Besides tools, many of the displays included pieces of work in progress, and a few even had wood shavings scattered about the work space, to add a sense of authenticity. Yet there was not a pencil to be seen.

While many early American craftsmen would have used sharp-pointed metal scribers to mark their work, pencils would also certainly have been used when they were available. And although there was no domestic pencil industry in America in the years immediately following the Revolution, that is not to say that pencils could not be gotten. A father, writing in 1774 from England to his daughter in what were still the colonies, sent her "one dozen Middleton's best Pencils," and in the last part of the century, even after the Revolution, English pencils like Middleton's were regularly advertised for sale in the larger cities. Imported pencils or homemade pencils fashioned from reclaimed pieces of broken lead would have been the proud possessions of woodworkers especially, for carpenters, cabinetmakers, and joiners possessed the craft skill to work wood into a form that could hold pieces of graphite in a comfortable and useful way. Not only would early American woodworkers have known about, admired, wanted to possess, and tried to imitate European pencils, but also they would have prized and cared for them as they prized and cared for the kinds of tools displayed two centuries later in the Smithsonian.

These stories of absence are interesting not so much because of what they say about the lowly status of the wood-cased pencil as an artifact as because of what they say about our awareness of and our attitudes toward common things, processes, events, or even ideas that appear to have little intrinsic, permanent, or special value. An object like the pencil is generally considered unremarkable, and it is taken for granted. It is taken for granted because it is abundant, inexpensive, and as familiar as speech.

Yet the pencil need be no cliché. It can be as powerful a metaphor as the pen, as rich a symbol as the flag. Artists have long counted the pencil among the tools of their trade, and have even identified with the drawing medium. Andrew Wyeth described his pencil as a fencer's foil; Toulouse-Lautrec said of himself, "I am a pencil"; and the Moscow-born Paris illustrator and caricaturist Emmanuel Poiré took his pseudonym from the Russian word for pencil, *karandash*. In turn, the Swiss pencil-making firm of Caran d'Ache was named after this artist, and a stylized version of his signature is now used as a company logo.

The pencil, the tool of doodlers, stands for thinking and creativity, but at the same time, as the toy of children, it symbolizes spontaneity and immaturity. Yet the pencil's graphite is also the ephemeral medium of thinkers, planners, drafters, architects, and engineers, the medium to be erased, revised, smudged, obliterated, lost—or inked over. Ink, on the other hand, whether in a book or on plans or on a contract, signifies finality and supersedes the pencil drafts and sketches. If early pencilings interest collectors, it is often because of their association with the permanent success written or drawn in ink. Unlike graphite, to which paper is like sandpaper, ink flows smoothly and fills in the nooks and crannies of creation. Ink is the cosmetic that ideas will wear when they go out in public. Graphite is their dirty truth.

A glance at the index to any book of familiar quotations will corroborate the fact that there are scores of quotations extolling the pen for every one, if that, mentioning the pencil. Yet, while the conventional wisdom may be that the pen is mightier than the sword, the pencil has come to be the weapon of choice of those wishing to make better pens as well as better swords. It is often said that "everything begins with a pencil," and indeed it is the preferred medium of designers. In one recent study of the nature of the design process, engineers balked when they were asked to record their thought processes with a pen. While the directors of the study did not want the subjects to be able to erase their false starts or alter their records of creativity, the engineers did not feel comfortable or natural without a pencil in their hands when asked to comment on designing a new bridge or a better mousetrap.

Leonardo da Vinci seems to have wished to make a better everything, as his notebooks demonstrate. And when he wanted to set down his ideas for some new device, or when he merely wanted to record the state of the art of Renaissance engineering, he employed a drawing. Leonardo also used drawings to preserve his observations of natural facts, artifacts, and assorted phenomena, and he even sketched his own hand sketching. This sketch is usually identified as Leonardo's left hand, consistent with the widely held belief that the genius was left-handed. This trait in turn has been given as a reason for his mirror writing. However, it has also been convincingly argued that Leonardo was basically right-handed and was forced to use his left hand because his right was crippled in an accident. Thus Leonardo's sketch may really be of his maimed right hand as seen in a mirror by the artist drawing with his fully functioning left hand. The shortened and twisted middle finger in the sketch supports this view.

The precise nature of the drawing instrument in Leonardo's hand may also be open to some interpretation, but it appears most likely to be a small brush known from Roman times as a pencil. The lead pencil as we know it today does not seem to have existed in Leonardo's lifetime (1452–1519). Some of his sketches were done in metal point, but drawing with a pointed rod of silver or some alloy usually had to done on specially coated paper so that an otherwise faint mark would be enhanced. Some drawings were first outlined in metalpoint and then more or less traced over with a pen or a fine-pointed brush dipped in ink. This was the only kind of pencil Leonardo knew.

Nevertheless, even in their complex medium, Leonardo's notebooks were almost lost to posterity. Their author never published their contents, and after he died the thirty-odd volumes almost passed into oblivion. He left them all to his friend

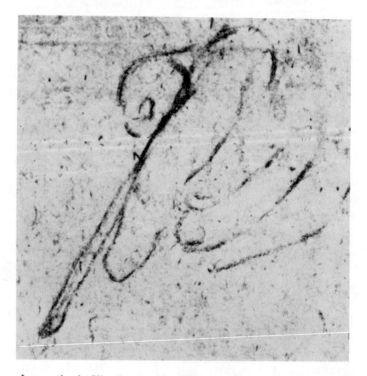

Leonardo da Vinci's sketch of his own hand sketching, either his left hand or his right seen in a mirror

and pupil Francesco Melzi, with an injunction: "In order that this advantage which I am giving to men shall not be lost, I am setting out a way of proper printing and I beg you, my successors, not to allow avarice to induce you to leave the printing un . . ." But the sentence seems never to have been finished, and the proper printing took longer than Leonardo must have hoped. Melzi kept the notebooks locked away for fifty years, so, except for a treatise on painting, which was extracted for publication in 1551, the bulk of Leonardo's engineering remained private, and by the time the notebooks were published in 1880, virtually all of the inventions were either rediscovered or superseded.

Engineers throughout history have tended to work out their plans in less permanent media and have suffered the obscurity that Leonardo escaped only through the sheer mechanical and artistic brilliance of his notebooks. Because they are the subjects of manuscripts and books, we know much more about the wrongheaded theories of the universe and the unrealistic utopias of dreamers than we do about the ingenious and successful engineering achievements of the ages. And this is due at least in part to the fact that, long before the time of Leo-

nardo, drawing rather than writing was the medium of thinking and planning for the engineer. But plans and drawings
were not the subject of scholarship. Lynn White, Jr., was
especially aware of the need to look beyond the written record.
In the preface to his brilliant study of the role of artifacts such
as the stirrup in the story of civilization, he wrote:

> If historians are to attempt to write the history of man-
> kind, and not simply the history of mankind as it was
> viewed by the small and specialized segments of our race
> which have had the habit of scribbling, they must take a
> fresh view of the records, ask new questions of them, and
> use all the resources of archaeology, iconography, and
> etymology to find answers when no answers can be dis-
> covered in contemporary writings.

The transient practice of engineering has been by and large
an invisible and unrecorded aspect of the history of civilization. While we do have artifacts from all ages that we recognize
as tools, structures, or machines, we tend to see them as dis-
crete pieces of material detritus in the context of cultural development. It is less easy to deal with the origins of those
artifacts as deliberate acts of invention and the evolution or
"perfection" of them as deliberate acts of engineering, especially since such interpretations depend ultimately upon
presumptions about the thought processes of our distant
ancestors. Did they really practice engineering or did they just
stumble upon happy accidents of nature in the form of fortuitously shaped rocks and fallen trees bridging streams? Have
we always been victims of circumstance or have we from the
start been conscious inventors and conscious engineers?

Marcus Vitruvius Pollio, whose *De Architectura* in ten
books is the main source for the history of engineering in
ancient Rome, argues that our ingenuity is innate. But Vitruvius did not believe that the advancement of civilization could
rest on innate qualities alone, and he listed skill with the pencil
—the fine-pointed brush that Leonardo used—only behind
education as one of the prerequisites for the architect, or engineer, of two millennia ago. Drawing was essential.

What the earliest engineers do not seem to have done, of
course, is to have *written down* much, if anything, about their
work. Vitruvius' twenty-centuries-old classic is generally considered to be the oldest surviving work on engineering, but it
is also about the aesthetics of building, and it seems to have

survived for that reason alone. One historian said of Vitruvius what many have implied: "He writes in atrocious Latin, but he knows his business." And a classicist said, "He has all the marks of one unused to composition, to whom writing is a painful task." But whether he deluded himself about his own writing ability or simply did not think skill with the pen to be as important as that with the pencil, beginning with Vitruvius and continuing to this day, writing about engineering has been generally less than poetry and often dominated by a labored description of artifacts, a prosaic prescription of rules for emulating those artifacts, and an overwhelming concentration on the technical "business" of making artifacts. There is a paucity of any kind of literature, either well articulated or written in forgettable and forgotten prose, on how the earliest engineers used "their natural gifts sharpened by emulation" to come up with the ideas for new and improved artifacts in the first place.

But whether it is recorded or not, the process of engineering, what is commonly referred to as the engineering method, is actually much older than Vitruvius—indeed, as old as civilization itself—and it has come down to us today essentially unchanged in its most basic characteristics. While engineering as a formal and distinct profession may be only a century or two old, engineering as a human activity has been, and is, virtually changeless and timeless.

Vitruvius propagated the myth that engineering is applied science. Yet there is an astonishing imagination in engineering, an imagination independent of science, but it has been realized in pictures and artifacts and not in words. And as the pictures are erased as the artifacts themselves remove the need for pictures, so the artifacts wear out because they are designed not as objets d'art but as things to be used, indeed things to be consumed in their very use. While every artifact embodies the methods of technology, the pencil is an especially appropriate one to study. Not only can the pencil serve as a symbol of engineering itself; the development of this artifact of remarkable ingenuity, complexity, and universality may also serve as a paradigm for the engineering process generally.

As there have always been engineers, so there have also always been philosophers. The artifacts of philosophers are, of course, their writings, and the survival of writings about matters philosophical has too often led to the facile conclusion that matters practical were somehow of lesser importance. This is not necessarily so, but as late as the Renaissance, it was still largely the case that "the social antithesis of mechanical and liberal arts, of hands and tongue, influenced all intellectual and professional activity," and well into modern times the artisans and craftsmen who helped advance the technology, albeit slowly, of everything from writing implements to ships were not educated and "probably often illiterate." And if even Leonardo's notebooks could remain unread for so many centuries, what expectation could there be that humanists would "read" the poetry and history embodied in artifacts? With the rise of what have been called "artist-engineers" like Leonardo, technological subjects came more and more to be recorded, but for the most part only in notebooks and manuscripts that circulated among other artist-engineers.

The business and technology of making pencils have obscure roots and have evolved in fits and starts out of the unwritten traditions of craftsmanship. The reasons for many of the physical characteristics of the pencil are as lost in those traditions as are the origins of the sizes and shapes of many a common object, but the relatively recent origin and short history of the modern pencil also makes it a manageable artifact to twirl about in the fingers and reflect upon in the mind. When we do this we also realize that for all its commonness and apparent cheapness, the pencil is a product of immense complexity and sophistication. Thus there is much to be learned from the pencil and the story of its development for illuminating the nature of engineering and engineers and, by extension, modern industry. Problems faced over the centuries by pencil makers and manufacturers are not without their lessons for today's international technological marketplace. Used like the Socratic method, the pencil can draw out of us realizations about things of which we might never have thought.

In the late twentieth century, when there are billions produced each year and sold for pennies, it is easy to forget how marvelous and dear an object the pencil once was. According to the prayer of an old Nubian, recorded in an 1822 journal of a visit to Ethiopia: "Praised be God, the Creator of the World, who has taught men to inclose ink in the centre of a bit of wood." A century later and an ocean away, the pencil could still evoke wonder, but the manufacture of the artifact was seen to involve a lot more than just "ink in the centre of a bit of wood." In order to manufacture a pencil, according to the early-twentieth-century account of a participant in the process:

the writer has had to become familiar with the nature of hundreds of dyestuffs, of shellac and many other resins, of clays of all kinds and from all parts of the world, of the many varieties and qualities of graphites, or many kinds of alcohols and other solvents, of hundreds of natural and artificial paint pigments, many varieties of woods, and general knowledge of the rubber industry, of the glue industry and of printing inks, of nearly all varieties of waxes, of the lacquer or soluble cotton industry, of many types of drying equipment, of impregnating processes, of high temperature furnaces, of abrasives and many phases of extrusion and mixing processes.

Looking at my career in the pencil industry, along a perspective of some eighteen years, I am dumbfounded at the many angles it takes, at its polyphase ramifications, at the difficulties in developing a trained staff of assistants, at the extreme accuracy required of the tools, and at the broad knowledge of practical chemistry necessary, as well as the expert knowledge of the proper sources of supplies of raw materials, required to get anywhere with pencil manufacture, so as to compete in the markets of the world.

This is an excellent summary of the many facets of engineering involved in making a modern pencil. "Practical chemistry" is, of course, today called chemical engineering, and knowledge of the various specialties of mechanical engineering, materials engineering, structural engineering, and even electrical engineering is invaluable for manufacturing attractive pencils that can be sharpened to fine points that are strong and will write smoothly. And the fruits of all this expert knowledge are made available for a fraction of what it would cost merely to assemble the materials. While one oft-repeated definition of an engineer is someone who does for one dollar what anyone can do for two, in the case of a mass-produced pencil, the economic advantage is even more pronounced. In the 1950s, it was estimated that a "do-it-yourself addict" would have to spend about fifty dollars to make a single pencil.

While the Smithsonian Institution neglected to include pencils on the worktables of late-eighteenth-century craftsmen, in an earlier show, "A Nation Among Nations," it acknowledged that "all the principles of mass production can be seen at work in the manufacture of the common wood-cased pencil," and a pencil-making machine built in Tennessee in 1975 was displayed. Now, in the Smithsonian's most recently installed permanent exhibition, "A Material World," which serves as "an introduction to the entire National Museum of American History," there is a display showing how "stuff" is transformed into "things," and the raw materials of a pencil serve as the paradigm. These are fitting acknowledgments of the importance of the pencil and other engineered artifacts in influencing and being influenced by our more general culture. However, there remains a strong intellectual tradition that generally ignores the fact that the art and literature we cherish would be of quite a different nature without such technological artifacts as pencils.

In the Concord, Massachusetts, Free Public Library there are shelves of editions of Thoreau's *Walden* and shelves of books on the author's times, writings, and thoughts. One catalogue of these Thoreau Society archives lists more than one thousand items, but the number of those dealing specifically with Thoreau as pencil maker and engineer of pencil-making machinery is nil. While a "nail picked up at the Thoreau cabin site" is included among the literary works, no pencil is. Only a Thoreau & Company pencil label (printed in ink, of course) gives any hint of the activity that provided the family income. One must learn of Thoreau the pencil engineer almost by inference from the few scanty references within more general works that the curator happens to recall. There are now a few pencils among the books and literary material in the Thoreau alcove in the library, but their method of manufacture seems to be more mysterious than that of any of Henry David Thoreau's literary works.

While it may be excusable that Thoreau's pencil engineering is seldom emphasized relative to his other achievements, there is no excuse for ignoring engineering in our culture generally. Yet it is rare to find generalizations about engineering qua engineering that are the equivalent of the scientific method or to find universal insights about engineering that have the ring of Archimedes' "Eureka!" Great engineers have seldom left articulate generalizations or insights in ink; they have usually only sketched them in pencil, to be fleshed out in state-of-the-art structures and machines. Yet even as the state of the art is constantly evolving and developing, there are deep underlying similarities in what the first engineers or those described by Vitruvius did and what today's engineers still do. And it is the timeless features of the creative process sometimes called the engineering method, with the curious attributes that make it possible for essentially the same method to coexist in both naïve and sophisticated minds, that are innate in all of us. These features are also the reason why engineering always has been and always will be more than mere applications of mathematical theorems and physical principles. It is high time to write in ink for publication what engineers have for so long only sketched in pencil in their notebooks. The history of the pencil itself provides an excellent opportunity to learn more about engineering.

2/ Of Names, Materials, and Things

What has come to be known as a pencil was named that because it resembled the brush known in Latin as a *penicillum*. This fine-pointed instrument, which was formed by inserting a carefully shaped tuft of animal hairs into a hollow reed, much as a piece of lead is inserted into a mechanical pencil today, in turn got its name as a diminutive form of the more general Latin term for brush, *peniculus,* itself a diminutive form of the word *penis,* which is Latin for tail. This word was used for the very first fine brushes because they were actually formed from the tails of animals. Thus a pencil is literally a "little tail," which can be used for writing or drawing fine lines.

A Roman *penicillum,* or pencil brush

While it might be possible to give all sorts of anachronistic, prurient, and sexist interpretations for the etymology of the word "pencil," our interests are better served by looking at the functional rather than the Freudian antecedents of the object. The name of an artifact may certainly depend upon symbolic and subliminal evocations, but artifacts themselves do not come from their names. Indeed, the modern pencil is called what it is because it, like all technological objects, is more likely than not the product of distinctly nonverbal thinking.

Made things come to exist before they are named as surely as they come to be drawn, at least on the palimpsest of the mind, before they are made.

A former president of the Newcomen Society for the Study of the History of Engineering and Technology, H. W. Dickinson, has found etymological and functional connections between precursors of the broom and the pencil through the continuity of the way in which the art of making these common objects evolved. He associates the need for brushes with a more settled agricultural as opposed to a nomadic and hunting form of life and thus traces the earliest brushes back about seven thousand years. Dickinson agrees that the words used for such common objects are "fossil poetry into which the imagination of human beings has been compressed by centuries of use," and finds the first English word for something brushlike to be "besom," which as early as the year 1000 designated an assemblage of branches, twigs, or shoots. Since after a time besoms were commonly made from the broom shrub, the word "broom" came to be used by way of metaphor for all besoms. The earliest brooms required the user to bend over while using them, and the compound word "broomstick" suggests that the addition of a long handle came well after the besom itself.

The word "brush," which etymologically signifies twigs, has come to be the generic term for all implements used to do such things as sweep, dust, clean, polish, paint, and write. The craft methods of making all sorts of brushes, including the pencil, suggest a technological continuity that parallels the etymological. Even the size of brushes used for writing demonstrates their continuity of development. Egyptian pencil brushes surviving from the XVIIIth Dynasty (ca. 1500 B.C.) vary in length from about six to nine inches, a size not unlike the seven-inch pencils that have come to be standard.

Engineering, which is nothing but the modern name for a practice as old as civilization, cannot be easily understood by too narrow a focus on its etymology. Just as a nameless pencil would still make a mark on this page, so even if ancient engineering was an ineffable activity, or one called architecture, that would not be to say that it did not exist as a distinct pursuit. And while it is true that engineers in the eighteenth and nineteenth centuries would increasingly call themselves *civil* engineers to separate themselves from the military traditions of engineering that had grown so formal and strong, the civilian roots of engineering are at least as old as the military.

The evidence of the earliest Greek and Roman writers on engineering subjects makes it clear that in ancient times the Greek term *techne* and the Roman *architectura* definitely included what we today call engineering.

People, professions, and artifacts all have roots that continue to feed their greenest shoots. And as they can with virtually every common object, precursors of the modern pencil can be found in antiquity. The Greeks and Romans apparently were aware that metallic lead could make a mark on papyrus, and still earlier peoples knew that the burnt coals or ends of sticks from a fire were natural implements for drawing pictures on the walls of caves.

Nonliterary artifacts cannot tell their own stories as explicitly as people and books can, however. Thus the significance for engineering history of the existence of this or that object—and its manufacture—from prehistory or antiquity can be very problematic at best and at least as difficult as the explication of a story from a fragment of an obscure text. Yet much incontrovertible evidence, such as contained in the "Treasures of Tutankhamen" exhibit, which toured America in the 1970s, makes clear what a high level of craftsmanship and engineering must have existed in a place like ancient Egypt in order to make such beautiful and ingenious objects and the structure in which they were entombed. If such a level of achievement already existed more than thirty centuries ago, then surely the roots of our technological know-how and achievements must reach back myriad generations.

Yet, perhaps in part because specialization was no doubt as common in ancient times as it is today, the written history of engineering is sparse. Even the most able and articulate of ancient engineers, whether they were known then as artisans, craftsmen, architects, or master builders, might have had no more time or inclination or reason to articulate what it is they did and how they did it than do some of the most able of today's engineers.

The lead pencil replaced both the metallic-lead stylus, which made a dry light mark, and the pencil brush, which made a fine dark line, by bringing together the two desirable qualities of dryness and darkness in a single instrument. Although it may incorporate dozens of raw materials, the lead pencil derives its specific name from the one material that it is least likely to contain. The "lead" of today's lead pencil is really a mixture of graphite, clay, and other ingredients, and even the paint used on the pencil's exterior is likely to be lead-free in response to concerns raised in the early 1970s. Thus we no longer risk getting plumbism, or lead poisoning, when we chew on the ends of our pencils the way Harold Ross, the legendary editor of *The New Yorker,* used to, at least according to the reminiscence of his onetime office boy: "Mr. Ross ate pencils. I think he ate them, for I never found anyone else in his office; and when I inspected them in the morning, I discovered that they had been chewed, bitten into, and possibly eaten."

The names of many common objects, edible or not, derive from the materials of which they were originally made. Thus the British today still call an eraser a "rubber," even when it is made of some petroleum-based substance. We also eat off paper plates that are really Styrofoam, open tin cans that are really aluminum, set our tables with silverware that is really stainless steel, wear glasses that are really plastic, use golf irons that are really titanium and woods that are really not.

The persistence of the names of things deriving from the materials of which they were first made suggests the intimate relationship that can exist between ingenious objects and their materials. This relationship may develop because the object and the material seem, at least at first thought, to be made for each other. Indeed, in the excitement of discovery, the possibility of another material being used, if indeed another could be used, might be inconceivable. Thus while form may not necessarily follow function, material certainly may. Or can it be the other way around? In order for something to function properly, it must have the proper proportions, the proper weight, the proper strength, the proper stiffness, the proper hardness, and all the other qualities necessary to perform the function for which it is conceived. And all of these qualities depend upon the properties of materials. Finding the right material for a pencil lead can be as difficult as finding the truth.

Imagine being confronted with a box full of marble-size balls all painted mat black. Imagine that the balls are indistinguishable to the eye, but that each is made of a different material. There are balls of iron, wood, lead, stone, plastic, rubber, glass, foam, sand, graphite, and even balls of tin filled with such liquids as oil and water and nitroglycerine. Since the materials in the balls have different densities, it will be clear by hefting them that they have different contents. Since the balls will have different stiffnesses, squeezing them will further distinguish one from the other. Since the balls will have different resiliencies, bouncing them on the floor will give different rebounds. But without scratching off the uniformly black paint, it might take quite a few creative combinations of hefting, squeezing, and bouncing to determine what material is in what ball. And if we suspected, were told of, or once were shaken by the accidental discovery of a ball of nitroglycerine, we might be reluctant to be too aggressive in our hefting, squeezing, and bouncing. However, if we were adventuresome and yet deliberate, and if we were to pick out one of the balls and were lucky enough to choose one that could be melted, molded, or beaten into different shapes without exploding, we might fabricate something useful.

If we were challenged to produce a writing instrument out of the box of balls, some of us might dismiss the challenge as crazy or mischievous, some of us might become impatient and blow ourselves up by hastily dumping the contents out on the concrete, and some of us might take the time to think or tinker in a careful and methodical way. While a ball of wood would not work, a ball of lead might be just the right size to be held in the hand, not too heavy to be lifted and maneuvered, strong enough to be pressed down on a piece of paper or parchment and pulled along without breaking, stiff enough so as not to change its shape under this action, soft enough so as not to tear the writing surface, and, finally, of such a nature as to leave a visible line on the paper or parchment. Having found the lead to be so surprisingly serviceable a writing implement, we might curtail our search and put the rest of the balls aside, thinking that a ball of anything else in the box could not act nearly as well as a piece of metallic lead in fulfilling the requirements.

However, we might also reason that if the ball of lead had made such a good marker, perhaps another ball of another material might be even better. Continuing our trials of the balls, we might then come across the graphite and be astounded at how much blacker a mark it made. Needless to say, it would displace the lead. On the other hand, if the box had contained no ball of graphite, then even infinite patience might not produce something superior to the lead, but we might also recognize that a better material might exist somewhere outside the box. In short, whether one finds or invents something as good as one hopes to when setting out on a quest, whether one is satisfied with what one does come up with, and whether one continues a quest at all really depend not only on whether a suitable material exists but also on what one believes to be possible.

While the ancients did not have a convenient box of balls available for discovering a writing instrument, somehow they eventually did chance upon what seemed then to be just the right material in the right size and shape. And so the name of the ideal material itself, lead (the Latin *plumbum*), became the name of the seemingly ideal object that had been made from it. The "lead pencil" can trace its name back to the elemental *plumbum,* and also to the independently developed writing brush called a *penicillum,* precisely because the modern object is a direct descendant of the marriage of those two ancient objects. Such marriages are made not in heaven but here on earth out of the materials of the earth, and they are most likely to come out of the minds and hands of independent-thinking craftsmen, inventors, and engineers. And the marriages can be of chance as well as of convenience.

It is difficult to be certain how this or that better design for a brush, plow, house, or sword evolved from its predecessors, for the process was at best sketched metaphorically in pencil and seldom if ever copied in pen. It is because of this that the ideas and artifacts of technology—the processes and products of engineering—are so very different from the creations and theories of literature, philosophy, and science. For these latter activities of civilization, the preservation of the past can be a natural end in itself. The classics of formalized thinking have been treasured as finished entities, and they have been copied and later printed, translated, and referenced throughout the modern era. The classics, even if superseded in factual or theoretical sophistication, are considered models of thinking from which one can today still benefit by emulation, or at least by inspiration. Sometimes it is the rhetoric, or form of argument employed, that is the strength of the classics. But in the same way the engineering method, which might be called the rhetoric of technology, can be the very reason for preserving the history of engineering achievements long superseded. While it may be difficult, if not impossible, to recover with any degree of certainty the unrecorded method of the past, we have begun to preserve the next-best thing—the artifacts produced by that method.

The artifacts of technology, especially those that were not suitable to be buried with a king, were commonly thought to be made obsolete when they were improved through evolu-

tion. Hence there was not believed to be much reason to remember, through preservation or reproduction, an old plow
(or how to make one) that was no longer used because a new,
improved version, perhaps employing a new material, had
been developed or acquired. Similarly, old tools and old construction methods would eventually disappear with the craftsmen who were too aged to learn new ones. And any old swords
that were not already abandoned would be destroyed by the
harder, newer models. Curators of technological artifacts, industrial archaeologists, and historians of technology represent
rather new careers that have only developed with the realization that the products of our technological past may have an
intellectual and cultural value, and indeed may hold some lessons that are as irreplaceable as are the thoughts and writings
of Plato and Shakespeare.

Artifacts, the products of engineering, do replace artifacts.
However, the rhetoric of technology by which those artifacts
are articulated in timber or stone or steel or any material, pure
or composite, is more or less a constant of history. But since
the actual process of engineering is elusive, becoming tangible
only by being embodied in the artifact, the process of engineering itself can seem almost ineffable. Nevertheless, by looking at the continuity of artifacts and at the evolution of the use
of materials, we can begin to appreciate what it is that is behind the technological environment.

Although the artifactual landscape can appear to change
rather dramatically, if not chaotically, with time, in general
the evolutionary process is really quite slow and deliberate.
There tend to be three broad areas into which technological
developments fall: new concepts, new magnitudes, new materials. Truly revolutionary innovations tend to involve extrap-
olations within two or three of these categories simultaneously.
Thus the creation of the first bed, if it is meaningful to speak
of such an event, would have had to involve first the very
concept of a bed, then a conscious or unconscious decision
about its size, and finally the choice of materials out of which
to make it. Once the first bed existed, then it would be possible
to ask, with the Peripatetics, whether the size was right,
whether a different kind of wood could better withstand the
strain imposed by the cords supporting the mattress, and
whether the cords themselves could be arranged in a more
economical way.

The creation of the first writing implement required the
recognition of the fact or the concept that one can deliberately

and repeatedly make a mark with an object, the identification of a material that is capable of making the mark, and the determination of whether to use the material in the lump in which it was found or whether to fashion it into a probably more comfortable and convenient size and shape. Once this kind of discovery and determination has been made, the subsequent development of the concept is more likely than not to involve a change of magnitude or scale of the implement or a change of the materials that are used for making the mark and the materials on which the mark is made. The former class of changes, those of magnitude, tend to be associated with seeking the optimal or maximum size of artifacts or the multiplication of their numbers. The latter class of changes, those involving materials, tend to be associated with seeking economic advantages or improved qualities of performance or function. Engineers seldom talk of all classes of innovation at the same time.

Most engineering projects do not involve wholesale innovation. And the choice of material can often be an arbitrary one, determined more by aesthetic preference or notion of status than by any technological imperative. But more often than not, getting an idea and trying to realize it in the wrong material can be disastrous. For example, a wooden raft works fine, but a stone one would not, and even some woods work better for rafts than others. Thomas Edison certainly knew that not every material was suited for the filament of an incandescent light bulb, but he did know in his technological heart that his idea should work in principle—if he could only locate a material with the correct properties. Edison tried numerous different materials until he hit upon the right one, and when asked if the long quest ever discouraged him, he reportedly replied that it had not, for every failed filament taught him something —namely, one more material to exclude from further consideration. Charles Batchelor, Edison's co-worker at Menlo Park, described the failure of one such experiment: "Made hemp fibres with clamps of plumbago, graphite such as used in lead pencils—they have got too much stuff mixed with them for us —seem to swell up and form gases or arcs which bust up the lamps."

The lead pencils that Batchelor found too adulterated as a source of graphite were dear to Edison for other reasons, however, and they were not beneath his attention, for "Edison liked short pencils and he persuaded a pencil factory to turn out short pencils especially for him." Perhaps he settled on his ideal pencil in much the same trial-and-error manner in which he selected a filament for his lamp, but once he did find his ideal, he stopped looking. Edison's pencils, which he ordered in lots of one thousand and always carried in his lower vest pocket, had very soft lead, were thicker than average, and were only about three inches long. Once when his order was not filled to his liking, Edison wrote to the Eagle Pencil Company that the "last batch was too short." He complained of the pencils: "They twist and stick in the pocket lining."

While the story of Edison's preferences for pencils is understandably not so oft-repeated as that of Edison and the electric light bulb filament, the history of the lead pencil itself does involve quests as memorable. For example, the search for suitable materials for pencil leads and the means of processing them into good alternatives to a diminishing supply of ideal graphite was one that ranged all over the globe and spanned centuries. But once such a quest is ended and once an artifact like the pencil or the light bulb has evolved to a certain advanced state, the artifact and its materials can seem to be made for each other. It is at that stage of technological advancement that novelty items like Edison's short pencils and square light bulbs come to be produced.

But to substitute materials in a perfected artifact can be at its worst disastrous, as when inferior steel was discovered in millions of bolts imported into America not very long ago, or at its best nothing but a practical joke, as when I tried to write with a pencil that my son had purchased in a London gag shop. The pencil looked like any other one that had been sharpened several times, and I only got the punch line of its unfamiliar name, "Tryrite," when I pressed on its rubber point.

3/ Before the Pencil

How does a lump of lead that draws a creditable line evolve into a modern pencil? How does a rounded rock turn into a wheel? How does a dream become a flying machine? The process by which ideas and artifacts come into being and mature is essentially what is now known as the engineering method, and the method, like engineering itself, is really as old as *Homo sapiens*—or at least *Homo faber*—and the process is about as hard to pin down and as idiosyncratic in each of its peculiar applications as is the individual of the species. But while each invention and artifact has its unique aspects, there is also a certain sameness about the evolutionary way in which a stylus develops into a pencil, a sketch into a palace, or an arrow into a rocket. And this observation itself is as old as Ecclesiastes, who may have been the first to record, but probably was not the first to observe, that "what has been is what will be, and has been done is what will be done; and there is nothing new under the sun."

Anything as old as civilization can be associated with the different professions but cannot be claimed to be exclusively theirs. Thus drugs predate medicine, belief religion, conflict law, and artifacts formal engineering. The essence of engineering, like that of medicine or religion or law or any other codified human endeavor, is only one manifestation of the human mind, and what distinguishes engineering by modern engineers from the creative processes of artisans in antiquity is merely a matter of the self-consciousness, intensity, circumspection, efficiency, science, and probability of success with which the engineering method is applied by engineers to generally more and more complex artifacts and processes. Just as we all have a sense of health without being physicians and a sense of justice without being jurists, so we all have a sense of design without being engineers. What distinguishes the modern engineer from our earliest ancestors is not the basic intuitions of the engineer but the development of analytical and synthetic powers and devices that at the same time both maximize and minimize the chances of exploding nitroglycerine to discover a writing instrument or of fashioning a sword out of a plowshare. But even the complexity of modern life is nothing new.

Today artifacts evolve and displace their own precursors much the same way they did in ancient times. Cicero gives clear evidence that the way we do things today would not be unfamiliar to a Roman—whether architect, engineer, or layman—for what the Roman wrote is not unfamiliar to us. In a letter to his friend, adviser, and confidant Atticus, who apparently had some criticism of the size of the windows in the country house he was coming to visit, Cicero wrote that in finding fault with the narrowness of the house's windows, Atticus was finding fault with the education of Cyrus, the house's architect. And Cicero gave to Atticus, complete with a geometrical argument that implies the accompaniment of some kind of diagram, the explanation Cyrus might have given as to why the windows were just right for an appreciation of the garden. After relating the argument, Cicero continued, "If you find anything else in my house to criticize, I shall always be supplied with reasons quite as good to give you in reply, unless, indeed, I can remedy the difficulty at small expense."

It is the ideal of design to make and furnish the best artifact for the money by using the best of available resources, where resources include style, time, and energy, as well as hard cash and materials. Because there are always the constraints of economy and possibility, any product of engineering can always be criticized because it will never be totally efficient or flawlessly made or perfectly strong or absolutely safe, if indeed it can be made at all safe, at all strong, or at all, period, and still perform adequately the function that is its raison d'être.

Whenever something is criticized and an improvement can be realized, the engineer, like Cicero's architect, wishes to respond with the improvement or be able to defend the imperfect product as arguably the best that can be done at the time with the given materials and resources, within or even a bit over budget. But the true engineer also will recognize, without being prompted, the desirability of any incremental improvement that is possible without being too expensive, too demanding of existing technology, or too time-consuming to achieve. If, however, new financial resources become available, or if new technologies or new materials develop, or if there is a less urgent need for an improved design, thus allowing time for more development, then the conscientious engineer does not counter criticism but embraces it as a new initiative. Sometimes, however, criticism calls attention to a risk that is too great to let stand at even a great expense. The case of the space shuttle *Challenger* is a now familiar example that has affected everyone's perception of risk and expense.

The history and practice of engineering is often thought of in terms of increasingly large and complex machines and structures and devices and systems and processes, all of which tend to be ever more technical and esoteric and threatening. The primitive bridge consisting of a fallen log evolves into steel and concrete bridges that fall out from under interstate highway traffic. Television and newspapers find it easier to tell us exactly how many people were killed than to inform us exactly how such an accident can happen. Initial reports tend to offer conflicting theories that seem to change daily as different engineers give their opinions, and the public (and perhaps even the media) can hardly be blamed for giving up trying to understand what the experts themselves cannot seem to agree upon. And as the technical debate often continues in the courts, it tends to become even less and less visible an issue. Thus the conventional wisdom is reinforced that the workings and the works of engineers are unfathomable to all but, if not to all including, the engineers.

But the basic ideas of engineering and the fundamental principles of the engineering method are not really as complex as some of the more involved products and personalities of engineers and engineer-managers. It is by trying to understand simple ideas and principles in terms of the most complex of examples and issues that we tend to feel overwhelmed. If we can capture the essence of engineers and engineering through the most elementary and least abstract of examples, then we can more easily get to the heart of the matter when confronted with something so large and unfamiliar that we can barely conceive what it really looks like, let alone hold it in our hands and think about it. What might seem to be the secrets of engineering are in the common as well as in the uncommon, in the small as in the large, in the seemingly simple as in the indubitably complex. But on closer inspection, even what can appear to be the commonest, smallest, and simplest of objects can reveal itself to be on its own terms as complex and as grand as a space shuttle or a great suspension bridge. So to scrutinize the trivial can be to discover the monumental. Almost any object can serve to unveil the mysteries of engineering and its relation to art, business, and all other aspects of our culture.

The example of the pencil's evolution may serve as a paradigm. While a lump of lead or charcoal could certainly be serviceable as a primitive pencil, it could also be easily criticized. Writing or drawing with a lump of anything for an extended period of time can cramp the fingers, thus cramping one's style and perhaps even cramping one's mind. And the relative bulkiness of a lump would hide from the view of the writer or drawer the very thing being written or drawn, thus making fine and detailed work difficult at best. And the line made by a lump of lead might not be as dark as one might like, while the line made by a lump of charcoal might be too dark and smudgy on the parchment or paper, not to mention on the hands. The ancients were no doubt complainers as chronic as we are today, and they no doubt articulated such criticisms about the sorry state of their writing implements. Couldn't anyone do better than a lump of charcoal or lead?

Well into modern times, the act of writing was occasioned by much preparation and inconvenience. Reed pens were known in antiquity, and quill pens have been used for well over a thousand years. But both required preparation of their points and repeated dipping in ink, which was at risk of being spilled and smeared. The paraphernalia of writing with a quill, for example, included not only pen and ink, but also a penknife to shape the writing point and an absorbent substance such as pounce, a powder used to dry the ink and prevent it from spreading. Albrecht Dürer's familiar portraits of St. Jerome with a quill and Erasmus with a reed pen show them writing on sloping surfaces and with an ink bottle either in hand or on a level surface nearby.

For a long time, the principal alternative to pen and ink on papyrus, vellum, or parchment was the metal style and wax tablet. *Codex,* the Latin word for tree trunk, came to be used for the wax-covered wooden tablet that would evolve into the modern book. The tablet's hard frame, later also made of ivory, contained a surface of wax that could be incised by the sharp end of the style or stylus, whose other end was often flattened or otherwise shaped so that it could smooth out the wax for alterations or reuse and thus effectively function as an eraser. Some scholars believe that Chaucer may have composed such works as *The Canterbury Tales* on wax tablets, writing only his final draft in ink on vellum or parchment. "The Summoner's Tale" relates the use of a wax tablet, as well as its erasability, by an unscrupulous friar who had no intention of keeping track of those to whom he had promised favors:

> *His comrade carried a staff tipped with horn,*
> *Waxed tablets backed with ivory to write on,*
> *A beautifully polished stylus pen,*
> *And always wrote the name down on the spot. . . .*
> *No sooner had he'd got outside the door,*
> *The friar would smooth out every single name*
> *He'd written on the tablets earlier.*

Thomas Astle, whose first edition of *The Origin and Progress of Writing* appeared in 1784, described how early epistles were written on multiple tablets of wood, known also as table books, held together with string whose knot was sealed with wax. It is from breaking this brittle seal to read the epistle that the expression "to break open a letter" presumably comes. Astle's treatise goes on to note that by the late eighteenth century books of ivory "written upon with black lead pencils" were used in his day for "memoranda." But since the principal object of Astle's book was to illustrate what was called the "diplomatic science," which was used to determine the age and authenticity of important written documents, especially those under dispute, it is not surprising that he tells as asides a few tales of violence. Astle dramatizes his point about the size and weight of wooden and ivory tablets by noting that "in Plautus a school boy of seven years old is represented breaking his master's head with his table book."

Astle also contends that sharp-pointed "iron styles were dangerous weapons, and were prohibited by the Romans, and those of ivory were used in their stead," and he relates stories of Roman violence stemming from the misuse of styles that accompanied *pugillares,* as the table books were called in Latin, presumably because the smaller ones at least could be held open in one hand while the other held the style to inscribe the wax. He tells of Caesar using a style to pierce Cassius' arm

A young girl from Pompeii with a style and *pugillares,* or book of wax-surfaced tablets

"in full senate," of Caligula inciting the massacre by styles of a senator, and of the torture to which "Cassianus was put to by his scholars, who killed him with their *pugillares* and styles."

Styluses of metal, or metalpoints, used for peaceful purposes, were long known to be capable of making a faint mark on many surfaces, but in the Middle Ages they came to be used, especially by merchants and others who needed to keep easily readable lists, on surfaces coated with special chalklike preparations that enhanced the mark. The alternatives to pen and ink have persisted, and the slate tablet, written on with slate pencil or chalk, is still used in some primitive schools in third world countries. Slate pencils were still sold in America in the latter part of the nineteenth century, and some of the more curiously fluted sticks were no doubt preserved because of their pleasing design and the attractiveness of their gray stone that does not show its age. Even though sharpening slate pencils made a noise that no one wanted to remember, they can be found among the trinkets of antique dealers who would discard a wood-cased pencil of the same era.

For want of a writing instrument many an individual has

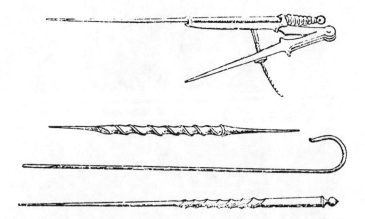

Some lead and silverpoint styluses, including one inserted in a compass

been known to have resorted to unconventional means of recording his thoughts, and the Scottish poet Robert Burns is said to have composed some of his verses by scratching the words into the glass of a windowpane with his diamond ring. While some innkeepers deplored this practice, others welcomed the popular poet, and to this day one hotel in Stirling reminds its visitors that Burns once etched a verse on one of its windows and proudly maintains a Robert Burns Suite, where the poet is said to have been inspired.

Although in time the graphite used in pencil lead would be found to be chemically the same as diamond and no less valuable, to replace a clearly awkward means of writing with a device that was as portable as a ring and yet not nearly as expensive must have been the dream of many a writer through the ages. That dream was answered by an invention that not only needed no liquid ink but also could make a relatively clear and smudge-proof and yet erasable mark on untreated paper that was itself much more portable than a wax tablet, a slate, or a glass windowpane.

Many times the mere articulation of a shortcoming in an artifact implies the means of correcting the shortcoming in a new or improved artifact. If a lump of lead is uncomfortable in the hand, then it should be reshaped to fit comfortably in the hand. If the lump is blocking the writer's view of what is being written, then the lump should be made smaller. But just making a writing implement smaller could make it even more likely to cramp the hand that holds it. So the lump should perhaps be pounded into something that is small at the writing end and yet not so thin elsewhere as to be uncomfortable to hold. Acting on engineering observations as simple as these, an artisan could re-form the lump of lead into the stylus shape of a plummet for finer writing and drawing and into the disk shape of a *plumbum* for scribing lines without blunting easily. And since these changes for the better could be had at little expense, whoever made or supplied the ancestors to our own pencils could not easily have resisted the changes, especially in a free-enterprise society where not to respond to the short-comings of one's product is to invite the competition to capture the market with an improved version of that product.

As for the darkness of the line left by a piece of metal, lead is lead, and its mark cannot easily be darkened by a mere reshaping. One might try to heat the lead or mix it with other materials in an attempt to achieve a darker mark, but such changes involve research and development (primitive as it may seem to be in this example) and can involve an unpredictable expense that may never pay off. Thus the ancient stylus engineer could reply to the ancient stylus user that the mark could not without expense be made darker (unless one wanted to use crumbly charcoal) because no better pencil material than pure metallic lead was known. And perhaps in an attempt to excuse himself for the material's flaw, the engineer might remind the critics of lead of the advantage that its mark could be erased with bread crumbs.

The shape of the stylus could thus evolve much more easily from a lump of lead than could the quality of its mark. With time, however, through thinking and tinkering, through search and serendipity, alloys of lead, perhaps with tin and bismuth and maybe mercury, would be developed that would produce a better mark and, incidentally, give a point that would wear better and be less likely to scratch. As early as the twelfth century, the German monk Theophilus wrote of an alloy of lead and tin being used, in conjunction with ruler and compasses, to lay down a design on a wooden board. Such styluses, consisting of perhaps two parts of lead to one of tin, also came to be known as plummets, regardless of their shape or precise metallic content. However, since Theophilus was, according to Cyril Stanley Smith, "the first man in all history to record in words anything approaching circumstantial detail of a technique based on his own experience," it is not easy to know how long alloyed plummets were around before they were first described.

While implements made of lead, even when alloyed with

other metals, might still be grossly inferior to pen and ink for writing, they would continue to be used into the Middle Ages for ruling lines and margins on papyrus, parchment, and paper, so that the text of the manuscript could more easily be kept straight and uniformly spaced. Apparently the custom became less common sometime in the early fifteenth century, and the lines of manuscripts after that period became "crooked and oblique." But the practice was not entirely forgotten.

Edward Cocker was the seventeenth-century schoolmaster who wrote, among other books, the first English arithmetic directed to commercial applications. The book was so widely known that "according to Cocker" came to be a phrase signifying accuracy. But it was not just account books that Cocker wished his readers to keep straight. Under the title "How to manage and use the Pen," he began his instructions on writing: "Having a Book to write in, or a sheet of paper to write on, which must be ruled with lines with a black-lead Pen or a pair of Compasses . . ."

The pair of compasses, or dividers, which would faintly indent the paper in two parallel lines, had the advantage that the lines could guide the penmanship so that "every Letter would be kept even at head and foot." But at the time Cocker was writing, "black lead" or graphite was still such a recent replacement for the plummet that the name pencil was not yet standard. Yet its mark could be made so much darker'than that of the plummet that the new instrument took not that name but the name of the pen, whose ink the mark of black lead more closely suggested. However dark a line it could make, the idea still was to use the "black-lead Pen" with just a little pressure, so that the guidelines it left were as faint as the "parallel lines marked with the tin stylus or a pen-knife so as not to show black," as one sixteenth-century writing manual put it. The guidelines were not to distract from the text, and it was a long time before black lead could entirely displace plummets and other more primitive writing instruments.

While old lead pencils may be virtually nonexistent, the practice of using faint lines to guide the margins of penmanship may not ever disappear entirely. Wood-cased pencils and plummets coexisted into the nineteenth century, and one writer recalled that in upstate New York in the 1820s "goose-quills were used for copy-book writing, and plummets for ruling; *pencils,* though common, were not necessary to school life." The plummets were often homemade, and a boy could

cut a mold in a pine block to cast molten lead in the form of
axes and tomahawks. These plummets were carried in the
pocket, their sharp edges always at the ready to rule lines to
guide the boy's penmanship. Besides their appeal to a young
boy, the shapes of axes and tomahawks had distinct advantages
over pointed styles of lead. The latter would bend easily under
the pressure needed to draw a line, while the flatter shapes
would better hold their edge. The ancients knew this, and that
is why plummets used to draw guidelines were made in the
shape of the flattened, sharp-edged disk called a *plumbum*.
The Romans also called such a disk a *productal*, meaning
something to draw ahead, and the Greeks called it a *paragraphos*, meaning to write beside, from which the English word
"paragraph" derives.

While the plummet may now have disappeared, in mid-
twentieth-century America, when school desks had inkwells
and penmanship was taught by the Palmer method, pupils still
used a pencil and ruler to mark off margins in their copybooks
before even hoping to dip their pen nibs in the ink. Not so
long ago, among the earliest lessons learned by engineering
students in their first mechanical-drawing class was to use a
hard and sharp lead pencil to rule almost invisible lines to
guide block lettering. Even today tablets of lineless writing
paper can be found with a single lined sheet included. This
sheet not only keeps the impression of the pen from going
through to the next blank sheet but also provides uniformly
spaced guidelines for the pen. According to an Italian writing
manual published in 1540, "the sheet with the black lines is
called by some the 'blind line' or 'show-through,'" and its use
by the beginner was supposed to strengthen the hand until one
could "write well and very confidently" without any lines.

But even if practices change slowly, improvements in hardware do not necessarily wait for discoveries of new and better
pure materials. Long before black lead was generally known
as a marking substance, plummets were altered in response to
actual or anticipated criticism, if not to the whims and imagi-
nations of schoolboys. Some styluses were even made in the
form of thin rods pointed at one end for writing and flattened
at the other end for ruling lines, thus obviating the need for a
separate *plumbum*, or *productal*, or *paragraphos*. Perhaps this
refinement evolved from the idea of a wax-tablet stylus with
one end flattened for erasing, or perhaps it evolved because
monks complained about having to put down one piece of lead

to pick up another, or because it provided an economical use of material. These same reasons have certainly been given to twentieth-century engineers in engineering drawing textbooks recommending the sharpening of pencils differently at each end to have different points for different uses.

Regardless of how it was shaped, as the pure-lead stylus evolved into the alloyed stylus, different stylus makers would unquestionably employ different degrees of alloying with varying degrees of impurities, perhaps depending upon the quality and availability of local raw materials. Thus writers and artists could still complain and argue about the relative merits of this stylus over that. But, generally speaking, by ancient times the stylus had already evolved into an instrument of a convenient shape that could make a reasonably good mark, one that must have looked all the better to those who had never known a modern pencil. And once its mark had been "perfected," less essential features of the writing implement could be the focus of improvement. Thus lead-alloy styluses eventually came to be wrapped in paper holders, perhaps to answer with little expense the complaint that bare metal made the fingers dirty. And this was indeed a complaint as old as the time of Pliny the Elder, for in discussing in his *Natural History* the popularity of gold, the Roman scholar said of the precious metal: "Another more important reason for its value is that it gets extremely little worn by use; whereas, with silver, copper and lead, lines may be drawn, and stuff that comes off them dirties the hand." Paper-cased metallic lead pencils that did not dirty the hand remained in common use well into the eighteenth century, and they did not become extinct before the early part of the present century.

While five small pieces of graphite dating back to 1400 B.C. were discovered in an Egyptian excavation, they are of an impure nature and are believed to have been used for pigment rather than writing or drawing. Less ancient reports of graphite, dating from the year 1400 of the modern era, apparently refer to isolated and inferior European supplies. What finally caused graphite to replace lead alloy as the preferred substance for a dry writing and drawing medium that required no specially prepared writing surface was, quite simply, the discovery around the middle of the sixteenth century of an easily mined abundance of the material that made a superior mark. While it would be centuries before its true chemical nature was known and it was thus properly named, its exceptional properties were noticed shortly after the substance was discovered near the town of Keswick in Cumberland, in northwestern England, in the middle of what is commonly known as the Lake District. The discovery of graphite itself appears to be unrecorded, but it has definitely left its mark on scholarship, engineering, art, and all of civilization.

4/ Noting a New Technology

he modern history of the familiar wood-encased pencil goes back at least four centuries, because a clearly recognizable ancestor of today's pencil is described in a book on fossils written by the German-Swiss physician and naturalist Konrad Gesner and published in Zurich in 1565. Like virtually all scholarly treatises of the time, the book is in Latin, bearing the ponderous title that begins *De Rerum Fossilium Lapidum et Gemmarum Maxime, Figuris et Similitudinibus Liber,* meaning that this is a book on the shapes and images of fossils, especially those in stone and rock. But unlike most other contemporary treatments of natural history, this book is illustrated. And among the illustrations is one showing not a fossil but what Gesner describes as a new kind of stylus or writing instrument, pictured beside a piece of the mineral from which its marking point was made. Not surprisingly, we know much more about the person who used this first known graphite pencil than we do about the pencil itself, its antecedents, or its artificer.

Konrad Gesner was born in Zurich, Switzerland, in 1516, and his precociousness led his father to send him to school in the household of a relative who grew and collected medicinal herbs. The boy learned to read Greek and Latin, and at the age of twenty-one he prepared a Greek-Latin dictionary. His ability in Greek enabled him to earn enough money to study medicine, and he lectured on Aristotelian physics even after becoming a practicing physician. It was quite natural for Gesner to be as interested in a new kind of writing implement as

Aitij puto, quod aliquos Stimmi An-
glicum voca-
re audio) ge-
nere, in mu-
cronem dera
si, in manubri
um ligneum
inferto.

L. Lateres
è luto finguntur & coquunt, ad ædi-
ficiorum parietes, pauimenta, cami-
nos: item ad furnos, aliosq̃ vsus.

Lithostrota dicuntur loca lapidibus strata: vt apud Varronem pauimenta nobilia lithostrota. fiebant au-
tem è crustis paruis, marmoreis præ-
cipuè, quibus folum pauiméti incru-
stabatur. Vide Agricolam libro 7. de
nat. fossilium.

M. Mensæ fiunt nó folùm è ligno:
fed etiam lapidibus & marmore, fiue
folidæ: fiue marmore aut lapide fissili
incrustatæ duntaxat.

Molaris lapidis icon posita est Ca-
pite

The first known illustration of a lead pencil, from Konrad Gesner's 1565 book on fossils

he was in fossils, for his curiosity was boundless. He was an
omnivorous reader and an equally wide-ranging writer and
editor, having about seventy books to his credit, including
works that have earned him sobriquets ranging from "father
of bibliography" through "German Pliny" to "father of zoology." He was said to have been "born with a pen in his hand,"
and he seems to have put it down only long enough to take up
a pencil to make notes for a new book. Gesner died of the

plague in 1565, the year his illustration of the pencil was published.

His other works include a medical tract on the virtues of milk, an account of about 130 languages known in his time, surveys of plant and animal life (heavily illustrated with woodcuts), and a critical bibliography of over 1,800 items, an encyclopedic work in which the recorded knowledge of the world was supposed to be surveyed. Needless to say, works explicitly on engineering were not prominent in Gesner's sixteenth-century bibliography. But for someone whose life must have been so fully occupied with reading and writing, coming across a new and convenient writing instrument must indeed have been as exciting as happening upon a new plant while mountain climbing.

The instrument pictured in Gesner's book looks like a tube of wood with a point of lead inserted in one end and a fancy knob on the other end, where we now expect an eraser. Another of Gesner's illustrations makes it clear that such a knob was used to provide a means of securing a piece of string to a stylus so that it might be tied to a naturalist's field book of tablets, which Gesner referred to as *pugillares,* and the vestiges of such a practice have continued into more recent times. Pencils with knobs and rings on their ends have long been manufactured, including, for example, program pencils for dances or ballot pencils for voting booths. In Victorian times, wooden pencils could be retracted into gold and silver cases to protect the lead as well as their owners' clothes, and the cases invariably had a ring for attaching them to a chain. Workmen and note takers have frequently been known to cut a notch around one end of their pencil, thus making it possible to attach a string and tie the pencil to a desk or clipboard. The modern ball-point pen, which has been described as "no more than an inky pencil," can still be found chained to the counters in post offices and banks.

However, to Gesner it was not the already familiar knob that was really remarkable, but the marking substance inserted in the business end of the tube, thus eliminating the need for any specially prepared surface on which to write or sketch. Gesner says of the object he illustrates only that:

> The stylus shown below is made for writing, from a sort
> of lead (which I have heard some call English antimony),
> shaved to a point and inserted in a wooden handle.

Gesner's illustration of a stylus attached to a set of bound wax tablets, also known as a table book

As common as the lead pencil is today, one can still find modern books on sketching and on engineering and architectural drawing that include illustrations of pencils among introductory discussions of equipment. But these books do not show lumps of graphite and only rarely do they describe the origins of the stuff. It must have been the novelty of the use of that substance when Gesner was writing that led him to describe not so much the pencil as its point. And being accustomed to including woodcuts of new species of plants and animals in his books, it was natural for Gesner to include an illustration of the new writing instrument and substance. Exactly what Gesner's illustration shows is somewhat subject to interpretation, but consistent with the idea that it is the material of which the stylus's point is made that is really novel, the lower nondescript object in Gesner's illustration must be a piece of "English antimony."

Although Gesner's stylus appears to be like a modern mechanical pencil, it really is much more primitive than that. The point of "lead" presumably has been sawn or shaved off the larger piece illustrated and appears to be inserted in a tube that is perhaps in turn inserted into another, larger tube, probably not unlike the way tufts of hair were to make pencil brushes. Alternatively, Gesner's illustration can be interpreted to have some kind of ring compressing what might be a slit tube of wood around the piece of graphite. A small pointed piece of "lead" could easily be held in a tube in either fashion. A properly tapered tube press-fitted into another one and related types of connection to which tightening devices are usually added are in use today. Such devices are used in things as diverse as the telescoping legs of a photographer's tripod, the bit assembly of an electric drill, and, appropriately, the clutch mechanism of a mechanical pencil. The familiar Chinese-finger prison, in which the fingers are trapped when they are pulled apart, is a related device, though one that tightens when it is pulled rather than pushed, and the old penholder, into which the nib is forced and held by friction, is still another variation on the same basic idea.

But whatever the exact mechanism for holding the point in the invention illustrated by Gesner, this represented a considerable improvement over a piece of metallic lead or lead alloy wrapped in paper. Now one could not only produce a darker line but also have a clean, convenient, and comfortable instrument (which could be tied to a traveler's or a rock climber's sketchbook or notebook) in which to hold "leads" of different shapes and sizes, even leads too small to be held in the fingers. This would be an important consideration if the supply of the wonderful new marking substance were to become scarce and expensive and available only in small pieces, as it would from time to time.

In form and function, the marvel that Gesner described is clearly what we today call a lead pencil, and he treated it as a curiosity presumably because, at least to him, it was a new, improved portable writing instrument "from a sort of lead (which . . . some call English antimony)." This stuff would make a good mark on common paper and thus obviate the need for any naturalist to take into the field either *pugillares* in which to scratch with a metal-pointed stylus or a cumbersome and messy pen and inkwell—and the related paraphernalia—with which to record the fossils, flora, and fauna found in all sorts of inconvenient rock formations and other natural settings.

While Gesner's appears to be the first illustration of a modern pencil, it is not the very first reference to one. In 1564, a year before Gesner's book was published, Johann Mathesius wrote of a then new discovery for writing: "I remember . . . how one used to write with silverpoint . . . and now one writes on paper with a new unrefined mineral." But this unspecific reference falls far short of the thousand words needed to equal Gesner's picture, and so neither it nor its author is remembered the way Gesner's illustration is. And neither Mathesius nor Gesner exactly dates the first appearance of the pencil, for they are only reporting on something already in use.

Gesner's illustration was reprinted, enlarged, in a book published in 1648. This was a posthumous continuation of the encyclopedic *Musaeum Metallicum* of Ulisse Aldrovandi, the sixteenth-century naturalist whose work was a "more complete but less critical compilation than that of Gesner." Aldrovandi, also writing in Latin, did not call the pencil's essential ingredient *"stimmi Anglicanum,"* but rather *"lapis plumbarius,"* or lead stone, but the inclusion of an illustration of the pencil indicates that it could still be considered remarkable—and the names for its marking material still multifarious—in the mid-seventeenth century.

While references to the pencil do document when it already existed, they do not reveal the precise date of its first appearance. But while the absence of references to the pencil does not prove its nonexistence, the silence of certain books can suggest it. For example, a book published in 1540 by the Italian writing master Giovambattista Palatino contains a description and illustration of what he claimed to be "all the tools that a good scribe must have." While a pair of compasses and a metal stylus are included for their use in marking guidelines for the pen, there is not a hint that the author had even dreamed of a piece of graphite, by any name, or a pencil made from it. Thus it may be assumed with some confidence that, at least in Italy in 1540, neither the graphite pencil nor its marking substance was known.

Exactly when and where pencils containing graphite were first made and used appears to be unrecorded, as are so many technological milestones. There are undocumented assertions that place the discovery of the graphite that Gesner refers to as early as about 1500 and as late as 1565, the date of his book. But the scanty evidence generally points to the unearthing of the pencil's marking substance—the "new unrefined mineral" or "English antimony"—as sometime in the early 1560s in Cumberland. However, in his *History of Inventions and Dis-*

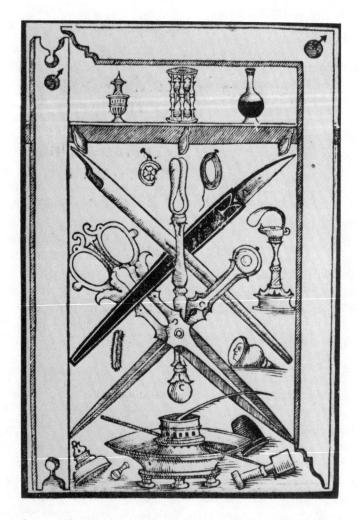

A 1540 illustration showing "all the tools that a good scribe must have," but showing no lead pencil

coveries, John Beckmann wrote near the end of the eighteenth century that he was "unacquainted with the time when the pits in Cumberland, which, as is well known, produce the best plumbago, were discovered." The Latin word *plumbago,* which means that which acts like lead, was only one of many names for the curious material, which Beckmann noted was also "called *black lead, kellow* or *killow, wad* or *wadt,* which words properly mean black."

As could be expected to be the case in naming a material whose value lay in its replacing other materials, many other names, including "black-cowke," "kish," and "crayon noir," were commonly used before the functionally descriptive and scientifically accurate name "graphite," derived from the Greek *graphein,* which means "to write," was suggested by A. G. Werner a full decade after the true chemical nature of the substance was finally determined by K. W. Scheele in 1779. The earliest names for the material, some of which appear in Beckmann's account, ranged from the inexplicable to the obvious. The traditional local name dating from the sixteenth century for the then new material found in the Cumberland hills was "wad" or "wadd," and in dialectal English that word also came to mean a graphite pencil, with the term "wad pencil" being used well into the twentieth century in the vicinity of the plumbago pits. Wadd was also called *nigrica fabrilis,* "for its use in scoring," as late as 1667 because it then still had no universally agreed-upon Latin name. A short communication to the *Philosophical Transactions* for May 1698, entitled "Some Observations Concerning the Substance Commonly Called, Black Lead," which mentions some of the above names, shows that well over a century after its discovery there was still much uncertainty about the nature of graphite:

> The Mineral Substance, called, Black Lead (our common Lead being the true Black Lead, and so called, in Opposition to Tin, which is the White Lead) found only in *Keswick* and *Cumberland,* and there called *Wadt* or Kellow . . . is certainly far from having any thing of Mettal in it, that it has nothing of Fusion, much less Ductility; nor can it be reckoned amongst the Stones, for want of hardness; it remains therefore that it must have Place amongst the Earths, tho' it dissolve not in water . . .

Being uncertain about how to classify graphite, the author of the note concludes, tentatively, with a confusing justification that "the most Proper Name that can be given it, perhaps, may be *Ochra Nigra,* or Black Ochre."

However, since the wadd behaved so much like metallic lead, it eventually came to be more widely called by the Latin term *plumbago.* And, of course, the name black lead was a natural description in English for something that made a much blacker mark than real metallic lead, though the color of the shiny new substance itself was not so unambiguous. The German word adopted for black lead was *Bleiweiss,* which translates literally as "white lead," and this word stems from an early "misconception of graphite being a shiny white lead-type metal," perhaps akin to tin. Today, of course, "white lead" is used to designate a poisonous paint pigment containing metallic lead carbonate. But by whatever name, the history of discoveries and uses of black lead was of more than academic interest for the "diplomatic science," at least according to Beckmann:

> To ascertain how old the use of black lead is for writing might be of some importance in diplomatics, as the antiquity of manuscripts ruled or written with this substance, or of drawings made with it, could then be determined. What little I know on this subject I shall here communicate, in order that others may be induced to collect more.
>
> I allude here to pencils formed of that mineral called, in common, *plumbago* and *molybdaena,* though a distinction is made between these names by the new mineralogists. The mineral used for black-lead pencils they call *reissbley, plumbago,* or *graphites.* . . . Plumbago . . . contains no lead; and the names *reissbley* and *bleystift* have no other foundation than the lead-coloured traces which it leaves upon paper. These lines are durable, and do not readily fade; but when one chooses, they may be totally rubbed out. Black lead, therefore, can be used with more convenience and speed than any coloured earth, charcoal, or even ink.

While Beckmann's pioneering scholarship may have induced others to collect more, there does not seem to have been very much more to collect. At least, not much more has been collected in the ensuing two centuries. In the early twentieth century debates did arise about the earliest pencil markings on manuscripts, with C. T. Schönemann claiming that lines drawn in black lead appeared on an eleventh- or twelfth-century codex in a German library. This was disputed by C. A. Mitchell, whose pioneering microscopic investigations of the nature of pencil marks turned up none in British museums produced earlier than the seventeenth century, thus defending the dating of the discovery of pure graphite in Cumberland.

Among the most recent histories of the area that became the center of English pencil making and the source of raw material for early pencil factories throughout Europe is Molly Lefebure's book *Cumberland Heritage.* In the introduction she is blunt about the task of tracing the origins of the principal ingredient of the lead pencil to a fell, or hill, where shepherds once roamed:

> The recorded history of wadd is patchy and largely unreliable, as is the case, too, with Keswick's famous pencil industry. Seatoller Fell [where the ancient wadd holes are], both literally and figuratively, is mist-distorted: distant rocks, looming through the fog like Peruvian pinnacles, dwindle, when approached, into stones a few feet high; Greenland bears some sheep; wadd, an established commercial commodity for over 300 years, proves amazingly elusive. The quest for its story . . . developed for the author into an almost Alice-like adventure; the nearer she drew to wadd, the further did wadd recede.

In her chapter entitled "In Quest of Wadd," Lefebure writes quite frankly about her frustrating scholarly search for the history of the discovery and the early exploitation of the material:

> Reading up wadd one discovers that most of the authorities are merely repeating the words of a previous writer; thus one digs one's way downwards through a slag-heap of endless (and sometimes erroneous) repetition.
>
> The wadd, according to legend, was discovered originally by shepherds, after a large ash-tree on the fellside (an alternative version of the tale gives it as an oak) had been uprooted by a gale. The date of the discovery is unknown. When first found the substance was simply used by the local people for marking their sheep (continues the legend).

One alternative version of the story of how the first graphite mine was discovered in the Cumberland manor of Borrowdale does not claim the scholarly "reading up" of Lefebure's work but does make a good story that is not quite as tentative. "The Pencil," an essay by Clarence Fleming published in a booklet originally issued in 1936 by the Koh-I-Noor Pencil Company, begins:

> The uprooting of a large oak tree during a storm, led, it is said, to the discovery of the famous graphite mine of Borrowdale, England. This was in 1565, in the time of Queen Elizabeth. A wandering mountaineer, attracted by

the particles of a strange, black substance clinging to the roots of the fallen tree, soon had the people of the countryside excitedly discussing the mysterious mineral.

The year 1565 is actually, of course, the date of publication of Konrad Gesner's book. Even Lewis Mumford, in the list of inventions appended to his famous *Technics and Civilization,* also dates the introduction of the lead pencil itself as 1565 and appears to credit Gesner with its invention. But this is certainly more than the naturalist claimed for himself, and all that seems certain is that plumbago was available and widely appreciated, especially by naturalists and artists. In 1586, for example, the English antiquary and historian William Camden could write of Borrowdale: "Here also is found abundance of that mineral earth, or hard shining stone, which we call Black-lead, used by painters in drawing their lines and shading."

Queen Elizabeth had encouraged new industries during her reign, and the help of experienced Germans was sought to develop the mining and smelting of various ores in several English counties, including Cumberland. The Germans were involved with mining in Keswick and its environs by the late 1560s, and it is very possible that it was through them that English graphite found its way to the Continent. It may also be a result of their sudden exposure to an abundance of metals in England that the Germans confused the English names for tin, "white lead," and wadd, "black lead," or developed their own reasons for calling graphite *Bleiweiss,* but the true reason, like the true date of the discovery of graphite, may never be known. Flemish traders and Italian artists have also been credited with introducing English graphite to Europe, and in Italy it was known as "Flemish Stone" or "Flanders Stone" because it came to southern Europe via Belgium and the Netherlands. Whatever its route to getting there, graphite was well known throughout Europe by the end of the sixteenth century, and in 1599 the Italian natural historian Ferrante Imperanti wrote of *grafio piombino* that "it is much more convenient for drawing than pen and ink, because the marks made with it appear not only on a white ground, but, in consequence of their brightness, show themselves also on black; because they can be preserved or rubbed out at pleasure; and because one can retrace them with a pen, which drawings made with lead or charcoal will not admit."

As Gesner's reference made clear decades earlier, however, it was not only artists who were interested in using black lead, and there soon developed various means of holding pieces of it for writing as well as for drawing. Just as metallic lead had been wrapped in paper, so rough pieces of wadd, perhaps straight from the mine, were wrapped in sheepskin, and stylus- or pod-shaped pieces of the raw material were wrapped in paper or string, thus keeping the fingers clean. Rod-shaped pieces of graphite could also be pushed into the end of a hollow twig or reed. A series of short pieces could be inserted into a piece of straw bound around with string, which could be unwound and peeled away as the point wore down in much the way we peel the wrapper off a diminishing pack of mints today. Such natural cases for black lead as vine twigs must not have been uncommon, for even in the present century the term "vine" was still in use for a pencil in parts of Cumberland and County Durham.

A pointed piece of wadd wrapped in string

By the seventeenth century, Borrowdale lead was widely exported. In Germany, where it was regarded as a mixture fused with antimony, it was also known as "bismuth." And with the true chemical understanding of Borrowdale wadd still almost two centuries away, in 1602 Andrea Cesalpino, the Italian physiologist and innovative botanist, wrote of it by still another name, apparently confusing its places of application, where "pointed pencils were made of it for the use of painters and draftsmen," with its place of origin: "I think also that molybdenum is a certain stone shining with black color like lead, so slippery to the touch that it seems to have been polished, which comes off on the hands of anyone who touches it with an ashen stain shining like lead: painters use it in little sticks put into tubes; it comes from Belgium." Cesalpino also noted that some "say they find it in Germany, where they call it bismuth."

By 1610 black lead was sold regularly in the streets of London, for artists and others to fit into their wooden pencil cases, or perhaps for those who did their fieldwork in books in their studies to wrap in paper or string or to insert in twigs. By 1612 not only the blackness of the mark but also its removability were oft-repeated features of Borrowdale lead, and one writer, commenting on making notes in printed books, recommended for those "which you would have faire againe at your pleasure" to "note them with a pensil of black lead; for that you may rub out againe when you will, with the crums of new wheate bred."

The reputation of Borrowdale wadd continued to spread throughout the seventeenth century. Black lead was in great demand everywhere, and as its use grew so did the development of devices for holding it in a clean and convenient way. A holder, called by its French name of *porte-crayon,* had claw-like grips that could be locked in place to hold unrefined pieces of the black lead (or chalk or charcoal, for that matter), and M. C. Escher's haunting *Drawing Hands* are sketching each other with pencils held in metal *porte-crayons*. Since *crayon* unqualified is the French (and English) word for virtually any dry drawing or writing medium, the name *crayons d'Angleterre* came to distinguish black lead from other media.

A wooden *porte-crayon*

As the means to use it became more common and the popularity of the unique product of the Cumberland deposit spread, so did word that the readily stolen and disposed-of commodity was a source of quick money. Wadd was a strategic resource to be protected, and when stockpiles were adequate for England's purposes, the Borrowdale holes would be ordered closed for years at a time, even being flooded to ensure that no wadd could be removed. This precaution was taken because for a period the mine was being worked only six or so weeks every five or six years, for that concentrated effort was sufficient to dig out what graphite was needed. The Borrowdale mine remained closed from 1678, when it was believed to be almost worked out, until 1710, when new lodes were sought. Upon opening the mine it was discovered that pilferers had been at work during that period. Toward the end of the eighteenth century the mine would again be producing poorly, with only about five tons of inferior graphite mined in 1791.

It is not known exactly when the easily concealed wadd began to be smuggled wholesale from the mine, but the practice eventually grew to such proportions that elaborate security and legal steps had to be taken. Military and other uses of graphite, such as for casting "bomb shell, shot and cannon balls," must have been in part responsible for causing a bill to be introduced in the House of Commons entitled "An Act for the more effectual securing Mines of Black Lead from Theft and Robbery," and making it a "felony to break into any mine or wad-hole of wad . . . and steal any." The bill received three readings, was considered by a committee of the whole, and was passed by the House of Lords, to be made law on March 26, 1752, when His Majesty George II, in full regalia on his throne and with the robed Duke of Cumberland at his side, pronounced, "Le Roy le veult."

5/ Of Traditions and Transitions

*T*he history of the pencil, when it has been written down at all, is full of erasures and revisions. Perhaps this is inevitable, for history begins in storytelling, and storytellers seem to want mostly to relate what seem to be the most interesting things in the most interesting way. This is not to say that storytellers deliberately mislead, but they certainly do select their subjects and their words, and they can reject not only those subjects that do not seem right but also those that do not seem at the time to be necessary or important or apt to some larger story being told. Storytelling is also writing, and most writers must recognize the truth about revising manuscripts that Truman Capote alluded to when he said, "I believe more in the scissors than I do in the pencil." And Vladimir Nabokov articulated the same idea with a different image: "I have written—often several times—every word I have ever published. My pencils outlast their erasures."

But for all the writing with a pencil, there is little written about it. Who has read about the pencil what has been written about the pen? Who has read about engineering what has been written about science? Who has read about factories what has been written about cathedrals? But that is not to say that what is seldom celebrated is not relevant to the history of civilization. In one prosaic account, a German pencil firm wrote of its product: "Few articles have contributed more to the spread of the arts and sciences. Not one can be named so nearly universal in its daily use. Our very familiarity with it tends to make us regard it with indifference." Yet the sketch goes on to claim, perhaps more for self-justification than historical perspective, that "like most articles that depend for their existence upon mechanical skill, the Lead-Pencil is entirely a product of modern times."

The long, slow development of the pencil into the seventeenth century parallels the story of contemporaneous engineering. Much of what was achieved up to the century before the Industrial Revolution was ostensibly an emulation of ancient practice. Although the great Gothic cathedrals had been erected, the Roman arch remained the norm for stone bridges. Although Stonehenge and the pyramids had been standing for millennia as monuments to mechanical advantage, Galileo was just asking anew questions that the Peripatetic philosophers had raised but not fully answered. By the end of the seventeenth century, not only would Galileo and Newton have laid the foundations for modern science and engineering, but the lead pencil would achieve its present form. While science and engineering may not have needed the pencil to advance, its development would eventually benefit greatly from increasingly scientific approaches to the practice of engineering. But that would not occur before late in the eighteenth century.

There are several reasons why it took so long for such a common and seemingly simple artifact as the modern pencil to evolve from the lead stylus. While it was easy to criticize the stylus, it was not all obvious to the early craftsmen and artists who produced styluses how to make them better—even barring the expense. Before the discovery of a truly suitable material for the marking medium, attempts to improve the stylus by the use of alloys were frustrated, and the basic instrument was only gradually modified by the development of specialized forms for specialized purposes. The primitive, prescientific understanding of the principles of chemistry that prevailed meant that there was little theoretical foundation on which to build or formalized engineering practice within which to make fundamental changes in the materials of the pencil. While trial-and-error methods would have served as well as they do even today, without the availability of suitable materials or methods to work them there were not even opportunities to try.

Once black lead was discovered in Cumberland, however, the development of the pencil could accelerate. Even though the true chemical nature of graphite was not known, the makers and users of pencils in the sixteenth and seventeenth centuries could concentrate on how to adapt the ideal material in the most convenient and efficient ways for writing and drawing. Since the plumbago was initially in the form of lumps taken from the Borrowdale mines, all efforts were essentially devoted to devising means of shaping pencil leads out of the lumps of wadd and enclosing the leads in holders. As the product became known and demand for it grew, pencil makers and vendors naturally multiplied and competition intensified.

With growing competition, criticism of pencils was easier not only among competitors but also among consumers, for there were now examples of different styles of pencils to contrast and compare. As artists and draftsmen who would use pencils became more discriminating, their criticisms especially had to be answered or, if the expense was not too great, removed. This dynamic process is essentially self-propagating and self-regulating and results in the development of better pencils, or can at least produce less expensive ones. More or less the same process is followed in all modern engineering development, but in the early days of the pencil it was left largely to artisans and craftsmen to evolve changes in response to forces of imagination and creativity, of art and craft, of supply and demand, of criticism and economy. And as long as the ideal material remained readily available, the ideal pencil should naturally have been expected slowly to emerge out of the tradition of the craftsmen who would come to specialize in pencil making.

While pencil makers and engineers, as opposed to historians, may have chosen now and then to write about their processes and products, the makers and doers have seldom written history, or at least creditable and objective history that has been also a good story well written and therefore memorable. Rather, what we have generally inherited from the early days of pencil making and the early days of modern engineering, at least up to the nineteenth century, are snatches of stories, snippets from the scissors that have escaped the eraser. This is due in part, of course, to the fact that engineers of the Industrial Revolution, like early pencil makers, were more craftsmen than scholars. They thought of and with their pencils, and more often than not what they thought became unimportant to them once they had realized it in the form of artifacts. Their scissors and erasers might have been metaphorical, but most likely it was the improved artifacts that were the new objects of thinking with a pencil that displaced the old and rendered them forgettable. On many backs and bottoms of old pieces of furniture we can find the pencil calculations and sketches of their craftsmen, but the pencilings are usually incomplete, elliptic, enigmatic. The pencil work of the engineers on their great structures and machines, the furniture of civilization, is no less generally hidden from view, if it is preserved at all.

It is certainly not just the invention and development of the physical pencil that are unrecorded. While the actual use of Borrowdale graphite grew throughout the seventeenth century, exactly how the black lead was held for writing and drawing and what it was called by those who used it are subjects of much speculation and little documentation. What passing references there are do not give a complete picture, as when a character in Ben Jonson's play *Epicoene,* which was produced in 1609, describes the contents of someone's box of mathematical instruments: "his square, his compasses, his brass pens, and black-lead to draw maps." Some references later in the century, including one in 1644 by the English diarist John Evelyn, refer to the use of a "black-lead pen" for drawing. In 1668 a distinction between instruments was made in a book entitled *The Excellency of the Pen and Pencil,* and it named "black-lead pencils" among "the necessary instruments pertaining to drawing." But while the pen could be used for either drawing or writing, the pencil by and large was an instrument of drawing, thinking, and noting. Among artists and writers, the pencil could be a guide for the pen, but more often it came also to be seen by readers as a follower, being used to jot erasable notes in the margins of books.

Getting at the history of the pencil has not been made any easier by the proliferation of confusing names for its marking substance that persisted even into the nineteenth century, and even what was called a pen and what a pencil at any given time is not easily pinned down. It was not only in English that the pencil and the names for it were multifarious and confusing, of course, and in his book *The Mastery of Drawing* the art historian Joseph Meder heads his chapter on graphite with a list of the names by which that medium has been referred to in German, Italian, Dutch, French, and English—the languages of the masters: *"Bleistift, Blay-Erst, Wasserbley, Blei, Bleifeder, Englisch Bleyweiss, Reissbley—Grafio piombino, Lapis piombino—Potlot, Potloykens—Mine de plomb d'Angleterre, Crayon de Mine de plomb, Crayon de mine, Crayon— Black-lead Pencil, Plumbago."*

Meder goes on to lament the fact that "in catalogues, dictionaries, and other art publications there is no other drawing medium that has occasioned so many errors and misunder-

standings as the 'lead' pencil." While Meder's concerns were
with the use of the pencil for drawing rather than for writing,
there has been no less confusion in distinguishing metallic lead
and its alloys from the true graphite pencil used on manuscripts. Such confusion is no wonder, considering that it remained for the early-twentieth-century microscopic and chemical investigations of C. A. Mitchell to explain how one might
scientifically distinguish between the line of a metallic-lead
point and the line of a piece of graphite. According to Mitchell, when viewed in side illumination under the microscope,
lead and its alloys leave a line with a characteristic luster and
striations that are absent from the uniformly distributed
masses of black pigment left by ordinary graphite. Furthermore, different types and mixtures of graphite may be distinguished by such close examination, thus enabling pieces of
suspected pencil writing or drawing not only to be confirmed
but also to be dated. Mitchell found graphite writing in notebooks dating from 1630, but his analysis certainly could not
tell how the black lead was held.

The marks of, top to bottom, modern graphite pencil, lead
point, lead-tin point, lead-bismuth point

Whatever the circumstances of the discovery of graphite and
its use in the earliest pencils, the idea of sticking a piece of
black lead into the end of a tube, as illustrated in Gesner's
work on fossils, is akin to that of putting some hairs or bristles
into the handle of a brush. And since "pencil" was principally

the name for a brush of a pointed sort, it was natural for that word to replace "stylus" as the name for Gesner's remarkable object. But the evolutionary process from what Gesner illustrated to what today is so common that it is virtually an invisible and ignored object was long and arduous in name and deed. In fact, as late as 1771 the *Encyclopaedia Britannica,* which claimed in its subtitle to be a "dictionary of the arts and sciences," could still totally ignore the black-lead instrument in its definition of "pencil," which reads in its entirety:

> PENCIL, an instrument used by painters for laying on their colours. Pencils are of various kinds, and made of various materials; the larger sorts are made of boars bristles, the thick ends of which are bound to a stick, bigger or less according to the uses they are designed for: these, when large are called brushes. The finer sorts of pencils are made of camels, badgers, and squirrels hair, and of the down of swans; these are tied at the upper end with a piece of strong thread, and inclosed in the barrel of a quill.

Why there is no mention of the fact that the term "pencil" means more than this is as confusing as why there is no entry for "pen." "Wadd," which by that time also had important uses in making cannonballs, was defined only in terms of a wad of paper stuffed in a gun barrel to keep the shot from rolling out, but the dialectal nature of the old term might excuse its omission. No such explanation can excuse the sole identification of "plumbago" as, "in botany, a genus of the pentandria monogynia class." It appears that the increasingly common materials and instruments of drawing and writing were as neglected as was the growing importance of engineering outside the military, for that same encyclopedia allows an engineer to be only "in the military art, an able and expert man."

The word "pencil" continued into the nineteenth century to designate an artist's brush, as it does to this day. The pioneer photographer Fox Talbot was referring to that meaning when he called his early camera work the "pencil of nature," and today's dictionaries still give the first meaning (chronologically) of "pencil" as "a brush of hair or bristles used by artists to lay on colors." So, like all revolutionary artifacts, the earliest black-lead pencil had ahead of it a long period not only of physical but also of etymological evolution. Both would go largely unrecorded and generally unobserved except among pencil makers and special users.

Whatever the writing instruments were to be called, in the seventeenth century there apparently were already pencil makers, as opposed to dealers in uncased Borrowdale black lead, in the Bavarian town of Nuremberg. But whether these were the first pencil makers, or whether, as the Cumberland Pencil Company claims, that distinction belongs to producers closer to the Borrowdale mine itself, in the English town of Keswick, may never be known for certain. According to one writer, local legend has it that Keswick was "making pencils" in Elizabethan times, which would be consistent with the ready supply of material. But that is not to say that all pencils were bought already fabricated. As late as 1714 a London broadsheet could depict an itinerant hawker whose cry suggests that he did not sell fabricated pencils as we know them, but rather still peddled the marking substance itself, suitable for the purchaser to insert in an appropriate holder:

> Buy marking stones, marking stones buy
> Much profit in their use doth lie;
> I've marking stones of colour red,
> Passing good, or else black Lead.

While pieces of black lead may have been sold directly to the citizens of London, to be wrapped in string or held in porte-crayons, wood-cased pencils seem also to have been available before the end of the seventeenth century. Exactly what the earliest cottage-industry pencils looked like is not certain, but it appears that toward the end of the century the word and the instrument itself were beginning to take on a modern aspect. In a 1683 book on metallurgy, Sir John Pettus wrote under the heading "lead":

There is also a MINERAL LEAD, which we call BLACK-LEAD, something like ANTIMONY, but not so shining or solid, of which sort I know but of one MINE in England, and this yields plenty, both for ourselves and other nations, and this mine is in Cumberland, which they open but once in seven years, (I suppose the reason is, least they should dig more than they can vend), this also is used by PAINTERS and CHYRURGEONS &, etc., with good success, especially being mixed with

Buy marking ſtones, marking ſtones buy
Much profit in their uſe doth lie ;
I've marking ſtones of colour red,
Paſſing good, or elſe black Lead.

A London hawker of black lead, from a 1714 broadsheet

the products of metals; and of late, it is curiously formed
into cases of DEAL or CEDAR, and so sold as dry PEN-
CILS, something more useful than PEN and INK.

So by the end of the seventeenth century the idea of the
wood-cased pencil of black lead had definitely been formed
and apparently black lead was the central ingredient in an
artifact that was sold as a product of manufacture rather than
as something one assembled oneself from raw materials. Sir
John also makes it clear that at the time he was writing there
were at least three distinct users of black lead: painters, surgeons, and writers. Painters used it, presumably in place of
charcoal, most frequently to prepare preliminary sketches
whose lines were erased or inked or painted over. Surgeons
and others who treated the sick presumably used it for medicinal purposes, as minerals and mixtures of minerals had been
applied since ancient times, as reported extensively in Pliny's
Natural History. In his 1704 *Essay Towards a Natural History of Westmorland and Cumberland,* Thomas Robinson
makes more explicit this second use of wadd:

> Its natural Uses are both *Medicinal* and *Mechanical.* It's
> a present *Remedy* for the *Cholick;* it easeth the Pain of
> [the urinary disorders] *Gravel, Stone,* and *Strangury;*
> and for these and the like uses, it's much bought up by
> *Apothecaries* and *Physicians. . . .* The manner of the
> Country People's using it, is thus: First, they beat it small
> into *Meal,* and then take as much of it in white Wine, or
> Ale, as will lie upon a *Sixpence,* or more, if the distemper
> require it.
>   It operates by *Urine, Sweat,* and *Vomiting.*
>   At the first discovery of it, the Neighbourhood made
> no other use of it, but for marking their *Sheep;* but it's
> now made use of to glazen and harden *Crucibles,* and
> other *Vessels* made of *Earth* or *Clay,* that are to endure
> the hottest *Fire* [and] by rubbing it upon iron arms, as
> guns, pistols, and the like, and tinging them with its
> color, it preserves them from rusting.

Thus instead of worrying about "lead poisoning," as some
in the twentieth century still wrongly do about the leadless
lead pencil, people two centuries ago thought the central ingredient to have curative powers. It was, according to Robin-

son, even "bought up at great Prices by the Hollanders" ostensibly to make dyes, but Beckmann believed that purpose to be only "a pretense," and he was "inclined to think that they prepare from it black-lead pencils."

Whether ruses or not, medicinal and textile uses are less relevant for the history of pencil making than mechanical applications of wadd. Its importance in glazing and hardening crucibles would continue to be an important industrial application as the manufacture of iron and steel grew throughout the Industrial Revolution, and the name of at least one famous pencil manufacturer, the Joseph Dixon Crucible Company, would remind us that the same raw material, by then known to be graphite, could be an essential ingredient in products as diverse as crucibles for molten metal and pencils for molding thoughts.

But back in the late seventeenth century, it was the third use of black lead, in a "curiously formed" writing instrument, that Sir John singled out as then "of late" and "something more useful than pen and ink." Apparently the "curiously formed" wood casing had evolved over the previous century. And just as Gesner found the lead pencil a wonderful instrument to use for sketching in the field, so did others regard "pens of Spanish lead" useful for freeing horsemen from juggling pen and ink in order to make notes while in the saddle. Whether this late-sixteenth-century reference to "Spanish lead" indicates a Continental source of graphite or merely another location for pencil making does not seem to be recorded.

As great as were the advantages over pen and ink of the earliest lead pencil, consisting of a piece of graphite stuck in a tube, it no doubt had some disadvantages. Depending upon exactly how the case held the black lead in place, it might have easily worked itself loose or slid back into the case or fallen out —all frustrations we can still experience today with a cheap wood-cased or a malfunctioning mechanical pencil. Or the piece of lead may have been too thick to make a fine enough line. To recognize such shortcomings of the new device would not have seriously diminished contemporary appreciation of it, for any faults of the first black-lead pencil, whose marking qualities were so great an improvement over those of metallic lead, must certainly have been accepted as the best available under the circumstances. Cicero's rule of reason governed the pencil Gesner praised in the mid-sixteenth century no less than it governs technological gadgets in the late twentieth century. We would all like a better mousetrap, but that is not to say that we do not appreciate the ones we possess in the meantime.

While technology has not always worked with the speed of the electronic age, the pencil a naturalist could rave about in 1565 was dramatically improved upon by the end of the seventeenth century by a product that called attention to the faults of its predecessor, perhaps having pencil users go so far as to ask how they ever got along with the first primitive product. As much of a marvel as the first lead pencils must have been, cutting off a sliver or slice from a block of graphite must have been no easy task for the novice pencil user.

By having rods of graphite enclosed in pieces of pine or cedar, with the help of some glue, the graphite could be held firmly within the casing and exposed as the wood was whittled away with a knife. (Since artists and writers were used to shaping their quill and reed pens with a penknife, sharpening a pencil in this manner was nothing new and would not have been seen as a particular disadvantage of the pencil.) And the wood casing, which gave the real strength to the composite shaft, allowed the graphite to be slender enough to be formed into a fine point.

According to local tradition, a Keswick joiner first developed the idea of enclosing rods of graphite in wood. The product appealed to a clergyman, and he had some of the wood-cased pencils made for his friends, thus spreading the invention abroad. Other tradition places the invention of the "technique of glueing rectangular graphite rods in wood" in Nuremberg, where at first this was "an exclusive right of the carpenter's guild," but as early as 1662 Friedrich Staedtler was identified as a specialist, a "pencil maker." Wherever the idea originated, it seems easy to believe that its realization emerged from the woodworking craft of joiners, for the ability to shape and assemble rather small pieces of wood, not to mention cutting or sawing small pieces of graphite from odd-shaped chunks, was a necessary skill in executing the idea. Indeed, an early-nineteenth-century practical treatise on mechanical arts distinguished joinery from carpentry principally as the "art of working in wood, or of fitting various pieces of timber together, for the ornamenting of certain parts of edifices," and also noted that the French called joinery *menuiserie,* or "small work." But the execution of the idea of fitting graphite in wood must have been far from obvious, except perhaps to the imaginative joiner who first tried it. Even today, after we have grown up with the wood-cased pencil, one of the most commonly asked questions about it remains: "How do they get the lead into the pencil?"

The original process appears to have been as follows. Pure graphite was cut into thin slices of a roughly rectangular shape from the lumps of wadd taken directly from the mine. A desirable slice might be about one-eighth inch thick, one inch wide, and as long as possible. (Some of the edges might even have retained the irregular shape of the lump of wadd as found in the mine.) A strip of wood, about one-half inch wide, three-eighths inch thick, and six or seven inches long—approximately the length of and almost as thick as the finished pencil—was grooved lengthwise with a saw, with the width of the groove matching the thickness of the sheets of graphite. The longest straight side of a piece of graphite was dipped into glue and then inserted into one end of the groove. The graphite protruding was then sawn off or, more likely, scored like a piece of glass and broken off where it projected out of the groove. Since the graphite did not fill the length of the groove, a second piece of graphite was then inserted, butted up against the first, and also sawn or broken off. The process continued until a line of black lead almost filled the case. The piece of wood and exposed graphite were then planed flat. Glue was spread on this surface, and a strip of wood about one-quarter inch thick and one-half inch wide was clamped on to cover the lead. When the glue was dry, the square assembly could be used in that form or shaped into a finished pencil more comfortable to hold. Some of the first pencils were believed to

Steps in making an early wood-cased pencil from natural graphite

have been in the shape of an octagonal shaft of wood surrounding a square lead, but a good joiner could just as easily have
formed the first pencils into a hexagonal, round, or any other
shape.

An early pencil, with square lead in an octagonal wooden case

This method of enclosing pure plumbago in a wood case
was a valuable development because it gave pencil makers a
relatively efficient means of using in each pencil only small
slips of the Borrowdale material, which was becoming increasingly dear, but all the sawing and planing also left a considerable amount of unusable small pieces and lots of graphite dust.
No other black-lead deposit had been found to contain such
high-quality material, and it was becoming obvious that the
Borrowdale mines were not an inexhaustible source. The production of the mines was controlled once pilfering was made
punishable as a felony, but since wadd was well worth stealing
and easily could be sold, mine workers were searched as they
exited the mine to minimize unauthorized removal of the raw
material. Oral tradition has it that "a mouthful [of wadd] was
as good as a day's wage."

The export of black lead from England was discouraged,
except in the form of pencils. This gave a considerable advantage to British manufacturers, many of which were located
within about ten miles of the Borrowdale mines, in Keswick
and its environs, even though they had to buy the material at
auction in London, where it had been transported to under
armed guard.

Continental producers were forced to look for alternatives
to cutting their pencil leads directly out of pure Borrowdale
plumbago. The cutting process had actually always been
wasteful, but as long as the raw material was abundant this
had been accepted. However, when English black lead became
increasingly difficult to obtain at any price, and with European
supplies far inferior in quality and purity, Continental manufacturers were forced to engage in research and development
to find alternative ways of making pencil leads.

As early as 1726 the need to conserve Borrowdale wadd
was felt, and graphite dust and the smallest pieces of poorer-

quality plumbago were being used in a reconstituted form. This practice is implicit in a contemporary account of his process by a pencil maker in Berlin:

> The lead cutter pounds the graphite in a mortar and by sifting two or three times, frees it of all earthen particles such as sand. In a crucible, to every pound of graphite a fourth or a half pound of sulphur is added, melted and thoroughly mixed. After cooling and before it is quite dry, the mass is placed on a board and kneaded just as one would knead bread. This must then become fully cooled before it can be further worked. With a fine saw, the pencil-maker divides this cake into small plates from which he saws the four cornered leads of desired size. The workman cuts the wood to the requisite size, and forms the groove for the lead either with a grooving plane or by burning it out with a red hot iron tool. The lead is glued in the groove and a piece of wood glued over it to complete the pencil. The end of the pencil showing the lead is shaped to a neat point with a file. The entire surface of the pencil is then carefully finished by scraping with glass. It is evident that the lead pencil-maker, to show any profit, must complete the pencils in a short time, for a dozen costs but eight groschen.

This description also makes it clear that, unlike the pencils of today, the lead of an eighteenth-century pencil did not necessarily extend through the entire length of the wood case, for there was an "end of the pencil showing the lead" and, therefore, by implication, an end that did not show any lead. This was, of course, another way of conserving black lead with no real inconvenience to the pencil user. After the pencil had been filed or sharpened down to a very short stub it would have become difficult to hold anyway, and so the leadless wooden end would have been discarded for a new pencil. This practice even continued into the nineteenth century, as a passage from Jane Austen's novel *Emma* makes clear. Harriet is showing Emma secret treasures she has kept because of their association with a man she admired. First Harriet shows Emma a piece of court plaster (something like an old Band-Aid), which she had saved because the man had used another part of it to dress the finger he had cut while using Emma's penknife. And then she shows another treasure, "the end of an old pencil,—the part without any lead." It was a "superior treasure," because it had actually belonged to the man, who had wanted to make some notes on brewing spruce beer, but "when he took out his pencil, there was so little lead that he soon cut it all away." Emma had lent him another pencil, and the prized empty stub "was left upon the table as good for nothing." Harriet wanted to burn the treasures, but Emma distinguished between the two, confirming the uselessness of the leadless stub of wood: "I have not a word to say for the bit of old pencil, but the court plaister might be useful."

Although the ideal wood used to encase pencil lead would itself become scarce in the twentieth century, it was the graphite that was the focus of economy two hundred years ago. European graphite contained such a high level of impurities, which would scratch and tear the writing surface, that the mineral had to be pulverized first to separate impurities from it. In attempts to use graphite dust and powder to stretch the available supplies, they were mixed with binding agents such as gum, shellac, wax, and isinglass (a kind of gelatin made from fish bladder membranes and used in adhesives), in addition to sulphur, but these produced leads that were difficult to sharpen or use without breaking, and they did not produce a mark of very good quality. There was thus a revival of the use of metallic lead alloyed with various amounts of tin, silver, zinc, bismuth, antimony, and mercury, but the success of such efforts appears to have been only fair when compared with pencils made of pure Borrowdale graphite. Nevertheless, pencils, whether inferior or not, had become well established in the marketplace.

The second edition of the *Encyclopaedia Britannica,* which was published serially between 1777 and 1784, repeated verbatim the first's definition of a pencil as a painter's brush, adding only: "All good pencils, on being drawn between the lips, come to a fine point." But then the *Britannica* went on to record a newer meaning of the term:

PENCIL, is also an instrument used in drawing, writing, &c. made of long pieces of black-lead, or red-chalk, placed in a groove cut in a slip of cedar; on which other pieces of cedar being glued, the whole is planed round, and one of the ends being cut to a point, it is fit for use.

Black-lead in fine powder, stirred into melted sulphur, unites with it so uniformly, and in such quantities, . . . that though the compound remains fluid enough to be poured into molds, it looks nearly like the coarser sorts of black-lead itself. . . .

On this principle the German black-lead pencils are said to be made; and many of those which are hawked about by certain persons among us are prepared in the same manner: their melting or softening, when held in a candle, or applied to a red-hot iron, and yielding a bluish flame, with a strong smell like that of burning brimstone, betrays their composition; for black-lead itself yields no smell or fume, and suffers no apparent alteration in that heat. Pencils made with such additions are of a very bad kind; they are hard, brittle, and do not cast or make a mark freely either on paper or wood, rather cutting or scratching them than leaving a coloured stroke.

The true English pencils . . . are made of black-lead alone, sawed into slips, which are fitted into a groove made in a piece of wood, and another slip of wood glued over them: the softest wood, as cedar, is made choice of, that the pencil may be the easier cut; and a part of one end, too short to be conveniently used after the rest has been worn and cut away, is left unfilled with the black-lead, that there may be no waste of so valuable a commodity. These pencils are greatly preferrable to the others, though seldom so perfect as could be wished, being accompanied with some degree of the same inconveniences, and being very unequal in their quality, on account of being fraudulently joined together in one pencil, the forepart being commonly pretty good, and the rest of an inferior kind. Some, to avoid these imperfections, take the finer pieces of black-lead itself, which they saw into slips, and fix for use in portcrayons: this is doubtless the surest way of obtaining black-lead crayons, whose goodness can be depended upon.

Since this article was composed for the second edition of the *Britannica,* we can assume that it reflects the state of the pencil-making business around the 1770s. Yet while the article does contain explicit descriptions of the ingredients and methods of assembly of pencils, there is not enough detail for anyone without some prior experience to set up from scratch a pencil factory using inferior graphite. There are no precise formulas, for example, for mixing black lead and other substances to produce the best leads. In modern terminology, no trade secrets were revealed; nor were any likely to be found in such a publication.

But what the encyclopedia article does make incontrovertibly clear is what was wrong with the pencils of the day. Indeed, it is almost a catalogue of faults, and any reader, pencil maker or pencil user, could have derived a considerable set of criteria against which to judge the quality of a product or a purchase. The criticisms made explicit in the *Britannica* would no doubt be familiar ones to the pencil makers of the time, both scrupulous and unscrupulous, and the scrupulous ones no doubt would love to have answered the criticisms either by giving reasons why no better pencil could be made or by improving the pencils they did produce, if that could be done without making the pencils so expensive that no one would want to purchase them.

While wood-cased pencil making had begun as an offshoot of carpentry and cabinetmaking, by the time shortages of "so valuable a commodity" as black lead had begun to be felt, specialization even within this already specialized business was emerging, as the Berliner's clear distinction between the operations of cutting the lead and shaping the wood suggests. Indeed, in the early eighteenth century, pencil making was considered a distinct industry from general cabinetmaking, and the pencil continued to be developed into something produced more and more by a division of labor and of a relatively high technology for its time. What started as a cottage industry was to develop into a broadly commercial endeavor, with all the attendant difficulties associated with competing interests from guilds, governments, and foreign competitors. Toward the end of the eighteenth century, pencil making, like a lot of contemporary industries, was definitely emerging from the era when a pencil maker could "carry the week's production to town in a basket."

6/ Does One Find or Make a Better Pencil?

A lead pencil today might seem to be but a piece of graphite cleverly enclosed in a case of wood, but it actually involves an exacting process employing a multiplicity of raw materials. And the materials required for its manufacture make the pencil an object that depends on the most modern and cosmopolitan of political, economic, and technological systems. The lead in a single American-made pencil in the late twentieth century, for example, might be a proprietary mixture of two kinds of graphite, from Sri Lanka and Mexico, clay from Mississippi, gums from the Orient, and water from Pennsylvania. The wooden case would most likely be made of western incense cedar from California, the ferrule possibly of brass or aluminum from the American West, and the eraser perhaps of a mixture of South American rubber and Italian pumice stone.

While the world political climate can, as it has at times, severely affect the supply of necessary exotic ingredients, pencil manufacturers long have been resourceful in finding alternatives. Some have gone so far as to anticipate shortages by stockpiling graphite, or by gaining sole rights to the contents of a newly discovered mine or an island of cedar, or by planting whole forests of trees for their pencil wood alone. But even with guaranteed and unlimited supplies of raw materials, it is no trivial matter to process and assemble them so that they still provide a lot of pencil at a little price, for domestic and foreign competitors are a constant threat. The sophisticated use of materials is at the heart of all modern technology, whether it involves pencil or automobile or computer manufacturing. It is an ever-changing use that is dependent upon supply and demand, and it is nothing new.

Toward the end of the eighteenth century, pencil making, like much other technological activity, was becoming polarized between the British and the Continental approaches, and the differences were largely due to the availability and quality of graphite. This raw material was easily mined and could meet all demands once a pure and plentiful source was found, for as an article on "Borrodale" in the 1816 edition of the *Encyclopaedia Perthensis* stated, it was the black lead found in the valley's Derwentwater Fells "wherewith all the world is supplied."

Yet while all the world may have been supplied by the Borrowdale mines, they were by no means the only source of graphite at the opening of the nineteenth century, as another article in the same encyclopedia concisely states:

LEAD, BLACK, OF PLUMBAGO, . . . is found in different countries, as Germany, France, Spain, the Cape of Good Hope, and America; but generally in small quantities, and of very different qualities. The best sort, however, and the fittest for making pencils, is found in Cumberland, at a place called *Borrowdale,* where it abounds so much, that hence not only the whole island of Britain, but the whole continent of Europe, may be said to be supplied.

What seems to have caused Britain to shift back and forth, in the decades before and after 1800, between hoarding and exporting Borrowdale plumbago, was at least in part a matter of the exhaustion of old and the discovery of new finds. The article on black lead goes on to elaborate on the reasons the Borrowdale kind was superior, thus no doubt commanding a premium price on the world market, and we get a clue as to how there can be sudden changes in the supply. According to one Magellan, who in the article is quoted on the qualities of various sources of black lead,

I have seen various specimens from different countries; but their coarse texture and bad quality cannot bear any comparison with that of Borrowdale; though it sometimes, but seldom, contains pyritaceous particles of iron. It is but a few years ago, that this mine seemed to be almost exhausted; but by digging some few yards through

the strata underneath, according to the advice of an experienced miner, whose opinion had been long unat-
tended to, a very thick and rich vein of the best black lead
has been discovered, to the great joy of the proprietors
and advantage to the public.

Throughout the ebb and flow of the world supply, the very
best of English pencils could continue to be made with una-
dulterated native Borrowdale graphite, as long as the supply
of it was protected. But some extreme measures had to be
taken to ensure against the chronic problem of thievery at the
mine. According to an 1846 amendment to Beckmann's history of black lead:

> At present, the treasure is protected by a strong building,
> consisting of four rooms upon the ground floor; and im-
> mediately under one of them is the opening, secured by a
> trap-door, through which workmen alone can enter the
> interior of the mountain. In this apartment, called the
> dressing-room, the miners change their ordinary clothes
> for their working-dress as they come in; and after their
> six hours, post or journey, they again change their dress,
> under the superintendence of the steward, before they are
> allowed to go out. In the innermost of the four rooms two
> men are seated at a large table, sorting and dressing the
> plumbago, who are locked in while at work, and watched
> by the steward from an adjoining room, who is armed
> with two loaded blunderbusses. In some years the net
> produce of the *six weeks'* annual working of the mine has,
> it is said, amounted to from £30,000 to £40,000.

As a result of such protective measures, there was little
incentive among local pencil makers to develop the process
beyond the more or less static craft practice that it had become. Indeed, there was little incentive to modify the process
of producing pencils at nearby Keswick until toward the middle of the nineteenth century, when the mountain was clearly
fast becoming empty once and for all. But on the Continent,
where uncertain supplies of expensive Borrowdale and inferior
graphite from other sources were the rule long before they
were in England, a revolutionary process for making lead was
to be developed that is used to this day. And this technological
innovation was essentially forced by external social and political conditions, as well as the economic pressures of the supply

and demand of good graphite, rather than by any special initiatives of the growing but still craft-based pencil industries in England and Germany. Necessity was indeed to bring forth invention, but there were also technical factors that made successful innovation possible—and an accident that chose the time.

What forced the situation for pencil making in the 1790s was the unavailability in France of Borrowdale graphite, or *plombagine*. War broke out between, among others, France and Britain in 1793. Not only could France not obtain high-quality British plumbago; it could not even get the inferior but serviceable German pencils then made out of graphite dust mixed with sulphur and glue. Since war, revolution, education, and day-to-day commerce cannot get along easily without pencils, the French Minister of War, Lazare Carnot, sought a substitute method to produce them in France. At the time, Nicolas-Jacques Conté was a thirty-nine-year-old engineer and inventor whom the Revolution had led to abandon a career as a popular portrait painter for one in science. By the early 1790s he had established a solid reputation in many fields, and Gaspard Monge, the first director of the École Polytechnique, said of Conté that he had "every science in his head and every art in his hands." Thus it should not be surprising that someone like Conté was commissioned by Carnot to develop an alternative to using pure Borrowdale graphite for pencil leads, but it remains difficult to disentangle the facts from the fancy in the story of the crash research and development program itself. Conté promoted the military use of balloons, and he was apparently working on some experiments with hydrogen when an explosion injured his left eye. His experiments with graphite proved to be less dangerous. It was supposedly within a matter of days in 1794 that Conté found success, and within a year, in early 1795, he was granted a patent for his new process.

Conté's innovative process appears to have grown out of his familiarity with using plumbago to make crucibles in which to melt metals. The new way of making pencil leads consisted of mixing finely powdered graphite, from which impurities were removed, with potter's clay and water and rubbing the wet paste into long rectangular molds. When the leads were dry, they were taken from the molds, packed in charcoal, sealed in a ceramic box, and fired at a high temperature. Since the brittle ceramic leads could not easily be planed flat, they were inserted in wooden cases of a modified design, one used by some early German pencil makers to encase their sulphur-and-graphite leads. The piece of wood into which the leads were placed had a groove about twice as deep as the thickness of the rod of lead. A slat of wood was then glued in over the lead to completely fill the groove, and the pencil was ready to be finished to the desired exterior shape. Fine artists' pencils made by the new French process came to be known as *crayons Conté*.

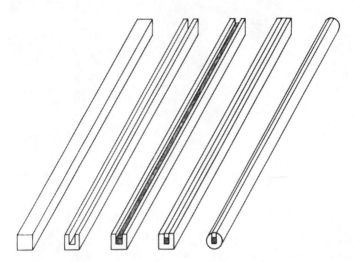

Steps in assembling an early Conté pencil

It has sometimes been claimed that the Conté process was discovered independently by Josef Hardtmuth, a Viennese mechanic, as early as 1790, but that appears merely to have been the date at which his pencil factory was founded. Hardtmuth himself claimed to have invented the process only in 1798, three years after the date of Conté's patent, but other sources state that the new process was not exploited in Vienna until it was carried there much later by Conté's son-in-law, Arnould Humblot, who also made several improvements in his father-in-law's "primitive process," and eventually became head of a pencil factory in Paris.

However and whenever discovered, once the process could be mastered, it had all the potential of freeing Continental pencil makers from their dependence on high-quality English graphite. At first applied only in France and Austria, and then later at a state-owned pencil factory in Passau, Bavaria, by the middle decades of the nineteenth century it was to be widely employed and remains the basis for pencil-lead making today.

The Conté process was apparently even used to some extent in England early in the nineteenth century, but the traditional way of making English pencils was not to be displaced completely until the Borrowdale graphite finally did run out, some say in 1869, though the exact date seems to be as difficult to pinpoint as the date of discovery.

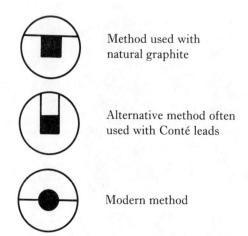

Method used with
natural graphite

Alternative method often
used with Conté leads

Modern method

Different ways of enclosing pencil lead in wood, with the present method for encasing round leads

Even though the Conté process gave European pencil makers a certain independence from the diminishing supply of Borrowdale graphite, it was generally believed that the new pencil lead could not write nearly as well as the best of English graphite. However, when the choice is between a very good pencil and no pencil at all, or between a prohibitively expensive pencil and an affordable pencil, concessions can be made. Besides, *crayons Conté* were far superior to those made with sulphur, and by varying the proportions of clay and graphite, pencil leads with different but uniform degrees of blackness could be produced, something for which English pencils could not always be counted on.

What distinguished Conté's improvement from the original way of making pencils out of pure graphite was its deliberately innovative nature, something that could not be said with so much certainty for the earlier stages in the development of the lead pencil. The traditional story of how the wonderfully pure Borrowdale graphite was accidentally discovered when a storm uprooted a tree, and how the hard lumps of strange black earth at first were used for marking sheep, is an account not of purpose but of receptiveness to whatever unrefined gifts the good earth might be willing to give up. Whether or not the story of the discovery of graphite is literally true, its insertion into the oral and written tradition of the pencil-making industry itself indicates that it was consistent with a climate of passive acceptance of things as they were and not of active prospecting for a writing material with the best properties.

But whether looked for or not, the wadd mined at Borrowdale became increasingly important and valuable as its unique qualities became evident. And those qualities were not only that it made a better mark on paper than metallic lead did but also that it was strong enough to be fashioned into convenient slips or rods that could be pointed and yet would not break so easily under the pressure of writing and drawing. Furthermore, while the traditional beginnings of the modern lead pencil are traced to the discovery of Borrowdale wadd, the fact that chunks of graphite, albeit not very pure ones, have been found in Egyptian excavations makes one wonder why the use of graphite as a writing material does not seem to have occurred before the middle of the sixteenth century. Indeed, it has even been said that graphite was actually discovered near Passau quite some time prior to its use in pencils. Apparently the material's ability to make a mark was recognized, and so the name "plumbago" may even have been first applied here. But it appears not to have been used as a marker because the graphite did not occur in abundant and convenient lumps of very high purity, and so it could not easily be formed into good sticks for writing. The material thus seems to have remained little more than a local curiosity. Apparently there were no deliberate attempts, and indeed there may not have been any significant attempts at all, to refine and re-form the Passau plumbago into an efficacious composition or shape for writing and drawing.

What was it that enabled Conté in France in 1794 to achieve success on a grand scale whereas others elsewhere at other times had not? There was plenty of intellectual prospecting going on during the scientific revolution, of course, but there was little claim jumping between material and philosophical works. Often the same prospector divided his efforts among the separate tunnels he drove into an unmapped mountain, yet barely getting under the surface of things and more often than not trying to infer the mountain out of his mole hole. Even the great Newton, whose interests were not without a practical dimension, could not always get beyond his tunnel vision. But he did not think of himself as a miner, for he likened his life to a walk along the seashore, where he was overjoyed to find a prettier shell now and then. He was even saddened by the fact that, as for the shell seeker strolling on the shore, "the great ocean of truth lay all undiscovered before him." When he did prospect, Newton looked not for a better material with which to make a pencil but for a philosopher's stone. While his library contained an astonishing variety of books, in fields as diverse as medicine and mathematics, he seems not to have greatly interleaved their contents. Although he had some practical interests, at heart Newton was not an engineer, and the bulk of those who would stand on his shoulders appear to have had little inclination to be engineers.

The Newtonian paradigm, the ideal of the seeker of truth standing on the shoulders of giants and therefore being able to see farther than any of the giants beneath him on the pyramid, is a paradigm of an objective and fully formed truth just over the horizon, a truth that one can see only by getting a higher vantage point. Achieving this presumably also redefines the horizon, and then one lends one's own shoulders to the next generation. The idea is consistent with Newton's metaphor of the seashore, in which the goal of mental activity is again suggested to be something objective and already out there, just waiting to be discovered. The seashore metaphor allows that one shell (theory) may be prettier (more elegant) than another, and perhaps that the searcher becomes less fond of the old shells as prettier ones are found, but the implication is still that the truths are whole in the ocean and it is just a matter of time before they are found thrown up upon the shore.

While pencils may be helpful in formulating abstract theories of motion and gravitation, abstract theories do not make pencils. And one will not discover black earth that writes and draws if one is kicking sand on it on one's way to fetch a prettier shell. In short, while the likes of Newtonian thought would contribute to the scientific basis of modern engineering, it alone was of little use in perfecting artifacts that were far from being regarded as proper objects of consideration for shell collectors.

In its native state, plumbago is not nearly so pretty as a shell lying on the beach or a planet rising above the horizon. Indeed, it only became an object of interest and value when its potential for making better pencils and crucibles and cannonballs became evident, just as black earth can be but dirt to all but the farmers who know of the nutrients it contains and how they are essential to feed the seeds of the next crop.

While it certainly seems that the modern pencil depends upon the accidental discovery of highly pure plumbago, it is not inconceivable that the idea for a pencil could have preceded any discovery of a material with the necessary qualities to realize the idea, just as manned flight was a dream long before it was a reality. Even had it turned out that there was no plumbago to be found, it might have been the case that the pencil could have been developed from some substitute material. Indeed, the fact that alloys of lead were used for writing and drawing suggests that there was some purposeful quest for an alternative to pen and ink. However, it does not seem that such quests were necessarily conducted with all the vigor we can imagine today, for through the sixteenth and even through most of the seventeenth century the paradigm that Newton finally made explicit and even personified seems to have been dominant.

But by the eighteenth century—when the foundations of modern chemistry were being laid, when plumbago was being properly identified as a form of carbon to be forever after known as graphite, and when the idea of the pencil and the need for the pencil were well established—it was also the case that a more pragmatic paradigm was gaining currency among practical engineers as opposed to theoretical scientists. In such a climate it was not only possible but also probable for an engineer like Conté to be asked what kind of material could be used as a substitute for pure but unavailable Borrowdale graphite. The theretofore impracticality of using impure graphite and the theretofore inferior means of mixing graphite dust and sulphur with wax to make pencil leads were challenges to Conté to prospect not in the earth but in his mind for the means to make a pencil with what was overlooked. And what Conté did also fit a non-Newtonian paradigm that would be epitomized in Edison trying numerous materials until he found the right one for the lamp filament that had lit up as an idea in his mind long before it was a reality. The fragile incandescent lamp was certainly not something Edison expected to find thrown whole upon the seashore.

Innovation, ingenuity, and inventiveness have always existed, as witnessed by the technologial advances of the oldest civilizations and by the machines and engines described by such ancient writers as Vitruvius and Hero of Alexandria, and later in the drawings of Leonardo, but a systematic approach to engineering, uniting craft and science to develop artifacts from dreams, took centuries to develop. While designing a pencil may seem to be easier than designing a bridge, in fact it can be even more difficult and uncertain. The principal requirements of a bridge are that it be strong enough and stiff enough to carry a particular load and that it do so economically and reliably for as long as the bridge is kept in good repair. How long that may be depends upon various decisions about the trade-offs between strength and cost, between durability and cost, between maintenance and cost—in short, on economic decisions made throughout the bridge's history as an idea and as an artifact.

Like a bridge, a pencil must have economy, strength, and durability. But these demands are of a more subtle and less calculable kind. The lead of a pencil must be made of a material that not only has the qualities of strength and durability but also leaves a good mark on paper. Borrowdale graphite, fortuitously and fortunately, occurred in pieces large enough to be cut into leads that had the proper combination of strength and marking qualities. But Borrowdale graphite dust mixed with sulphur and wax did not, for pencil points made of it tended to be soft in warm weather and did not make for very smooth writing.

Conté's great contribution to pencil making was in keeping with the French scientific engineering spirit that had been emerging in the widening and overlapping wakes of Galileo and Newton. This spirit recognized that the craft tradition could not be relied upon as the sole source of inspiration for its own new ideas. While the crafts were believed to be common, they also held indispensable information; it was from investigating their products and processes systematically and analytically that new benefits to mankind would come. The spirit was articulated in Denis Diderot's monumental *Encyclopédie,* the first volume of which appeared in 1751. Expressing a view reminiscent of Galileo's opening of his *Dialogues Concerning Two New Sciences,* the entry on "craft" was written by Diderot himself:

> CRAFT . . . This name is given to any profession that requires the use of the hands, and is limited to a certain number of mechanical operations to produce the same piece of work, made over and over again. I do not know why people have a low opinion of what this word implies; for we depend on the *crafts* for all the necessary things of

life. Anyone who has taken the trouble to visit casually the workshops will see in all places utility allied with the greatest evidence of intelligence: antiquity made gods of those who invented the *crafts;* the following centuries threw into the mud those who perfected the same work. I leave to those who have some principle of equity to judge if it is reason or prejudice that makes us look with such a disdainful eye on such indispensable men. The poet, the philosopher, the orator, the minister, the warrior, the hero would all be nude, and lack bread without this craftsman, the object of their cruel scorn.

At least one subsequent poet, D. H. Lawrence, would not look with disdain on the things craftsmen once made with their "wakened hands." To him, artifacts made centuries earlier could be still warm with the life put into them by their forgotten artificers. But others, back in the eighteenth century, did not want merely to feel the warmth of forgotten craftsmen. Like Galileo, the Encyclopedists recognized that there were among those practicing the crafts a small number of the first rank from whom much was to be learned. In Jean d'Alembert's "Preliminary Discourse" to the *Encyclopédie,* we find an observation that might be taken, incidentally, to describe a still-uncommon breed in the eighteenth century, "artists who are at the same time men of letters." But since "most of those who pursue the mechanical arts, took up their particular trades out of necessity and practice them by instinct," and some of whom "worked for forty years without knowing anything about their machines," it was the uncommon craftsman who became an engineer.

Whether craftsmen "know" anything about their machines or can articulate what it is they do in the practice of their craft, the craft itself can involve a complexity that no amount of theory in itself can reveal. This late-eighteenth-century recognition of an implicit and nonverbal tradition in the crafts was as slow to be widely recognized as it was to be exploited for accelerating technological advancement. The odds were clearly against rapidly occurring innovation within the crafts or activities of "sordid toil," as Agricola had described the common view of the mining industries of the sixteenth century. Except for the "dozen in a thousand craftsmen" who would also be men of letters or the men of letters who might have made themselves apprentices, it would appear that the typical craftsman would make "over and over again" the same bridge or pencil or the typical miner would dig where miners had always dug. And, conversely, the typical man of letters, or modern scientist for that matter, wishing to innovate would be hopelessly theoretical and derivative of what was already in the library. Engineers, by being in the tradition of curious and articulate craftsmen, even if only apprentices, and at the same time practical and experienced scientists, are thus the natural innovators.

Conté was able to make a revolutionary change in the manufacture of pencil leads because of his joint interest, in the spirit of Galileo, in matters both practical and theoretical. Before he had been approached by Carnot, and perhaps Carnot approached him precisely for this reason, Conté had already shown interest in graphite—as a refractory material for use in making crucibles and cannonballs, not to mention as an artist's medium. The apparent separation of such narrow craft activities as crucible making and pencil making in virtually everyone before Conté afforded little opportunity for any applications of the emerging scientific method to developing improvements in the craft of the one based on experience in the other. Conté could make a quantum leap in thinking about how to fashion a pencil lead out of graphite dust and clay because he was already familiar with the way those materials combined to produce excellent crucibles, broken fragments of which incidentally might act as marking stones, or so Conté might have noted in his tinkering in the laboratory.

The laboratory is really the modern workshop. And modern engineering results when the scientific method is united with experience with the tools and products of craftsmen. While it would emerge more slowly in Britain and America, modern engineering, in spirit if not in name, would come to play a more and more active role in turning the craft tradition into modern technology, with its base of research and development. And in the century following Conté this transformation would take place in virtually every aspect of technological life from common pencil making to monumental bridge building.

7/ Of Old Ways
and Trade Secrets

The slow improvement of the pencil and of pencil-making processes through the eighteenth century, up to the revolutionary discoveries of Conté in 1794, was made through what might be termed at best prescientific and primitive engineering practices. The lead pencil illustrated by Gesner merely mimicked the ancient metal stylus and the brush by inserting "English antimony" into a wooden tube, the way a piece of metallic lead or a tuft of animal hairs was stuck into the end of a hollow reed or twig, a leap of the imagination no more modern than that required to fashion the wooden-handled steel ax once the principle of the tomahawk was known.

Inserting Borrowdale graphite into more sophisticated wood cases could be expected to be the joiner's natural inclination. Pieces of metal had long been inserted into wooden frames and holders to make basic woodworking tools, such as planes and chisels. The seventeenth- and eighteenth-century joiners were naturally familiar with even more fundamental specialized knowledge, such as the qualities of woods required for holding a piece of graphite firmly and yet allowing the case to be whittled away without splitting or splintering. It was no accident, nor should it be considered surprising, that some of the first wood-cased pencils were made of cedar, the wood still believed to be the ideal one for pencils. Experienced joiners knew the properties of woods, and they knew the requirements of a case for graphite that would make a good pencil. The joiners would also have known well the techniques of gluing pieces of wood together, and they would have been able to shape a pencil to a fine finish as easily as they could turn a spindle. In short, the joiners had all the skills and experience necessary to fashion as good and convenient a wood-cased pencil as could be made out of the available raw materials. What joiners and carpenters were not able to do easily, however, was to make good pencils when good graphite or good woods were not available.

While it may long have been the dream of carpenters to reconstitute timber phoenix-like out of sawdust, it is not likely that such an achievement could ever have reached beyond mythology or wishful thinking for any but the most imaginative of old woodworkers. Mixing graphite dust with sulphur, gum, or glue and kneading the mass to re-form chunks of solid graphite from which square leads could be cut for insertion into wood would be equally beyond the basics of joinery. When the exceptional joiner did venture beyond the routine, there was no guarantee that there would be any satisfactory results. Indeed, up until the last decade of the eighteenth century, the pencil was made within a craft tradition that was generally so stifling as to inhibit rather than to promote true innovation. To develop a new means of making pencil leads required ambition and experiment beyond joinery.

How technological innovation can be both born and stifled within the tradition of the crafts and trades can be illustrated by the early development of pencil making in Germany. English graphite may have been known there not long after its discovery at Borrowdale via the German miners who were working in the Lake District since 1564. Whether it was through this connection or through Flemish traders, encasing *Bleiweiss* in wood was an established activity in Nuremberg by the 1660s. But the prevailing guild system was so rigid that it encouraged neither competition nor innovation, as Friedrich Staedtler learned in 1662.

Staedtler was the son of an immigrant wire drawer, but perhaps because of confusion surrounding the Thirty Years' War the young man did not learn his father's trade and instead became a shopkeeper in Nuremberg. He married the daughter of a joiner, and from his father-in-law learned the elements of that trade, including the joiners' craft of mounting graphite in wood. At the time specialization was so narrow that the *Weiss-macher,* joiners who made small wooden articles like sewing boxes and toys, did not even cut their own graphite but bought it from a *Bleiweiss-Schneider,* or white-lead cutter.

It was Staedtler's entrepreneurial idea to engage in pencil

A Continental "white-lead cutter" in a 1711 illustration, showing that the German terminology derives from the early misconception that graphite was a shiny white-lead type of metal

making exclusively, and not to diffuse his efforts on novelties and toys. Over the objections of his father-in-law and other joiners, he applied to the Nuremberg Town Council for permission to make *Bleiweiss-Steffte,* but the application was turned down because the city's Trade Inspection Board held that mounting graphite rods in wood was the exclusive right of all joiners, and further specialization was not allowed. Apparently Staedtler persisted, however, for official records identify him as a pencil maker on the occasion of the birth of his first daughter. Subsequent church records make further references to pencil makers, and by 1675 Friedrich Staedtler and others of his occupation, *Bleistiftmacher,* were evidently well enough established in the community for him to be granted the citizenship he had been refused earlier. Perhaps Staedtler's independent and persistent personality also kept him from forever making pencils the way traditional Nuremberg joiners did. The young pencil maker began to cut his own graphite and thus carry out all the major operations associated with pencil making in a single workshop. As this activity grew, pencil making was officially recognized as a separate trade, and pencil makers would form their own guild in 1731.

Friedrich Staedtler has been credited with experimenting in the late seventeenth century with mixing pulverized graphite with molten sulphur, thus utilizing graphite dust to form what then would have been considered an artificial pencil lead. Such a practice not only would have employed otherwise wasted material but also would have allowed poorer-quality graphite to be refined before being used in pencil making, and both these practices would have made Staedtler less dependent upon an increasingly scarce imported material.

Staedtler's pencil-making expertise was handed down to his children, and then to his grandchildren, and thus a family pencil-making dynasty was begun. Other pencil-making families got started in Nuremberg in the late seventeenth and early eighteenth centuries, and in the records of the city's *Rugsamt,* the Trade Inspection Board, for 1706 all but two authorized pencil makers belonged to the families of Staedtler, Jenig, and Jäger. While these family businesses may have been started by imaginative and innovative entrepreneurs like Friedrich Staedtler, most seem to have been inherited by less imaginative and less forward-looking children and grandchildren, and their pencils continued to be made by and large the same way year after year, generally in disregard of the need for or the appearance of technological developments that might produce better or less expensive pencils. Staedtler was the oldest pencil-making family to survive in business into the nine-teenth century, when a new era of pencil making was to begin.

In the meantime, the Staedtler family business, like those that did not survive, struggled in large part not because of fair competition from better pencil makers but from the unfair practices of a number of renegade pencil makers, sometimes known as *Stümpler,* or "bunglers," who worked outside the city limits and thus outside the control of city statutes and guild codes. The bunglers generally could offer pencils at low prices because they did not contain expensive materials. Rather than use solid graphite of high purity or even finely pulverized graphite carelessly mixed with inferior binders, the bunglers made pencil leads only good enough to sell the pen-cils or put very small amounts of reasonable-quality pencil leads into wood cases. A good piece of pencil lead might ex-tend only an inch or so into the case, for example, with the balance filled with the poorest-quality lead, if it was filled at all. Some unscrupulous pencil makers, who apparently did not worry about their family's or their firm's name, sought to stretch the supplies by a trompe l'oeil, for "the trade was (sometimes) supplied with mere useless pieces of wood, simi-lar to a lead pencil in appearance, but simply blackened at each end with graphite to imitate the lead running through the wood."

The practice of imprinting registered trademarks on pencils would develop as a means of guaranteeing to the trade, the shopkeepers, and also to the individual purchaser, that the pencil being bought had some legitimate claim to quality. But for some time there was opposition to pencil makers putting their marks on pencils, lest the ultimate purchaser learn who was making the best pencils and thus skip the middleman.

Apart from the question of trademarks, the competition of the bunglers in Germany was made all the more difficult to combat by the restrictive practices imposed upon the guild members to protect themselves from each other. It was only when the imperial city of Nuremberg became Bavarian in 1806, with the subsequent dissolution of the Trade Inspection Board, that a measure of free enterprise allowed pencil makers like Staedtler to expand.

Paulus Staedtler, the great-great-grandson of Friedrich, was responsible for the business during this time, and he called himself a *Fabrikant,* or factory owner, and his workshop a factory. Paulus Staedtler was officially a master pencil maker, and he served as a foreman supervising many people who had no license to manufacture pencils themselves. While this system may have produced pencils in quantity, the need to manage a large force of very narrowly skilled workers did not encourage much experimentation with new products or processes, and thus virtually no research or development was being carried out. As children inherited the family business and even as new pencil-making dynasties began, German pencil making was making little progress in evolving the pencil beyond a brittle, scratchy stick of reconstituted graphite in wood.

Marriages among pencil-making families often consolidated businesses, but they seem to have done little to encourage new blood in the industry. The parish register in the little village of Stein, near Nuremberg, holds records of marriages between "pencil makers" and "black-lead cutters," for both men and women were involved in the manufacturing process, but not all new pencil-making concerns came of marriages. In 1760, Kaspar Faber, a craftsman, settled in Stein, and the following year hung out a shingle there announcing his new pencil-making business. Faber's pencils were produced in his cottage, and at first the weekly output was small enough to be carried in a hand basket to be sold in Nuremberg and Fürth, another nearby village. When Kaspar Faber died, in 1784, his son Anton Wilhelm became the sole owner of the business and gave it the name by which it would become known worldwide, A. W. Faber.

The discovery by Conté in France of the process of mixing graphite dust with clay produced a pencil far superior to any made in the late-eighteenth-century German way. The ceramic lead of the Conté process had writing qualities that approached those of English pencils in ways that the graphite-and-sulphur German ones could not match. Furthermore, the French pencils were as strong and smooth as the German ones were brittle and scratchy. And finally, the French pencils could be made in a variety of hardnesses by varying the proportions of graphite and clay, and the different grades could be made uniform throughout the pencil, something not easily found even in some pencils of pure English graphite.

In the meantime, English graphite was becoming harder and harder to obtain, and so even the better German pencils made of pure graphite could not be manufactured in any significant numbers. The German inability to react quickly to the new technological development in making pencil leads with clay caused the industry to fall upon bad times, and the inertia of the German system of manufacturing allowed things to go from bad to worse during the quarter century following Conté's discovery.

Among the contributing factors in the decline of the German pencil industry in the early part of the nineteenth century was the tradition of trade and guild practices, perhaps compounded by deep-seated family biases in favor of their inherited systems of manufacture. In such an environment too strict an adherence to trade secrets can play the paradoxical role of delaying the mastery and acceptance of beneficial new developments. A climate of secrecy also discourages the free exchange of ideas and leaves very little in the way of written documents for historians of technology. But such frustrations are neither new nor unique to pencil making. The American mining engineer who became President, Herbert Hoover, and his wife, Lou Henry Hoover, in their translators' introduction to Georgius Agricola's sixteenth-century treatise on mining, *De Re Metallica,* made the following observations:

> Considering the part which the metallic arts have played in human history, the paucity of their literature down to Agricola's time is amazing. No doubt the arts were jealously guarded by their practitioners as a sort of stock in trade, and it is also probable that those who had knowledge were not usually of a literary turn of mind; and, on the other hand, the small army of writers prior to this time were not much interested in the description of industrial pursuits.

Such practices did not end with Agricola, and such sentiments were not new with the Hoovers. In the preface to his treatise on the various trades, arts, and manufactures of the early nineteenth century, the civil engineer Thomas Martin wrote in the third person of his difficulties in seeking information for his book, echoing the problems d'Alembert had perceived in the craft system:

> In almost all instances he has found persons engaged in trade extremely unwilling to communicate the processes and manipulations which distinguish their several arts; and, in the course of his inquiries, he had frequently to regret that those who were most disposed to afford him assistance were, from want of all literary habits and practice, utterly incapable of rendering him that aid which he

could have hoped for by the communication of their ideas in writing. Many persons refused him help lest they should be thought to betray the secrets of their trade, and others were equally reluctant to enter into the nature of their profession, fearing that a free communication of their own thoughts would expose their ignorance of its principles, or would prove that its excellence did not depend upon any thing secret, or that could be concealed.

Martin's perception of the fears of mechanics is probably not exaggerated. James Watt, whose improvements in steam engines brought him fame and fortune, found that the production of copies of business letters was proving boring and time-consuming for him, and "yet their confidential and technical nature, coupled with Watt's thriftiness, probably precluded using a copy clerk." This situation prompted Watt to "discover a method of copying writing simultaneously," by pressing tissue paper moistened with special liquids on the original written in a special ink. In 1779, when Watt wished to establish a separate firm to market the process, he proposed to protect his interests and those of his partners by opening "a subscription for 1000 persons who are to be put in possession of the secret, a quantity of proper paper and materials with a press for taking off the impressions," but with "no one to be put in possession until the whole is subscribed, as the thing is so simple and easy that after divulging it to a number we might lose the rest."

Watt's copy press was a success, but it was a cumbersome process in comparison to the use of carbon paper, which was invented by Ralph Wedgewood in 1806. However, since a quill pen did not exert a very heavy pressure and a pencil did not produce an original letter that could not be altered, the earliest carbon paper was really paper saturated with printer's ink and was used in a curious reversal of the modern procedure. In Wedgewood's scheme, a good piece of writing paper was put under the inked sheet and a piece of tissue paper above that. The writing was done on the tissue paper with a metal stylus, the impression causing ink to be deposited on the face of the good sheet and on the rear of the tissue. This latter then became the file copy, and its reverse impression was read easily through the flimsy paper by holding it up to a light. By the 1820s the more modern kind of carbon paper was developed, and the discovery of aniline dyes made the indelible pencil possible, so that an original letter could be written directly with the nonerasable pencil, with a directly readable copy made from a piece of carbon paper inked on one side only.

The importance of confidentiality of correspondence and the protection of trade secrets have certainly become no less important with advancing technology, and the pencil industry has been no exception. In the late nineteenth century, *Scientific American* was a major source of information about the latest industrial devices and processes, and the editors of that journal appeared ready to reveal to their readers any secret they could wrest from manufacturers. Indeed, in the nineteenth century *Scientific American* often read like a *Reader's Digest* for tinkerers and inventors, condensing reports from trade and technical publications on how virtually everything worked and was made. One late-nineteeth-century item, entitled "Black Lead for Pencils," repeated how a correspondent to *The Pharmaceutical Era* reported having tried unsuccessfully to make black lead out of "ten parts by weight of plumbago, seven parts German pipe clay, made into a stiff paste and run through a mould," and finally baked. The editor was quoted as saying that he could not help the reader, for "the successful production of pencil leads is a very valuable trade secret to the manufacturers of pencils, and we cannot be confident of giving you any very satisfactory information on this subject," but he went on to describe "in a general way" three methods of making pencil leads. One consisted of the original process of sawing blocks of plumbago, and another was identified as the method "invented by M. Conté in 1795." Conté's process is described, with few specifics, as consisting of mixing graphite powder with "an equal or any other desired proportion of pure washed clay," adding water, forming the lead in molds, and when dry exposing them to "various degrees of heat." While the essence of the Conté process is certainly contained in this published description, some all-important details are lacking: How fine a graphite powder? How pure a clay? How much water? How dry? How much heat?

While someone interested in making pencil leads might have attempted to fashion a batch of leads out of graphite, clay, water, and heat, what resulted might have been wholly unsatisfactory, as the correspondent to *The Pharmaceutical Era* had discovered. His formula had all the right ingredients, but that obviously is not enough, for he could no more make a good pencil lead without the proper sifting and mixing and baking of the ingredients than he could make a good cake out of lumps of flour and sugar and a cold oven. Even in the present century secret recipes have been hard to come by, as one new American manufacturer learned when he decided to make his own leads instead of buying them from an established pencil company. The neophyte described the state of affairs in the early 1920s:

A countess . . . owns one of the important lead factories in Czechoslovakia. She goes into a private room at the factory, two or three times a week, and mixes up a batch of material by means of formulas which are known only to her. When she dies her son is scheduled to inherit the formulas and the business, and to continue mixing the materials in the same secret way. A great deal of this sort of mystery, we found, surrounded most of the lead factories in [America]. In some cases, I believe, even the owners of the plants did not understand the processes, but were compelled to rely on men who had learned their formulas in foreign lands.

Industrial formulas are kept secret today no less than they were earlier in the century, and they were kept secret then no less than they were from the time that Conté discovered how to make a modern pencil lead, and even before that. While the German pencil makers may have known that the superior Conté leads contained clay instead of sulphur, that would not have been enough knowledge. If they could not gain expertise by marriage, they would have had to experiment with different kinds and purities and finenesses of graphites and clays, with different amounts of water and different methods of mixing and molding and drying, with different manners and temperatures of baking. Juggling such a number of variables takes time and a method, and it is not possible to save time by writing a letter to another pencil manufacturer asking for the trade secret or writing a letter to a trade journal's advice column. Those who can answer are not likely to, and those who would like to answer are not able to with any certainty. The Germans in the early nineteenth century would have had to do their own research and development, but they seem not to have felt the urgency or need to do so. Or perhaps they were unwilling to invest their resources at the time.

Pencil making, like all modern industry, whether in the time of Conté or today, owes its continued health to sound scientific and engineering practice in the form of research and development. This timeless maxim was articulated in an article entitled "What Industry Owes to Science," which appeared in *The Engineer* in 1917 and of which a third was devoted to the pencil-making industry: "The industry is restricted to comparatively few firms, the majority being of long standing. Details of manufacture are largely kept secret, but enough has been said to indicate that the industry owes much to chemical science, in the selection, mixing, and general treatment of materials, and to mechanical science in the invention of labour-saving machinery for the processes involved."

The chemical and mechanical sciences are known today as chemical and mechanical engineering, and these and other engineering sciences provide the essential tools and methods that lie at the heart of successful research and development. And they are no less important foundations for the electronic, petrochemical, automobile, aerospace, and construction industries than they are and have been for the pencil industry. And even though much of the two centuries of pencil development since Conté's rather formal and explicit reliance on the engineering-scientific method may have been carried out in a preprofessional era of engineering and science, not to mention by self-taught engineers and scientists who are even less recognized than Conté, the modern pencil is very much a product of deliberate engineering.

As the eighteenth century was giving way to the nineteenth, the likes of Conté notwithstanding, the rules of the mechanical arts could nevertheless still be in opposition to the new sciences that since Galileo's time had been growing increasingly capable of extrapolating from first principles to practical matters—and thereby to innovative thinking. The contemporary dichotomy is documented in the first edition of the *Encyclopaedia Britannica,* and its alphabetized arrangement makes it very easy to check what the encyclopedia's authors, "a Society of Gentlemen in Scotland," considered important enough to be included in the three volumes, of about three thousand pages, published in Edinburgh in 1771. Entries for "art" and "science" are included in the encyclopedia, and they are defined by the gentlemen as follows:

ART, a system of rules serving to facilitate the performance of certain actions.

SCIENCE, in philosophy, denotes any doctrine, deduced from self-evident and certain principles, by a regular demonstration.

Although "engineering" is not defined, the *Britannica* makes it clear that there was a class of people, mostly associated with matters military, who practiced something, albeit in no professionally organized fashion, that was both art and science. These people were called "engineers."

The history of pencil making in nineteenth-century America, then a new republic whose entrepreneurs were unhampered by restrictive trade practices and unfettered by guilds and trade councils, may serve as a paradigm for the history of contemporary technology generally. The pioneering spirit was no less an impetus to developing new ways of making pencils than it was to developing new lands. And as there are few professional pioneers, and as pioneering has but the shallowest roots in experience and tradition, so engineering in early nineteenth-century America, as elsewhere, was done by the self-taught and self-motivated.

What the pencil pioneers in early nineteenth-century America sought was what Conté sought and what the Germans should have been seeking: a quality pencil from a diminishing supply of quality graphite; a pencil whose lead would not easily break; a pencil whose lead was not in short pieces that worked loose with writing; a pencil whose lead did not deteriorate with heat or age; a pencil encased in wood that would not split or splinter; a pencil that was good to look at and yet comfortable to hold; and, finally, a pencil that was worth whatever it would cost, for there was not then, as there is not now, any real economy in inferiority.

8/ *In America*

There was no organized business of pencil making in America in 1800, but that is not to say that black-lead pencils or alternatives were not being used or made in the New World at that time. Just as in the Old World, where metallic lead was used long before black lead for making marks on paper, so in young America metallic lead may have provided an alternative to the foreign-made cedar-and-graphite pencil well into the nineteenth century. According to one source, "the most active competitor" of the black-lead pencil around the turn of the century was the "quill pencil":

> It was an inferior article, and it is not apparent that it was ever factory-made, but each person that wanted such a pencil made it. The tools and materials consisted of a goose quill, a bullet, a melting ladle, and a turnip. The quill was cut to a length of a couple of inches, the end was thrust into the turnip and was thus held upright, the bullet was melted and poured into the quill, and the pencil was ready for use.

The quill pencil apparently made a "pale, dull mark," and it seems to have been "in almost universal use by old-time schoolteachers to rule copy books." It is almost as if the young American republic was reliving the history of the world, with its homemade pencils recapitulating the development of pencils everywhere, for the quill pencil was but a cased plummet.

In the early 1800s foreign-made black-lead pencils were

available to and used by artists and the self-taught American engineers and surveyors, and so there probably were examples of various European pencils made of graphite dust and binders and perhaps rumors, if not specimens, of the new French pencils with lead made by the Conté process. Furthermore, there would very likely have been an annoying abundance of small pieces of rare Borrowdale graphite that were left in pencil stubs or that had perhaps broken off of or fallen out of the best English pencils. This coincidence of circumstances might easily have provided raw material, incentive (if not necessity), and models of the first black-lead pencils to be made in the New World.

Horace Hosmer was born in 1830 in Concord, Massachusetts, and later lived just five miles away, in Acton, where among other occupations he was a pencil worker and salesman. Hosmer, who read Walt Whitman, came to be known as "the goat of Acton," and the personality that earned him that title can be imagined from the style of a short article on early pencil makers of New England that he wrote in about 1880 for *Leffel's Illustrated News*. He credits a Massachusetts schoolgirl with making the first lead pencil in America:

> In the beginning there was a woman. Before men dated their letters A.D. 1800, there was a school for young ladies kept in the ancient town of Medford, and one of the pupils was from Concord, Mass. Besides learning to sketch, paint, embroider, etc., she learned to utilize the bits and ends of Borrowdale lead used in drawing, by pounding them fine and mixing a solution of gum arabic or glue. The cases were made from twigs of elder, the pith being removed with a knitting needle. So far as the writer knows, this was the first pencil-making establishment in the country. Forty years ago [in 1840], the writer, then a boy of 10 years, helped the same lady to make similar pencils from plumbago and English red chalk.

Hosmer's attention to detail and his personal association with the woman give this story the ring of truth. However, since he does not sound like a misogynist, it is curious that Hosmer does not name the woman with whom he worked, especially since he goes on to name many a man who played a subsequent role in the development of pencil manufacturing in and around Concord. Perhaps she was a relative of his.

There is a slightly different version of the story, related in 1946 by Charles R. Nichols, Jr., then director of engineering of the Joseph Dixon Crucible Company, which claimed to have been the first firm to mass-produce the pencil. His story is as follows:

> The first pencil factory in America was founded by a school girl whose name is not known. She obtained a few pieces of graphite from the Barrowdale [*sic*] mine, crushed them to a powder either with a hammer or stone, and then employed gum, mixing the two together, and stuffed a hollowed-out alder twig with the mixture. This first lead pencil made in America was produced at Danvers, Mass. Later, a man by the name of Joseph W. Wade co-operated with this girl in producing a number of lead pencils by the same process.

Although Nichols seems to be careless with spelling, it may be that he was recording what he obtained from oral tradition, which might easily interchange such near-homonyms as "elder" and "alder," rather than from written sources. However, since the name "elder" can also designate the European alder, the names of the trees, like their twigs, may have been used interchangeably. Unfortunately, since neither of these pencil historians documented his claims, we cannot be sure whether the first pencils were made of elder or alder twigs in Medford or Danvers. The inclination may be to trust more in Hosmer, who lived and wrote closer to the event and who claimed to have worked with the maker of the first American pencil, but the issue is further confused by a British historian's version of the young inventor's achievement:

> She obtained a few pieces of Borrowdale graphite, crushed them to powder, added a gum, and filled a holly twig with the mixture. Later a man named J. W. Wade co-operated with this girl in producing a number of such pencils, which, although obviously unsatisfactory, doubtless supplied a temporary want in a country which had not yet received supplies from England.

A footnote identifies Nichols as the author's source, but Nichols's article makes no value judgments about the quality of the pencils made of twigs, nor does he identify further the man named Wade. And while a holly twig might indeed have served as a pencil case for Borrowdale graphite, Nichols does not appear to have been the source of that information.

Whatever their quality or materials, however, the first American black-lead pencils are consistently credited to a young woman in Massachusetts. But her name or the details of her invention may never be known with any more precision than are the circumstances surrounding the discovery of the Borrowdale mine or the origins of the first wood-cased pencil. And the same uncertainty will likely remain about the name of America's first bridge builder and the location of and the kind of wood used in the first American bridge. The origins of pencils, bridges, and other artifacts of craftsmanship and early engineering can be imagined to belong almost more to mythology than to history, for it is the nature of much tinkering, inventing, and even engineering to save little of what is superseded. Ideas that are improved upon are displaced in much the same way the old files on computer disks are written over with the new, and the artifacts that are improved upon are often discarded or cannibalized for a new, improved product. What remains of original thoughts or things will be as obliterated as the first pencil marks on a much-used palimpsest.

As inaccessible as the true origins of thoughts and things may be, it is as natural to want to know them as it is to wish to know the future. And it is through the history of something as elusive and as illusive as past innovation that we can gain some insight as to how to understand our constant assault on the future. Since the products of American engineering in general are so numerous and complex in their early history, and since their origins are as dependent upon oral tradition and mythology in the making as are the origins of the pencil, it is a more humble and perhaps achievable goal to understand first something we can hold in our hand and whose essence we can grasp. Even though its beginnings may be blurred, the early history of the pencil in America, being the history of something arguably more simple than that of bridges and buildings, say, may give us more accessible insights into the history of larger things, such as structures, and of engineering generally. And we can learn from the uncertainties of the pencil's roots that there are lessons even in ambiguity.

Horace Hosmer's article on early pencil makers of New England continues by recalling that another Concord man, a David Hubbard, "made the first cedar wood pencils for the New England trade; but they were of little value, and but few of them were manufactured." How many other primitive American craftsmen-cum-engineers designed and made pencils "of little value" can no more be expected to be known than how many made bridges of stone or wood or iron that were of little value. But just as William Edwards, the eighteenth-century Welsh country stonemason who saw his first three attempts fail, could by ingenuity and perseverance finally build the bridge called Pont-y-tu-prydd, which still stands, so the collective if separate determination of early American engineers and pencil makers would by and by amount to something.

While the mathematical and philosophical foundations of modern engineering were being laid in France, they were largely being ignored in Britain and America, where the old apprentice system could punish innovation and stifle creativity. A memoir by his son about William Munroe, who was to become one of America's first pencil makers in the early nineteenth century, illustrates the climate very forcefully. In 1795, when Conté was patenting the modern pencil in Paris, seventeen-year-old Munroe was beginning his apprenticeship in Roxbury, Massachusetts, to a cabinetmaker who also happened to be a deacon. Young Munroe was not at all given his head: "Boys then had to submit to be thought boys, whose rights as well as duties were to be defined by the masters, and with no little strictness."

But the spirit of the Industrial Revolution, perhaps fired by that of the American and French political revolutions, was also sparking a spirit of rebellion in the boys. As would Oliver Twist, young Munroe and his peers protested about how their meals were given them. They demanded more than just the bread and milk that, except for Sunday morning's chocolate, they were given every morning and every evening. And some of their requests were eventually acceded to: they received chocolate every morning, for example. But other demands were not met: the boys never did get to go to the theater in Boston more than once a year. While such social oppression may have been bad enough, it was the climate in which the work itself was done that really discouraged technological innovation.

Young Munroe showed early promise in the cabinetmaker's shop, and he became "the best workman in it," with "the finest and most difficult work being entrusted to his hands" by the time he left. That is not to say his progress was without opposition:

Before finishing his apprenticeship, he had felt conscious of having powers that were cramped, whenever he thought out new modes of making or ornamenting his work. But the rules of the shop were imperative and did not permit innovations. On one occasion, however, he defied them, by proceeding quietly, and, as he thought, unobserved, to hang a table leaf by its hinges on a plan not known to the rules. He got started pretty well, when some tell-tale informed the deacon. A breeze was soon raised that was nearly serious, but on begging to be allowed to finish what he had started, he, with many admonitions, was allowed to go on. The result was such that no other way of hanging a table leaf but his was the rule of the shop from that time.

Munroe's determination was the exception, however, and the rigidity of the apprentice system must have stifled many a budding innovator and kept improvements from being introduced in products and processes. One observer noted in 1873 that Joseph Gillott, "the man who made more steel pens than any other, and better ones," never wrote with one: "With all the men and machinery at his command he was never able to produce a pen that suited him so well as the time-honored plume of the old gray goose." This lack of achievement was attributed to a single thing: "Only the blinding effect of tradition and training can account for the failure of penmakers to discover and correct the radical and plainly apparent faults of their productions."

After finishing his turbulent apprenticeship in the deacon's cabinet shop, the innovative Munroe stayed on for six months as a journeyman, earning enough money to buy some tools of his own. He joined his older brothers, who were clockmakers, and for years made wooden cases for their clocks. In 1805 he married the daughter of Captain John Stone, the architect of the first bridge over the Charles River at Boston, and Munroe no doubt carried over that bridge the furniture that he made to sell in the city.

In 1810 Munroe bartered some clock cases he had made for clocks from Norfolk, Virginia, and he invested in corn and flour the proceeds from selling the clocks. In "a round-about trade," he eventually obtained a shop near the Mill Dam where he attempted to make a living as a regular cabinetmaker. But Munroe realized that he could not continue, for, "finding that I could make with my own hands more furniture than I could sell, business of every kind being dull . . . I should in a few years at most . . . be poor." The war with England had begun, and embargoes and other restrictions on trade had depressed the economy.

According to his son, Munroe reasoned that under the circumstances there should be a good demand for essential articles theretofore made only abroad. Furthermore, he believed with respect to those scarce items that "invention . . . was encouraged and well rewarded." He first made cabinet-makers' squares, but "the demand was necessarily limited and competitions easy." So Munroe looked for something else to produce, something that would not only be in continual demand but also be more difficult for competitors to copy:

> And seeing what a high price had to be paid for a leadpencil, and that the article could hardly be procured at all, he said to himself, "if I can but make leadpencils I shall have less fear of competition, and can accomplish something." He acted upon this idea at once, dropped his tools, and procured a few lumps of black lead. This he pulverized with a hammer, and separated the fine portions by their suspension in water in a tumbler. From this he made his first experimental mixture in a spoon. And from this was his first attempt to make a pencil. The result was not very encouraging.

It has been written that Munroe's first experiments employed clay and that he tried to master the Conté process in a room that only his wife could enter. But, since the son who wrote the memoir did not pursue pencil making, it is possible that even years later he did not know of or appreciate the importance of clay in the lead-making process, for he did not mention it. However, regardless of the ingredients, Munroe evidently did not hide the fact that he was discouraged by the inferior pencil he produced. While he managed to eke out a living by making more squares and doing some cabinet work, he did not stop experimenting. As Munroe's son's memoir continues:

> But his mind was principally occupied for two or three months in devising ways of making leadpencils; having access to no information which could assist him, fearing to consult his friends on the subject, and sometimes get-

ting discouraged with repeated failures. But finally, securing some better lead, and picking up a little cedar wood of wholly unsuitable quality from the neighboring hills, he was able, on the second day of July, 1812, to proceed to Boston with a modest sample of about thirty leadpencils, the first of American make, and naturally not of very good quality. These he sold to Benjamin Andrews, a hardware dealer in Union Street, to whom he had sold the cabinetmakers' squares. This Andrews was an active, enterprising man, who encouraged all such novelties, and he advised going on with leadpencils. This advice suited [Munroe's] intentions, and on the 14th of July he went to Boston with three gross of pencils. These, also, were readily taken by Andrews, who then made a contract, agreeing to take all that should be made up to a certain time at a fixed price.

This story of initial enthusiasm, early discouragement, repeated frustration, constant distraction, prolonged determination, total isolation, and, finally, a serviceable but far from perfect product has all the ring of an honest recollection of a real engineering endeavor, an odyssey from idea to crude prototype to artifact to improved artifact as full of adventure as Ulysses' travels. And this is a story of research and development that can be repeated, *mutatis mutandis,* with "leadpencil" erased and "light bulb," "steam engine," or "iron bridge" written in its place. That the true course of development can be so easily forgotten, or not easily appreciated, especially by those who reap the benefits of the pioneers, is shown clearly by the romantic picture painted by someone who himself did not innovate so much as help make the pencils developed by others:

In 1812 William Munroe, a cabinet maker by trade, pounded some plumbago with a hammer, mixed it in a spoon with some adhesive substance, and filled the compound into some cedar wood cases. Some of these pencils were shown to Benjamin Andrews of Boston, who was ready to buy, and encouraged Munroe to make more of them. Twelve days after he carried five [*sic*] gross, which were readily taken and paid for, and a new industry was fairly started. Munroe was very poor, but a first-class workman; so, every step in the new business was smoothed by his matchless skill in the old. Cautious, me-

thodical, exact, he made few mistakes and took the shortest and best way known of securing a competency. He
made the "water cement" or paste lead which was filled
into the grooves in a soft state, and after remaining a week
or more the surface of the pencil slab was planed to remove the composition which adhered to it, and to leave a
clean surface for gluing on a veneer of cedar. The pencil
slab was about ¼ inch thick, and the veneer ⅛ inch and
of varying widths from 4 to 10 pencils wide.

Although the process described here does not appear to be
related to Conté's, having, for example, no baking involved,
the use of the "pencil slab" method of assembly was what was
eventually to give American manufacturers the edge. While
Horace Hosmer may be confusing decades of development in
this single passage, it is interesting alone for its unrealistic
depiction of the research and development process. What Hosmer really describes are the mature fruits of Munroe's efforts
and not the process of their cultivation. While there have been
some notable exceptions where success was remarkable and
where an engineer's "every step in the new business was
smoothed by his matchless skill in the old," that is more myth
than reality. Indeed, if a project goes as smoothly as this account of Munroe's development of the pencil, it is usually
because of the lessons learned from mistakes made in earlier
projects. Just as all mistakes do not lead to disasters, neither
do all attempts lead to success, as the account of Munroe's son
makes so clear.

If an ultimately successful bridge builder does not learn
from his own mistakes, as William Edwards did in eighteenth-
century Wales, then he most likely learns from those of others,
as John Roebling would before building his great Niagara
Gorge and Brooklyn suspension bridges. In the case of the
development of the pencil, William Munroe's son is likely to
have heard of the agony straight from the craftsman-engineer's
mouth, while Horace Hosmer is likely to have inferred the
ecstasy from the businessman's and master's rules for making
a successful product decades after a decade of development
took place.

That pencil making continued to be a challenge even after
Munroe sold his first few gross in Boston is clear from his son's
continuation of the story. Although he then had a promising
future as a pencil manufacturer, Munroe's struggles were not
over in 1812:

He had great difficulty in procuring the required materials, and lost some time in devising proper methods of performing the various processes on a larger scale than his experiments required. But encouragement stimulated his energies, developed his faculties of invention, and soon enabled him to overcome all difficulties. All the mixing of the lead and putting it into pencils was done entirely by his own hands in a small room of his dwelling-house, thoroughly protected from curious eyes; no one but his wife being permitted to know anything about his secret methods. The processes of pulverizing the crude lead, and preparing the wood-work, were performed by assistants in his shop on the Mill-Dam, the finishing being completed by himself, or within the family, at home.

William Munroe's problems of getting the proper materials, of scaling up his experiments to commercial production, and of keeping secret (in lieu of patenting) his hard-earned technology in order to realize a financial reward for his research and development are certainly not unique to pencil making. The development of practically every product of modern engineering, from patented iron bridges to nuclear power, has faced the same obstacles. And the frantic pace of development of today's computer technology, while generally considered so much "higher" than that of the lowly pencil, is really but a twentieth-century retracing in silicon of what the nineteenth century saw scribbled in graphite.

After overcoming his start-up problems, Munroe did have a profitable business—but only for the eighteen months during which he could procure black lead. While war raged on between America and Britain, Munroe made toothbrushes and watchmaker's brushes and did some cabinet work until, at the end of the war, he could once again obtain the raw materials to resume producing pencils.

The end of the war in 1815 brought a new problem, however, one that Munroe had predicted and feared: the importation of a better pencil than he could make. Instead of deserting pencil making, which he had developed through his self-taught discipline from a craft to a viable cottage industry, Munroe practiced scientific engineering, by studying the work of pencil designers as closely as natural philosophers studied the works of the Great Designer. Rather than forget or neglect pencil making, Munroe studied it and "made himself master of what little information he could gather about foreign methods of preparing the lead." Since little information was available, in books or anywhere else, he resorted to experiments, "occasionally making a few pencils for sale of not a quite satisfactory quality." According to his son's story:

> This continued until 1819, when, having prepared himself with better experimental results, and obtaining better lead and cedar timber, he decided to abandon the cabinetmaking part of his business, and to devote himself wholly to the manufacture of leadpencils. . . . It was not without a struggle that his reputation as a manufacturer of leadpencils was established and recognized; not until more than ten years of persistency in pushing the sale of his goods, and of study in improving their quality, that he was able to say "that purchasers were at length as ready to seek him as he had hitherto been to seek them." From that time, so long as he was in business, he stood before the public as the best and principal maker of leadpencils, as he had been the first, supplying a large part of the demand for that article.

Thus it took William Munroe ten long years to feel that he had succeeded in "perfecting" a pencil whose development others might see as being "smoothed by his matchless skill." And in those ten years the craftsman who had rebelled as an apprentice had indeed become an engineer, for his art was no longer without science. But whether he remained "the best and principal maker of leadpencils" requires a more objective assessment than a son's memoir.

Of course, Munroe was not without help in developing his business. He certainly needed assistance in operating the two-man saw used to cut slats and veneers from cedar logs. The slats were brought to proper thickness by hand planing, and, at least in the beginning, grooves for lead were cut one at a time. These and other labor-intensive operations characterized early pencil making generally. When he decided to quit the cabinetmaking business in 1819, Munroe sold some of his tools to his journeymen, Ebenezer Wood and James Adams, who continued to work wood for Munroe's use.

While Wood developed into a cabinetmaker in his own right, he continued to be connected with pencil making. Whether Horace Hosmer was right in stating that Wood's "hand and brain largely helped to make Munroe's fortune,"

Wood apparently did develop the first machines used in pencil manufacturing. His wedge glue press could hold twelve gross of pencils for drying, and his pencil-trimming machine could "hardly be simplified or improved." He apparently set up the first circular saw in the business and from it developed a machine that would cut six grooves for pencil leads at a time. He also fitted a circular blade with a series of knives for shaping pencils, including hexagonal and octagonal ones. Duplicates of his unpatented machines would be built for and by New York pencil-making companies in the latter half of the nineteenth century. At the Paris Exposition in 1867, where American progress in machinery would be well recognized, gold medals were awarded to a sewing machine, to a machine that made buttonholes, and to a machine that made six pencils in one operation.

But in the first half of the nineteenth century the machine was far from taking over America, or even pencil making. Ebenezer Wood, "a gentleman in looks and behavior," recited poetry while at work, as a mellowed goat remembered warmly. That same observer of Concord society also noted approvingly that this gave some justification to the fact that at least one nineteenth-century census numbered pencil makers among those engaged in literary pursuits. But Ebenezer Wood was not to be the only American to fall into that category.

However American pencil makers were to be counted, they evidently had achieved some mastery of pencil making in the first quarter of the nineteenth century. Andrew J. Allen, a Boston stationer who asserted in his 1827 catalogue that he "had always encouraged American Manufactures, and will constantly keep every article in the Stationery Line where they are of good quality," offered American-made items in direct competition with foreign-made goods he seems to have been able to acquire from around the world. Allen's catalogue also shows that he was discriminating in what he represented, for quills from America, of "various prices and in great variety," were listed just below quills from England "of a superior quality" and just above "good" ones from Russia. One could also purchase Dutch quills and "beautiful large Swan Quills, for large Hand," as well as the means to sharpen them: "Penknives—a great variety of superior Silver steel, made by Rodgers." "Knives for Eracing"—scratching away the error made with the quill and ink—were also offered.

Penknives could also be used to sharpen pencils, of course, and their marks could much more easily be erased with Allen's

India rubber than those in ink. The 1827 catalogue offered pencils of several kinds, including camel's-hair pencils for writing and drawing and pencils accompanying ass-skin memorandum books. But most significantly for what it implies about the state of American pencil making at the time, Allen also sold both English and American lead pencils, each kind offered with black or red lead. Conspicuously absent from Allen's catalogue are pencils from France or Germany. The former may not have been readily available, and while the latter may have been available, they could have been rejected for their poor quality.

In keeping with the contemporary merchant's desire to conceal the exact source of his goods, neither the English nor the American pencil manufacturers are identified in Allen's catalogue, and so it is not possible to say which of the several pencil makers who were located in the Boston area in the mid-1820s came up to Allen's standards. But there were certainly some who were trying to make American pencils that could compete with the best English products.

9/ An American Pencil-Making Family

Since it was an age of self-reliance and self-education, not only in the nascent profession of engineering and the emerging industry of pencil making but also among citizens generally, what one was called for the purposes of a nineteenth-century census or what one called oneself for the purposes of answering a questionnaire could depend very much on what one was doing at the time of the inquiry. One citizen of Concord, Massachusetts, a member of Harvard's Class of 1837, in response to a letter from his class secretary asking about his life ten years after college, wrote, with little regard for conventional punctuation:

> I dont know whether mine is a profession, or a trade, or what not. It is not yet learned, and in every instance has been practised before being studied. . . .
>
> It is not one but legion. I will give you some of the monster's heads. I am a Schoolmaster—a Private Tutor, a Surveyor—a Gardener, a Farmer—a Painter, I mean a House Painter, a Carpenter, a Mason, a Day-Laborer, a Pencil-Maker, a Glass-paper Maker, a Writer, and sometimes a Poetaster. . . .
>
> For the last two or three years I have lived in Concord woods alone, something more than a mile from any neighbor, in a house built entirely by myself.

Later in life this particular alumnus would also identify himself as a civil engineer. And while he would have had little inclination to join a professional society, as he had little class spirit and less concern for what his neighbors thought of him, his story is as relevant for an understanding of nineteenth-century engineering as it is for an appreciation of American Transcendentalism. This Harvard alumnus was given the Christian names David Henry in 1817, and that is the order in which the names appeared on the college commencement program. However, he had always been called Henry by his family, and for no apparent reason other than preferring the way it sounded, shortly after leaving Harvard he began signing his name Henry David Thoreau.

Thoreau's story, especially his involvement in the manufacture of pencils, is helpful for understanding the nature of nine-

Henry David Thoreau in 1854, from a crayon portrait from life by Samuel Worcester Rowse

teenth-century engineers and engineering for several reasons. First, an engineer before midcentury, like the alumnus Thoreau, would not necessarily be certain that his activity was a profession, for it was not yet "learned." Furthermore, the story of Thoreau shows again that one did not have to study engineering to practice it. College education in his days prepared one for the ministry, law, medicine, or teaching. Those who practiced and advanced engineering in the first half of the nineteenth century had come to it largely through the crafts and the apprentice system. Indeed, participation in the construction of the Erie Canal, which was begun in 1817 and took eight years to build between Albany and Buffalo, was believed to be the best civilian engineering education then available, and the canal itself has been called "the first American school of civil engineering." While the Institution of Civil Engineers was founded in London in 1818, the American Society of Civil Engineers did not exist until 1852, and it is generally the beginnings of such professional societies that are considered to mark the beginnings of professionalism itself.

Second, the story of Thoreau is instructive because it is a reminder that innovative and creative engineering was done by those who were interested in a wide variety of subjects beyond the technical. Whether or not they had college degrees, influential early-nineteenth-century engineers could be a literate lot, mixing freely with the most prominent contemporary writers, artists, scientists, and politicians. And this interaction hardened rather than softened the ability of the engineers to solve tough engineering problems.

Third, like Thoreau, innovative engineers tended to be a bit iconoclastic and rebellious, rejecting traditions and rules. Not a few eighteenth- and nineteenth-century engineers came from professional families that did not always understand why a young man wanted to pursue an apprenticeship rather than go to college, or why he would want to practice engineering after attending college. The Englishman John Smeaton, of whom it was said that he could not touch anything without improving it, was the son of a lawyer. But young Smeaton decided against a legal career and opened his own instrument shop in 1750. On the other hand, John Rennie, responsible for three great London bridges, attended the University of Edinburgh in the early 1780s, studying natural philosophy, chemistry, modern languages, and literature. But his son, also named John, and also to be a distinguished engineer, did not go to college. Even those who rose out of more humble backgrounds stood out precisely because they could, like William Munroe in America, challenge the craft tradition for its own improvement.

Fourth, like Thoreau's involvement in pencil making, engineering was practiced with the tongue and the pencil, and there was very little written of it or about it before the middle decades of the nineteenth century. Thus there was little left to tell posterity the technical story of how and why certain designs or processes were developed or chosen over others. The truths of the theories of the pioneer engineers were demonstrated by the successful erection of a solid bridge or the efficacious process of producing a good pencil. Major contributions to technology could be incontrovertibly demonstrated without a single word being spoken outside the workshop or committed to paper.

Henry David Thoreau's eventual involvement with pencil engineering in such an environment can be traced to Joseph Dixon, whose own introduction to pencil making was indirect. Dixon had a meager education, but he possessed a mechanical ingenuity that enabled him while still a youth to invent a machine for cutting files. He then took up printing but did not have enough money to buy metal type and so taught himself to carve his own wooden type. As his resources and ambitions grew, he began to experiment with graphite in Salem in order to make crucibles in which to melt his own type metal. Since there was a limited market for the crucibles, he also began to use graphite to make stove polish and lead pencils. However, unlike William Munroe, when Dixon tried to peddle his pencils in Boston, he found little call for them, and "he was told he would have to put foreign labels on them if he expected to make sales."

Infuriated, Dixon ceased making pencils, but apparently not before Henry's father, John Thoreau, learned the rudiments of pencil making and, perhaps incidentally, those of chemistry from the self-taught Dixon. There is some indication that Dixon may have learned of Conté's use of clay in pencil leads from a chemist friend named Francis Peabody, but without sufficient experimenting with the process even that knowledge would not have made Dixon's early pencils remarkable. While John Thoreau may in turn have learned that clay mixed with graphite could make an excellent pencil, he also would have needed to experiment with the process. However, there is no firm evidence to indicate that the French process of pencil making was really known at all, much less mastered in America in the 1820s.

In 1821 Thoreau's brother-in-law, Charles Dunbar, discovered a deposit of plumbago while wandering around New England. He who had been the black sheep of the family apparently stumbled upon the graphite in Bristol, New Hampshire, and so decided to go into the pencil-manufacturing business. Dunbar found a partner in Cyrus Stow of Concord, and the firm of Dunbar & Stow was established to work the mine and manufacture lead pencils. Their graphite was certified as far superior to any then known to originate in the United States, and so the future of the business looked bright. However, when some legal details of establishing mineral rights left the partners with only a seven-year lease on the mine, they were advised to dig out all the plumbago they could before their lease expired.

A faster production of plumbago meant that pencils could be manufactured at a faster rate, and this, it appears, was why Charles Dunbar asked Thoreau to join the business in 1823. Soon Stow, who apparently had other means of income, and shortly thereafter Dunbar, for unknown reasons, dropped out of the pencil-making business, and the firm was renamed John Thoreau & Company.

Either John Thoreau had more suitable graphite or he was more persistent than Dixon in improving his pencil-making process, for Thoreau pencils evidently could be sold without foreign labels. By 1824 Thoreau's domestic pencils were even of good enough quality to win special notice at an exhibition of the Massachusetts Agricultural Society. As reported in the *New England Farmer*, "the Lead Pencils exhibited by J. Thorough [*sic*] & Co, were superiour to any specimens exhibited in past years." The misspelling of the family name lends support to the oral tradition in Concord that the Thoreaus pronounced their name "Thorough." Indeed, Henry Thoreau to this day is quoted as having punned on his name by saying of himself, "I do a thorough job." However, there is also contrary evidence, such as a letter addressed to "Mr. Henry D. Thoreaux," suggesting that the French pronunciation of the name was not unheard of.

Whichever way the name was correctly pronounced, Thoreau pencils found a steady market, with or without the family name imprinted, perhaps even being offered by Boston stationers. By the early 1830s the pencils were threatening William Munroe's business and competition became fierce. Since both firms were having their plumbago ground at Ebenezer Wood's mill, Munroe apparently tried to get Wood to stop grinding Thoreau's material. However, Wood evidently made more money from Thoreau and so stopped grinding Munroe's plumbago instead.

While the Munroe business faltered, the Thoreau pencil business prospered. But to prosper is not necessarily to be without worries. One could not make pencils without graphite, and when it could no longer be obtained from the Bristol mine, other sources had to be found. These were located in a mine in Sturbridge, Massachusetts, and later, when that was exhausted, in Canada. It is very likely that, by the time he went away to college, the young Thoreau had become familiar with and helped with the manufacture of pencils, which by then had been the family business for about ten years. Indeed, in 1834, Henry David Thoreau made a trip with his father to New York City in order to sell pencils to stores there, apparently because the money was needed for Henry's schooling.

One of the reasons Thoreau pencils could compete successfully with the Munroe variety was that all pencils made in America at the time were "greasy, gritty, brittle, inefficient," and users, especially artists and engineers, were always looking for a better product. The inferiority of American pencils was due in large part to the fact that, since pure Borrowdale graphite was not available and since the Conté formula for pencil lead was apparently either unknown or not perfected in America, firms like John Thoreau's continued to mix their inadequately purified and ground graphite with such substances as glue, adding a little bayberry wax or spermaceti, a waxy solid obtained from the oil of the sperm whale or dolphin and also used in making candles. The warm mixture was then applied with a brush to the grooved part of a cedar case, and another piece of cedar was glued on top of it. John Thoreau worked at improving his imperfect product, and he achieved some success in making it less imperfect than that of his competitors. Although his or any other American pencils still did not come anywhere near the quality of the best English or French pencils, by offering reasonably priced alternatives it was possible for Thoreau & Company to be well established by the mid-1830s.

When Henry David Thoreau graduated from college he had no intention of making pencils for a living. Following in the tradition of his grandfather, his father, his aunt, and his brother and sister, all of whom had taught for a time or were then teaching, Henry accepted an offer to teach in his own childhood institution, Center School in Concord. However, after only two weeks, he was called to task for not using corporal punishment to keep order and quiet in the classroom. Apparently overreacting to this criticism, Thoreau proceeded to ferule students for no apparent reason and that evening resigned from his position. This seemingly irrational behavior, coupled with his insistence on reversing his names, confused the residents of Concord, and from then on many looked askance at the young Thoreau and his unconventional ways.

Without a job, Thoreau went to work for his father. But, true to his nature, the young man did not want to be just another pencil maker, and so he sought to understand why American pencils were so much inferior to ones made in Europe. Since he knew that the graphite was of excellent quality, though apparently not pure enough or occurring in large enough pieces to be used without being ground and mixed with binding substances, Thoreau deduced that the problem was in the filler or in the lead-making process itself. Thoreau pencils at the time were still being made by pressing a mixture of graphite, wax, glue, and spermaceti into a paste, warming it, and brushing it or pouring it soft into the grooves of the wooden cases.

To identify and correct what is causing a product to fail to perform as hoped is the essence of engineering research and development, and whether he or anyone else called it that, that is exactly what Thoreau proceeded to engage in. Since the problem of identifying what was missing from the pencil-manufacturing process was so open-ended, Thoreau wondered if he could determine what was in good European pencil lead or what the European pencil manufacturers did differently.

While it has been said that German pencils made by the Faber family were the models that Thoreau was trying to emulate in the mid-1830s, there is some question whether many German pencils themselves were then being manufactured by the Conté process, which made possible the "polygrade" pencils whose hardness or softness depended upon the proportions of clay and graphite in the lead mixture. According to one historical sketch of the German industry, in a booklet published in 1893 by the Johann Faber pencil factory in Nuremberg:

> . . . the first Polygrade lead pencils of "Faber" were offered to the trade in Germany in the year 1837 (with

French labels) through "Pannier & Paillard" of Paris and represented to be a *French* article, whereas when Mr. Faber on his early journeys explained to his customers that the "Faber" pencils were of *German* and not of *French* origin, his statement was very often discredited.

While the literature of some pencil manufacturers, themselves descendants of the German industry, claims that clay was used in German pencil leads as early as the 1820s, it was certainly not used widely, if at all, for export. It was only after he took over his father's A. W. Faber pencil factory in 1839 that "Lothar Faber occupied himself with opening up business connections throughout the civilized world." Thus it is most probable that German pencils were not at all common in America when young Thoreau first sought to improve his father's product, and any German pencils that did exist may not even have been made by the superior Conté process. What Henry Thoreau may have been hoping to do was emulate a French pencil or perhaps just find out how the Germans mixed and processed their ingredients to make a good but far from perfect pencil.

Not being trained in chemistry, Thoreau could not easily analyze a specimen of pencil lead, so he evidently proceeded to look for clues in Harvard's library. The oft-repeated story is that in a Scottish encyclopedia published in Edinburgh, Thoreau found that German manufacturers combined graphite with Bavarian clay and then baked the mixture. The story appears to have its origins in what Thoreau himself is believed to have said years after the fact, when the Faber pencils were indeed being made according to the Conté process and were being pushed "throughout the civilized world." But at the time Henry is said to have used the Harvard library, in about 1838, it does not seem possible that a Scottish or any other encyclopedia could have described the use of Bavarian clay in German pencil making, for the Germans themselves apparently were not yet using that process to any considerable extent.

It has been generally assumed that it was the *Encyclopaedia Britannica,* whose thistle trademark still recalls the work's Scottish origins, from which Thoreau got the idea of mixing graphite with clay. But the "pencil" entry in any edition of that work available to Thoreau had not changed since the second edition, completed in 1784, which was before the clay-and-graphite process existed. While German pencil making is described in the article, it is the process of mixing sulphur with graphite that is discussed—and criticized, for such pencils were said to be inferior to English ones. The encyclopedia article also tells the reader how to detect an inferior German pencil—by the fact that the lead will melt and give off a "strong smell like that of burning brimstone" when held in a flame—and this may have given Thoreau a clue about how to make a better pencil. Or perhaps he got a clue elsewhere.

There were many encyclopedias published in Edinburgh in the late eighteenth and early nineteenth centuries, and Thoreau may also have consulted the *Encyclopaedia Perthensis,* whose second edition was issued in Edinburgh in 1816, in spite of its title's association with the nearby city of Perth. Or he may have read *The Edinburgh Encyclopaedia,* whose first American edition was published in 1832. But since German pencils were not then generally made with clay, it is not surprising that neither of these encyclopedias describes such a process. Why none refers to the French pencil-making process is more problematic, though it may have been merely a matter of national pride. The omission of any mention of the French industry may also have been due to the fact that Diderot's great *Encyclopédie,* completed in 1772, appeared before Conté made the discovery that put French pencil making in the forefront on the Continent. Given the derivative nature of encyclopedic works, it is perhaps not surprising that in the 1830s the secret of pencil making was not so readily available in print as has been assumed by some students of Thoreau's literature. But that is not to say that Thoreau did not look.

Most encyclopedias published in Thoreau's time seem to have relied heavily upon other encyclopedias for information, as a comparison of the "pencil" entries in near-contemporaneous works will demonstrate. The 1832 edition of the *Encyclopaedia Americana,* for example, repeats almost verbatim the earlier *Britannica*'s entry that defines the pencil as "an instrument used by painters for laying on their colours." This edition of the *Americana* is most likely the one that Thoreau was complaining about in an 1838 letter to his brother, John. According to Thoreau, who was evidently trying to learn from books how to form gunflints, the encyclopedia had "hardly two words on the subject." "So much for the 'Americana,' " he wrote to John, and then quoted from another source an explanation that he clearly found inadequate: "Gunflints are formed by a skillful workman, who breaks them out with a hammer, a roller, or steel chisel, with small, repeated strokes."

From such laconic written descriptions Thoreau could no more learn to knap gunflints than to bake pencil leads.

But if he did not read about combining clay and graphite to make an excellent pencil, then where did Thoreau come up with the idea? If he did not read it explicitly, it is still possible that he did find something in the Harvard library that made him put two and two together. For example, if Thoreau had looked up "black lead" in the *Encyclopaedia Perthensis,* he would have been referred to an entry where he could have read among other things about pencils:

> A coarser kind are made by working up the powder of black lead with sulphur, or some mucilaginous substance; but these answer only for carpenters, or some very coarse drawings. One part of plumbago with 3 of clay, and some cows hair, makes an excellent coating for retorts, as it keeps its form even after the retorts have melted. The famous crucibles of Ypsen are formed of plumbago mixed with clay.

In a first reading of this passage, one might anticipate finding, after the criticism of sulphur as an ingredient suitable only for the lead of a carpenter's pencil, an indication of an ingredient to be preferred for better pencils. Thus, "one part of plumbago with 3 of clay" might be expected to be followed with the phrase "makes a pencil suitable for the use of artists and engineers." Even if Thoreau did not anticipate words, and even if this encyclopedia entry did not tell Thoreau or anyone else exactly how to make a Conté pencil, it might have provided a catalyst to thought.

By juxtaposing the disadvantages of sulphur as an additive with the advantages of clay as a heat-resisting ingredient, albeit for retorts, such an article might have provided the climate for making a leap of invention—or reinvention. Even if it did not make a better mark, a pencil produced with some clay might have a point that would not melt or soften so easily as one containing sulphur. The further juxtaposition of the mention of crucibles may have also sparked an idea in Thoreau's mind, for he may have been aware of the Phoenix Crucible Company in Taunton, Massachusetts, and thus he would have known of a possible source of appropriate clay. Or he may have known that the New England Glass Company was also importing Bavarian clay at the time. And even if he was not familiar with these sources of supply, Thoreau might easily have found out about them once he had it in his head to experiment with a clay-and-graphite mixture for making pencil leads. Whatever his source, Thoreau apparently obtained some clay and proceeded to work with it. While he could immediately produce a harder and blacker pencil lead, it was still gritty, and he suspected that this fault could be corrected by grinding the graphite finer.

As with much of engineering, it seems to be unclear exactly how much Thoreau and his father interacted in developing a new grinding mill for graphite. The older Thoreau's habit of reading chemistry books and his early association with Joseph Dixon may also have provided the basic idea of mixing graphite with clay, but such details as how fine to grind the graphite and how to remove impurities that caused pencil leads to scratch would most likely have remained to be worked out. While it may have been at his father's suggestion that he focused on a new graphite mill, Henry Thoreau apparently worked out all the mechanical details. But whether a suggestion to work out the details is engineering or managing can depend on whether the suggestion is anything more than simply that—a suggestion. One thing is clear, and that is that Henry Thoreau, at least later in life, was capable of making what we would today call mechanical drawings or plans. He certainly designed and built his own cabin at Walden, and examples of a more mechanical bent in Thoreau exist in the Concord Free Public Library in his drawings for a barn and stanchion for cows and for a machine designed for making lead pipe. So it certainly seems that the younger Thoreau was not without the talents or inclination to "practice engineering" by working out the details of a solution for a machine to produce finer graphite. According to Ralph Waldo Emerson's son, Edward, who was a young friend of Thoreau, the solution consisted in having a "narrow churn-like chamber around the mill-stones prolonged some seven feet high, opening into a broad, close, flat box, a sort of shelf. Only lead-dust that was fine enough to rise to that height, carried by an upward draught of air, and lodge in the box was used, and the rest ground over." Walter Harding, in his biography of Thoreau, continues the story by describing the action: "The machine spun around inside a box set on a table and could be wound up to run itself so it could easily be operated by his sisters."

The demand for the quality pencils that the Thoreaus produced with refined graphite enabled them to expand the business. At the same time they restricted access to its premises

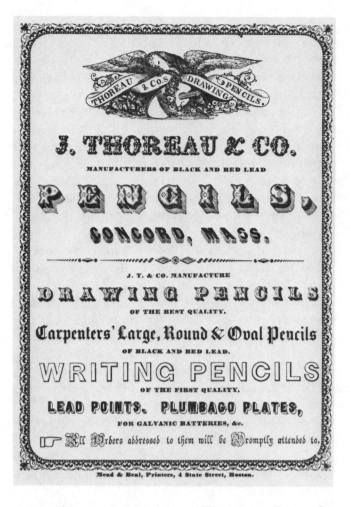

A broadside advertising a variety of Thoreau pencils, ca. 1845

because they did not want to spend money patenting their machines—or to reveal the process that was not precisely described in any encyclopedia. But apparently Henry Thoreau's personality was such that, once he had succeeded in making the best pencil in America, he found no challenge or satisfaction in the routine of doing so. What he wanted to do then was teach.

Just about the time he joined his father's pencil business, Henry Thoreau began his *Journal,* whose two million words were to comprise his major written work. In Thoreau's time, the journal, while a seemingly private form of writing, was actually a common means of communication among the Transcendentalists. They would exchange journal passages to supplement their more spontaneous forms of intercourse. Thoreau's first journal entry is dated "Oct 22nd 1837," but over the following decade, during which time he was engaged on and off in the business, he would mention pencil making rarely and then only in passing.

Thoreau grew restless when he did not find a teaching job, and he made plans to travel, setting out for Maine in 1838, but later in the year he was back in Concord running a private school with his brother. The brothers took their excursion on the Concord and Merrimack rivers in 1839, and Thoreau presumably carried his diary and pencil, even if he did not list the latter as a necessary part of anyone else's outfit for such an excursion.

John's health forced the Thoreau brothers to close their school in 1841, and shortly thereafter Henry moved into the Emerson household, where he would stay for two years, conversing with Ralph Waldo Emerson, doing odd jobs around the house, and entertaining the Emerson children. As Edward Waldo Emerson would recall later, after Thoreau told them stories, "He would make our pencils and knives disappear, and redeem them presently from our ears and noses." When Thoreau's father needed help in the pencil factory, Henry would go home for a time, and he would also put in a few days at the shop when he had to earn a few dollars. The younger John Thoreau died early in 1842 and it was a great loss for Henry, who would eventually write *A Week on the Concord and Merrimack Rivers* as a memorial tribute. Its dedicatory quatrain ends: "Be thou my Muse, my Brother—."

Henry David Thoreau spent about eight months in 1843 tutoring on Staten Island, writing home often to report on his reading in libraries and to inquire after "improvements in the pencil line." Thus the family business was out of sight but not out of his mind, and he may have been thinking of improvements of his own. He returned homesick to Concord late in the year, but soon he was in debt and so went back to work in the family factory—with renewed vigor and inventiveness. He apparently conceived of many ways to improve still further the processes and products of the factory, and according to Emerson could think of nothing else for a while (but with an engineer's characteristic literary silence about things technical).

Thoreau is reported to have developed many new approaches to fitting the lead in the wood casing, including a reputed method employing a machine to drill holes into solid pieces of wood into which the lead could be inserted. In the

Concord Free Public Library there is a pen-nib holder that Thoreau is believed to have made out of a round piece of wood, but which appears in fact to be a pencil case rejected for that use because the hole in it is very eccentric. While it might not be easy or efficient to insert and glue a brittle pencil lead into a close-fitting hole, and while the idea has even been the object of ridicule, one of the rare passages mentioning pencils in Thoreau's *Journal* suggests that a seamless pencil case is at least a dream he might have had. In describing his 1846 travels through Maine, after commenting with disdain on a shop full of frivolous toys, he continues: "I observed here pencils which are made in a bungling way by grooving a round piece of cedar then putting in the lead and filling up the cavity with a strip of wood."

While this differed from the usual American and British ways of making pencils, it was similar to the procedure used for encasing leads formed by the Conté process. Nevertheless, the passage does indicate that Thoreau certainly thought he knew the ideal or at least the proper way of making a pencil. And, after all, a round pencil lead should certainly be the preferred shape for sharpening to a point, and leads made by the Conté process could be extruded into round shapes as easily as any other. Conté himself apparently produced round leads, and they were made in England for mechanical pencils well before midcentury by passing square strips of plumbago successively through polygonal and round holes in rubies, as if drawing wire. So to insert a round lead in a round hole might have seemed to many to be the most rational of ideas, regardless of how difficult it might have been to execute, for it would have eliminated a lot of grooving and gluing operations. But whether it was even a dream of Thoreau's is not clear.

Apparently there was plenty of reason for Thoreau to believe that he knew what constituted good pencil making. Not only had he developed a fine pencil; he had also found that by varying the amount of clay in the mixture he could produce pencils of different hardness and blackness of mark, just as Conté had discovered. The more clay a pencil lead contained, the harder would be the pencil point, and that Thoreau did not realize this immediately suggests that he did not read about the Conté process explicitly. Thus Thoreau & Company could offer pencils in a variety of hardnesses, "graduated from 1 to 4," as claimed on the wrappers around the pencils, one of which advertised "IMPROVED DRAWING PENCILS, for the nicest uses of the Drawing Master, Surveyor, Engineer,

Architect, and Artists Generally." By 1844 Thoreau pencils were apparently as good as any to be had, whether of domestic or foreign manufacture, and Ralph Waldo Emerson thought enough of them to send some to his friend Caroline Sturgis in Boston. An exchange of letters in that year tells the tale:

Concord Sunday Eve, May 19

Dear Caroline,

[I] only write now to send you four pencils with different marks which I am very desirous that you should try as drawing pencils & find to be good. Henry Thoreau has made, as he thinks, great improvements in the manufacture, and believes he makes as good a pencil as the good English drawing pencil. You must tell me whether they be or not. They are for sale at Miss Peabody's, as I believe, for 75 cents the dozen . . .

Farewell.
Waldo

[22 May]

Dear Waldo,

The pencils are excellent,—worthy of Concord art & artists and indeed one of the best productions I ever saw from there—something substantial & useful about it. I shall certainly recommend them to all my friends who use such implements & hope to destroy great numbers of them myself—Is there one softer than S—a S.S. as well as H.H.? I have immediately put mine to use. . . .

[Caroline]

While there appear to be some discrepancies about exactly how much the improved Thoreau pencils did cost, with some reports that a single pencil cost as much as twenty-five cents, there seems to be little doubt that they were more expensive than other brands, some of which sold for about fifty cents a dozen. The discrepancies in price no doubt exist because over the years the Thoreaus made a variety of kinds, as surviving labels and broadsides document, and thus sold pencils at a variety of prices. Today, of course, any artifacts associated with Henry David Thoreau are prized possessions, and even as long ago as 1965 a dozen pencils offered by a Boston bookstore sold for $100 to a collector.

The variety of Thoreau pencils is further suggested by the

fact that some were "graduated from 1 to 4," which was the system adopted by Conté, while Caroline Sturgis's letter indicates that the ones Emerson sent her were graduated in terms of the letters S, presumably for "soft," and H, for "hard," with S.S. being softer than S and H.H. harder than H. While Thoreau's, by using antonyms, was a more consistent use of the language than the European system employing abbreviations for "black" and "hard," such dual systems of grading were used throughout the nineteenth century, and they continue to be used with some modifications to this day, with the numeric system now usually designating common writing pencils and the alphabetic one the more expensive drawing and drafting pencils.

Thoreau pencils also appear to have been packaged in a bewildering variety of ways, another practice that persists, presumably to make the buyer feel there is a pencil for every need. Still, all of the Thoreau pencil labels and advertisements that seem to survive, including one in a University of Florida library collection offering black- and red-lead pencils that has been dated as late as about 1845, read "Thoreau & Co.," as do the pencils in the same collection. Pencils in Concord collections, on the other hand, are imprinted "J. Thoreau & Son. Concord Mass."

While the changing designations and packagings of Thoreau pencils are difficult if not impossible to place in any incontrovertible chronological order, the confusion of undated artifacts only underscores the challenge for the historian of engineering and technology. As the Thoreaus introduced a greater variety of pencils and further improvements in their process from the late 1830s through the mid-1840s, it was no doubt desirable, if not necessary, for them to distinguish the newer and improved pencils from the older and superseded ones, but evidently they felt no need to chronicle their changes.

There was certainly no confusion among the Thoreaus about the fact that the latest new pencils they offered were at least different, if not their best, for otherwise there would be little reason to change labels and designations, and there is little doubt that before Henry David Thoreau was the literary celebrity he has come to be, the pencils he and his father made came to be without peer in this country. But the Thoreaus, like other pencil manufacturers, did not expect their word alone to sell pencils. Shortly after the Emerson-Sturgis correspondence, the family business was able to issue a circular which included a testimonial from Emerson's brother-in-law, Charles Jackson:

JOHN THOREAU & CO.,

CONCORD, MASS.

MANUFACTURE

A NEW AND SUPERIOR DRAWING PENCIL,

Expressly for ARTISTS AND CONNOISSEURS, possessing in an unusual degree the qualities of the pure lead, superior blackness, and firmness of point, as well as freedom of mark, and warranted not to be affected by changes of temperature. Among numerous other testimonials are the following.

Boston, . . . June, 1844

Dear Sir:—I have used a number of different kinds of Black-lead Pencils made by you, and find them to be of excellent quality. I would especially recommend to Engineers your fine hard pencils as capable of giving a very fine line, the points being remarkably even and firm, which is due to the peculiar manner in which the leads are prepared. The softer kinds I find to be of good quality, and much better than any American Pencils I have used,

Respectfully,

Your Obedient Servant,

C. T. Jackson

Boston, June, 1844

Sir:—Having made a trial of your pencils, I do not hesitate to pronounce them superior in every respect to any American Pencils I have yet met with, and equal to those of Rhodes, or Beekman & Langdon, London.

Respectfully yours,

D. C. Johnston

J. THOREAU & CO. also manufacture the various other kinds of BLACK-LEAD PENCILS; the Mammoth or Large Round, the Rulers or Flat, and the Common of every quality and price; also, Lead-points in any quantity, and plumbago plates for Galvanic Batteries. All orders addressed to them will be promptly attended to.

The use of English pencils as the epitome in the Johnston testimonial and in the Emerson-Sturgis correspondence adds further doubt that it was a German pencil that the Thoreaus set out to emulate. But whatever product they had improved upon, in the late summer of 1844 Henry's mother, Cynthia Dunbar Thoreau, felt the family business had earned them a house of their own, and he put more hours into pencil making to help earn the capital. Thus, contrary to the conventional wisdom then and still current around Concord and elsewhere, Henry David Thoreau was no slouch, even though in May 1845 he left home and the pencil business and began to build his cabin near Walden Pond, where he would live until 1847. Among the many activities he engaged in at Walden was a form of chemical engineering known as bread making, and among his innovations was the inclusion of raisins in some of his dough. This reputed invention of raisin bread is said to have shocked the housewives of Concord. But while he may not have won any ribbons for his cooking, in Thoreau's absence from the family pencil business, the Massachusetts Charitable Mechanic Association awarded a diploma to "John Thoreau & Son for lead pencils exhibited by them at the exhibition and fair of 1847" (perhaps reflecting that a pencil so imprinted was displayed in that year). However, in 1849, the Salem Charitable Mechanic Association awarded a silver medal to "J. Thoreau & Co. for the best lead pencils" at that year's exhibition, suggesting that the son's name even then was not consistently associated with the father's on their products.

There is no mention of pencil making in *Walden,* but there is plenty of economics and sound thinking about business, qualities not alien to good engineering. Thoreau's famous accounting of the cost of the materials of his cabin ($28.12½) and the profit he made from his "farm" ($8.71½) attests to his fondness and understanding of business as well as of engineering. As he wrote in *Walden:* "I have always endeavored to acquire strict business habits; they are indispensable to every man." Yet at the same time he recognized the absurdity of the economic system: "The farmer is endeavoring to solve the problem of a livelihood by a formula more complicated than the problem itself. To get his shoestrings he speculates in herds of cattle."

The Thoreaus had successfully speculated in pencils to get their shoelaces, and when Henry David went into debt in 1849 to publish his first book, *A Week on the Concord and Merrimack Rivers,* he manufactured a thousand dollars' worth of pencils to sell in New York. However, the market was becoming flooded with products of American and foreign manufacture, especially those of the world-market-conscious Germans, who by then had mastered the Conté process themselves, and Thoreau had to take a loss on his speculation, selling the lot for only one hundred dollars. While his book got favorable reviews, it did not sell, and he hauled hundreds of copies of it into his attic study. He is said to have remarked that his library there contained "nearly nine hundred volumes, over seven hundred of which I wrote myself."

While Thoreau was trying to sell his book, the pencil business was beginning to receive large orders, not for pencils, but for ground plumbago. The Boston printing firm of Smith & McDougal was secretive about why it wanted such quantities of the material, and the Thoreaus suspected that the firm desired to enter the pencil-manufacturing business. But after swearing the Thoreaus to secrecy, the firm explained that high-quality graphite was ideal for the recently invented process of electrotyping and the company wished to keep its competitive advantage. Selling the fine graphite powder was extremely lucrative, and the Thoreaus continued to manufacture pencils only as a front. Eventually, in 1853, they gave up the pencil business altogether, and Thoreau is said to have put off his friends, who asked why he was not continuing to make excellent pencils, with the response: "Why should I? I would not do again what I have done once."

Once pencil making was abandoned as a front, "John Thoreau, Pencil Maker," publicly announced his new product as "Plumbago, Prepared Expressly for Electrotyping," and the black-lead business continued to do well. When his father died in 1859, Henry took over the business, his conscientiousness indicated by his getting himself a copy of *Businessman's Assistant*. In the meantime the American pencil market had become overrun by German manufactures.

All the while he was dealing in fine plumbago, Henry Thoreau was also writing, publishing, and lecturing about slavery and other matters. But he always maintained a sense of the machine, even in his philosophizing. When he reflected on writing itself in his *Journal,* he wrote: "My pen is a lever which in proportion as the near end stirs me further within— the further end reaches to a greater depth in the reader." While Archimedes felt that, given a place on which to stand, he could move the earth with a mechanical lever, Thoreau apparently believed that, given a place to sit and think, he could move the soul with his metaphorical lever.

Another of Thoreau's professions was surveyor, and among

his surveys was that of Walden Pond, a model of quantification that arose out of debunking myth. He wrote in *Walden:*

> As I was desirous to recover the long lost bottom of Walden Pond, I surveyed it carefully, before the ice broke up, early in '46, with compass and chain and sounding line. There have been many stories told about the bottom, or rather no bottom, of this pond, which certainly had no foundation for themselves. It is remarkable how long men will believe in the bottomlessness of a pond without taking the trouble to sound it.

While his map of the pond in *Walden* has been considered a joke by some critics, who apparently do not wish to allow that Thoreau could seriously be both engineer and humanist, there is too much evidence to the contrary. Among the artifacts in the Concord Free Public Library is a leadless cedar pencil end with a pin projecting from it. Such a simple instrument was a means of copying drawings in the days before the blueprint and xerography. The original outline would be carefully pricked through to another piece of paper, and then the pin marks would be connected with a continuous line. Thoreau apparently not only copied but simplified his map of Walden Pond, not because the details he left out were unimportant, but because they were unnecessary for him to make his point and because they made the survey appear too cluttered. He was as critical of his drawing as he was of his words and his pencils.

Thoreau was no Sunday surveyor, for he goes on in *Walden* in true engineering fashion to specify how accurate his measurements are (three or four inches in a hundred feet). But after observing that the deepest part of the pond is at the intersection of the line of greatest breadth and that of greatest length, he reverts to philosophy and generalizes about the highest parts of mountains and morals:

> What I have observed of the pond is no less true in ethics. It is the law of average. Such a rule of the two diameters not only guides us toward the sun in the system and the heart in man, but draw lines through the length and breadth of the aggregate of a man's particular daily behaviors and waves of life into his coves and inlets, and where they intersect will be the height and depth of his character.

Thoreau always pursued a multiplicity of careers and ideas, and while he wrote his famous books he also practiced surveying throughout the 1850s. His pond surveys were even incorporated into the 1852 map of Concord, at the bottom of which he was credited as "H. D. Thoreau, Civil Engineer," a title he sometimes used. He even advertised his services, as follows:

LAND

SURVEYING

Of all kinds, according to the best methods known; the necessary data supplied, in order that the boundaries of Farms may be accurately described in Deeds; *Woods* lotted off distinctly and according to a regular plan; *Roads* laid out, &c., &c. Distinct and accurate Plans of Farms furnished, with the buildings thereon, of any size, and with a scale of feet attached, to accompany the Farm Book, so that the land may be laid out in a winter evening.

Areas warranted accurate within almost any degree of exactness, and the Variation of the Compass given, so that the lines can be run again. Apply to

HENRY D. THOREAU.

This side of Thoreau was as integral a part of his character as any other. According to Ralph Waldo Emerson, Thoreau became a land surveyor naturally because of "his habit of ascertaining the measures and distances of objects which interested him, the size of trees, the depth and extent of ponds and rivers, the height of mountains, and the air-line distance of his favorite summits." Furthermore, "he could pace sixteen rods more accurately than another man could measure them with rod and chain."

Thoreau's penchant for measurement and surveying is on display behind Plexiglas in an upstairs cul-de-sac at the end of the tour in the Concord Museum. Among the artifacts from his years at Walden are a T square and compasses and, of course, pencils. But while Emerson knew of Thoreau's pencil making, the fact that he made arguably the best pencil in America seems not to have been sufficient for the essayist, for near the end of the obituary he published in *The Atlantic*, Emerson wrote of his friend Thoreau:

I so much regret the loss of his rare powers of action, that I cannot help counting it a fault in him that he had no ambition. Wanting this, instead of engineering for all America, he was the captain of a huckleberry-party.

But Thoreau accomplished much more than Emerson seems willing to grant. Thoreau surveyed and built his own cabin on Emerson's land, and it was Emerson's pride that Thoreau fed with excellent domestic pencils, pencils made right in Concord. There are many kinds of engineering for America and for the world.

10/ When the Best Is Not Good Enough

*T*he *Complete Book of Trades* was published in London in 1837, and it contained "a copious table of every trade, profession, occupation, and calling" then pursued in England. This "parents' guide and youths' instructor" gave the usual fees for an apprenticeship and the amount of capital required to set oneself up in business. While the profession of civil engineer demanded among the highest of apprentice fees (between £150 and £400) and moderate capital (approximately £500 to £1,000), the trade of pencil maker required no apprentice fee and as little as £50 of capital investment. This was a graphic indication of the emergence of engineering per se from its traditional subsumption under specific craft and manufacturing rubrics. But the distillation of an engineering discipline from the fire and steam of foundries and factories did not at all mean there was no more engineering to be done under their roofs. It simply meant that there was a gap of specialization developing between the research and development and the production aspects of manufacturing. As this gap would widen, the necessity of bridging it would become more acute.

The widening gap was as important in pencil making as it was in any other industry of the time. In the middle third of the nineteenth century, critical questions of diminishing supplies of raw materials and increasing complexity of processes and distribution made engineering, whether or not considered a separate endeavor, more rather than less important to the maturing pencil industry. The state of the pencil-

making business in England in the mid-1830s was described
in *The Complete Book of Trades:*

> . . . plumbago, is a dark shining mineral, found on the
> Malvern hills, and in Cumberland; whence great quan-
> tities reach London; and the latter produce is sold by
> monthly exhibition, or vent, from the depot underneath
> the chapel, in Essex-street, at various prices. Its value is
> regulated by its evenness and solidity, qualities which are
> bettered by age, and which some makers extend to indef-
> inite periods. Formerly, Mr. John Middleton was the
> most celebrated maker in this respect; but at present
> Messrs. Brookman and Langdon manufacture the most
> desirable surveyors' pencils; and these necessarily com-
> mand astonishingly high prices. Needless, perhaps,
> would be the task of pointing out the numerous imposi-
> tions that are daily practised upon the public in this very
> necessary article; rank deceptions, which are also sought
> to be carried further home, by affixing to them the most
> respectable names—forged. A pencil of a penny price,
> and another value a shilling, have frequently the same
> appearance, externally.
>
> The Maker who should lay himself out for the superior
> trade, it is obvious, would require a large capital to buy
> in and mature his stock of lead . . .

As long as it was abundant and inexpensive, genuine Cumberland plumbago of the best quality was easily bought and
stored until it was "bettered by age," with new supplies added
to the pencil maker's inventory as the old was ready for use.
But as best-quality plumbago became unavailable or prohibitively expensive, and before other significant deposits were
discovered, the temptation must have been strong to make the
plumbago into pencils before it had been "bettered" and also
to label inferior pencils as containing "English lead" when they
had little if any of the precious stuff.

Other pencil makers, "of a deceptious character," used "sophisticated lead." This was not just improperly aged or inferior plumbago, but rather some substitute for the real stuff.
Such substitutes could include mixtures of graphite dust or
powder with sulphur, clay, and other additives, which naturally would be better or worse according to the ingredients and
process used. Thus, toward the middle of the nineteenth century, English pencil buyers could no more escape the problem

of inferior pencils or false claims than could their American or Continental counterparts. But as the diminishing supplies of high-quality graphite attracted unscrupulous hawkers who would promise a better pencil for a lesser price, the same scarcity would attract inventors and engineers, from within and without the pencil-making business, who would develop a better pencil, much as they would develop better bridges, but not always for a lesser price.

By midcentury, competition among pencil makers, like that in expanding industries, reached across continents and oceans, and it was in this climate that the Great Exhibition of the Works of Industry of All Nations was planned to be held in London's Hyde Park in the summer of 1851. All the world's attention was focused on the event, and among the many associated festivities held that year was the first race for what was to become known as the America's Cup. The race was only one of the many symbolic indications that the Great Exhibition marked a watershed in international technological development and competition and in the internationalization of large-scale industry and commerce. Just as all subsequent world's fairs have provided the occasion for industries and nations to display their achievements and culture with pride and to boast of their new and future products and dreams, so did this paradigm of world's fairs.

The Crystal Palace, the building made to house the Great Exhibition, outshone much of its contents. That such an enormous and magnificent and yet temporary structure should have been erected at midcentury was fitting. The Great Exhibition demonstrated by example after example how much the concept of engineering had united art and science in the eight decades since the first edition of the *Encyclopaedia Britannica* had appeared. While the Scottish gentlemen had seen no connection whatsoever between art and science, now the connections were so all-pervasive that society was assumed to view with equal interest a famous diamond and a lump of coal. One hundred thousand exhibits were spread over seventeen acres under the one remarkable roof of the Crystal Palace, and visitors were not expected to be able to take it all in at once.

Hunt's Hand-Book, which was really a Baedeker to the miles of aisles and galleries of the exhibition, actually begins by leading the visitor up to the South Entrance of the Crystal Palace, under the great semicircular electric clock, and into the transept. The printed guide draws the visitor's attention back and forth between art and science and comments on some highlights, among which was a model of the Britannia Bridge. It was actually during a visit to that bridge's construction site in northwestern Wales that Joseph Paxton conceived of his design for the Crystal Palace and sketched it on a blotter. And Robert Stephenson's daring use of new structural concepts and materials on an unprecedented scale no doubt gave Paxton heart when designing the enormous building that would have to be completely finished and occupied in record time in order to house the Great Exhibition. While the Crystal Palace itself and the artifacts it contained were no doubt a visual delight, the visitor was expected to be educated as well as entertained. For example, before embarking on a description of the numerous models and drawings of bridges on display, the handbook explains first that "for subjects so important we propose to enter into much greater detail, and shall endeavour . . . a general review of the subject of constructing bridges of great span."

In fewer than three hundred words the reader is then given the essence of a course in strength of materials that today's engineering students normally study in their sophomore year, after having taken prerequisite courses in physics, calculus, and engineering mechanics. Yet the Victorian guidebook author expected his reader to follow the argument, complete with such terms as "strain," and that was a reasonable expectation, for in the context it all must have seemed very understandable. The author was not expecting his readers to calculate or predict when or how a bridge would break, for that was left to engineers with the proper mathematics and formulas; the author was merely expecting his readers to imagine and observe what it is that engineers must calculate in order to obviate failure. The little experiment that the author directs his readers through is an excellent way of making the abstract concept of "cross-strain" tangible, and it is advisable to try it, for these same principles would become the subject of quality-assurance tests for pencil leads themselves as the international competition among pencil manufacturers intensified in the years following the Great Exhibition.

A square piece of clean, dry fir was suggested to represent one of the rectangular tubes of the Britannia Bridge, and it would be ideal for the experiment, but an unsharpened No. 2 pencil will do just as well. A typical pencil is about seven inches long and about a quarter of an inch across. This gives a ratio of length to thickness of almost 30 to 1, which is the ratio of length to depth of the average piece of wood specified in

Hunt's Hand-Book. Maintaining a constant length-to-depth ratio is important when substituting pencils for pieces of wood or models for prototypes, for otherwise different mechanical phenomena are liable to predominate and thus invalidate any analogies that one might wish to draw. One should also maintain the same cross section, but although square pencils are made, they are uncommon, and a hexagonal pencil is close enough for the purposes of the experiment. While pencil wood is likely to be cedar, and the pencil will have a piece of lead along its center, these factors are of little importance. As the handbook pointed out, materials as diverse as wood, iron, and stone behave pretty much the same under the circumstances.

To perform the experiment, first select convenient supports for the pencil bridge. These may consist of a pair of bricks or two books of equal thickness, preferably with spines out, just under seven inches apart. Place the pencil across the gap as if it were a bridge. (It will be helpful to have the imprinted face up so that the top can easily be identified after the pencil has been broken.) With the proper care that should be exercised in any experiment designed to break something, push down slowly in the middle of the pencil bridge with either a finger or a short pencil placed crosswise on the bridge. While pushing slowly, watch for the bending action and listen for the wood slowly giving way. The pencil will continue to bend and crack as long as the weight on its middle continues to be increased. According to *Hunt's Hand-Book,* "the upper surface presents the appearance of being compressed or shortened, whilst the under surface presents the appearance of being stretched or lengthened." The experiment can be stopped at any time to inspect the damage done, or the pencil can be bent until it breaks. When it is broken, the pencil will display crushing on the top and tearing on the bottom. "An attentive consideration of these appearances will teach you the nature of the forces called into play, when a beam of any material is subjected to what has been termed *cross-strain*"—that is, strain that changes from pushing to pulling from the top to the bottom.

This simple experiment also subjects the pencil to the same basic action that concerned Aristotle when he wondered about breaking sticks with the knee. In that case the hands are the supports and the knee is the means of applying the weight or force. Whether the supports move while the force remains stationary or vice versa makes no difference to the piece of wood being broken. This phenomenon must be understood in order to design pencils with leads more resistant to snapping, a goal that has been called the oldest problem in the industry.

The explication of technical matters related to bending sticks and breaking pencils was not occasioned merely by the Great Exhibition of 1851. It was rather the expectation and receptiveness of the general public for such details that ensured the success of the exhibition itself. The contemporary public was used to reading about engineering achievements in such journals as the *Illustrated London News* and, across the Atlantic, in the recently begun *Scientific American,* and one of the marvels of the age that people were reading about was the new kind of beam bridge, the Britannia Bridge across the Menai Strait, whose significance interested rather than escaped them. But while bridges could obviously be seen as unique and spectacular products of engineering, good pencils were no less an achievement.

Hunt's Hand-Book was only one among many of the publications offered for sale to the six million fairgoers, and to those who might not have gotten to the fair itself. Another popular offering was *Tallis's History and Description of the Crystal Palace,* whose three volumes sometimes went into considerable detail about everyday objects. Pencils made in England, France, Germany, and Austria were on display at the Great Exhibition, and the black-lead pencil was among the topics treated in *Tallis's History,* in a section on artists' implements, and the contemporary reader was reminded of the problems with pencils containing sulphur and similar additives. Among other drawbacks, the marks made by such pencils could not be erased and, even if apparently erased, could be restored by unauthorized persons through chemical reactions with the sulphur residue left on the paper. The degree of consumer consciousness being raised, albeit perhaps motivated by national pride, is demonstrated by the background and detail provided in the caveat:

> It is not generally known that lead dust, or inferior plumbago, is combined with sulphuret of antimony, or pure sulphur; and the greater the proportion of this ingredient, the harder the composition. When ground with the lead —generally that called Mexican—the compound is put into an iron pot, or frame, and subjected to the degree of heat required to semifuse the combining ingredients. It is then, whilst hot, put under a press, and kept there until it is cold; when it is turned out as a block, ready to be cut

into slices, and inserted in the cedars. . . . A good black-lead pencil . . . should work freely; be free from grit, yet without a greasy, soapy touch; bear moderate pressure; have a lustrous and intense black colour, and its marks easily erased. . . . The softer or darker degrees of lead are weaker, and yield more readily than the harder varieties.

The varieties of German pencils, with ornamental exteriors, which have recently been imported in large quantities, are, it appears, made of clay mixed with Bohemian lead. . . . All these pencils, however, are harsh in use, and their marks cannot be entirely erased.

Although by the middle of the nineteenth century, French, German, and even some American pencils were being made by the Conté process of mixing graphite dust with clay, those English pencils made with pure Cumberland graphite were still the world's standard. But the vicissitudes of the supply of the best graphite, due not only to wars and pilfering but also to the sheer exhaustion of the mines, presented an acute problem at the time the Great Exhibition opened. While English plumbago and pencils were proudly displayed by some British firms, *Tallis's History* made it clear that, even if the state of the industry was not what it had been, there were new developments in which to take heart:

Messrs. Reeves and Sons, of Cheapside, contributed a [display] case of some importance to artists, inasmuch as it contained the proofs of an efficient substitute for the far-famed black-lead mine of Cumberland, which is now thoroughly exhausted. It is well known, that, for all purposes having reference to art, this lead of Cumberland was unsurpassable; that no other could compare to it in quality of colour, absence of grit, nor was any so easy to erase; indeed, that no other yet found could be thus made use of in its natural state. That from the Balearic Islands is "cindery," that from Ceylon, though purer than any plumbago known, in the excess of its carbon, and the small portion of iron and earthy matter, is too soft and flaky; that termed Mexican is really produced from mines in Bohemia, and is also friable and earthy. Other varieties, from Sicily, from California, from Davis' Straits, and elsewhere, have been tried, but all have proved unfit for the use of the artist. Cumberland lead was the only black-lead that in its native state could be cut into slices; and thus be inserted into the channels of the cedars.

But after all this praise, instead of a description of the "efficient substitute," the article went on to acknowledge that not all Cumberland black lead was perfect and that some of it had been known to possess "uncertain temper and occasional grit." Thus, as pencil makers were using up the last and the best of their previously mined stocks of imperfect Cumberland lead, the artist was finding it difficult to get a top-quality pencil. Yet in spite of the material's near-extinction, a large lump of unprocessed Cumberland graphite was on display in the Crystal Palace, and *Hunt's Hand-Book* did not hesitate to make a natural philosophical connection between the dark plumbago and the sparkling object in a neighboring display:

> The diamond, as Sir Isaac Newton conjectured from its high refracting power, is a combustible body. The researches of Lavoisier, and others, have shown that this gem is nothing more than pure carbon, and under the influence of the voltaic battery diamonds have recently been converted into coke. The plumbago, which is in the adjoining bay, and the coke but a short distance from it, differ only in physical condition; in chemical constitution they are similar to the diamond. *The Koh-i-Noor* diamond is to the east of the Transept. . . .

Hunt's Hand-Book goes on to explain that Koh-i-Noor means "Mountain of Light" and to relate the Hindu legend of the diamond's discovery and its history. Pages later, after descriptions of such objects as corundum, turquoises, and a "case of pearls found in the deepest part of the river Strules," the reader reaches the adjoining bay in the Crystal Palace, where five English exhibitors are grouped under the heading "Plumbago—Black Lead—Pencil Manufacture." It is here that we learn that the output of the Cumberland mine fell from five hundred casks in 1803 to "about a half-a-dozen casks, weighing a hundred and a quarter each" in 1829. And we also learn from specimens exhibited that at midcentury England was importing plumbago from India, Ceylon, Greenland, Spain, Bohemia, and the Americas. A then recent small find in the north of Scotland is described, and its chemical analysis by weight is even given, showing it to contain 88.37 percent carbon.

One of the exhibitors in the Crystal Palace was William Brockedon, who had found the "efficient substitute" process whereby graphite powder could be re-formed into blocks without the use of a binder. Brockedon was the only son of a watchmaker, from whom the child acquired a taste for scientific and mechanical pursuits. He began to work as a watchmaker, devoting his spare time to drawing, which he also had come to love since childhood. With the encouragement of a patron, Brockedon studied at the Royal Academy and became an established painter. He was also an author of some reputation, having written and illustrated many travel books. Throughout his career as an artist and author, Brockedon maintained his interest in mechanical pursuits, and he patented inventions ranging from a method of drawing wire through holes pierced in sapphires, rubies, and other gems to a novel pen point. In 1843 he patented his method of producing "artificial plumbago for lead pencils purer than any that

An 1834 pencil sketch of Thomas Telford, first president of the Institution of Civil Engineers, done by William Brockedon, who invented a process to compress graphite dust

could then be obtained, in consequence of the exhaustion of the mines in Cumberland." The pencils made from Brockedon's plumbago were considered especially valuable to artists because they were free from grit.

While *Tallis's History* mentioned only in passing "the somewhat recent discovery by Mr. Brockedon of a process by which lead is made perfect," its readers were informed that the new process was being used and that pencils made by it for Reeves and Sons were "unquestionably what they affect to be." *Hunt's Hand-Book* went further and revealed how Brockedon accomplished his black magic. He first ground the graphite in water and made it as finely and uniformly divided as possible. Then he enclosed the powder in special paper packets, evacuated the air, and subjected the mass to a great pressure. Presumably the slippery graphite powder moves freely under pressure, and, according to the handbook, after the pressure is relieved and the graphite removed from the paper, "close examination shows that the particles have arranged themselves under the influence of the pressure in precisely the same manner as in the natural productions; and if one of these compressed masses is broken, the fracture is exactly similar to that exhibited by a piece of native plumbago."

The process was first worked commercially by Mordan & Company, but upon Brockedon's death in 1854 the plant and machinery were sold at auction to a Keswick merchant. According to one obituary, Brockedon's was the "best black-lead" then produced. Apparently the process continued to be used for a number of years in England and America, but it seems to have been expensive and to have become less commonly used after the Borrowdale mines were finally exhausted.

It was the Conté process that was to be almost universally used for pencil making, and *Hunt's Hand-Book* makes it clear that this process was common knowledge at the time of the Great Exhibition, at least in England. According to Hunt, "M. Conté, in 1795, was successful in combining plumbago with clay, and then calcinating the mass, so as to produce crayons of any shade." In this context "shade" refers to the degree of blackness of the lead and not the color. Colored Conté crayons, on the other hand, like colored pencils generally, contain wax and are not subjected to the heat that is so important in the creation of the ceramic lead containing graphite and clay. While pencils "for drawing, engineering, &c." were exhibited in the French section of the Crystal Palace, no actual pencils, French or English, black or colored, are described by Hunt.

Meanwhile, in America, an article in *Scientific American* for the year 1850 gave no indication of an awareness of the advantages of the Conté process, noting only that "pure Cumberland black-lead (plumbago) is of too soft and yielding a nature to enable an artist to make a fine clear line." As a remedy for this the journal reported a process by which shellac was mixed with plumbago in a repetitious process of melting and pulverizing, giving no hint that using clay accomplished the same end—namely, pencils "of various degrees of hardness . . . their blackness . . . in proportion to their softness."

Pencil makers in "Great Britain and the Islands in the Seas," as the 1851 Census Population Tables reported them, numbered 319, and while many were located in London a good number were concentrated in a small area around Keswick. This Cumberland town was historically the center of a pencil-making cottage industry because of its proximity to the source of plumbago, of course, but other minor industry had been located in the Lake District generally since the eighteenth century. The area was a natural host for such activity because of its abundance of timber and minerals and because of its ample rainfall, which provided for plenty of waterpower. According to an industrial archaeology guide to the district, which is commonly associated with poetry and beautiful scenery, its "streams, minerals and woodland resources have been turned to the uses of man for many centuries." Furthermore:

> Against the magnificent background of the high fells, altered only by the touch of the seasons, men have dug, transformed and contrived, winning minerals from a thousand twisting workplaces beneath the walker's feet, and leaving only the occasional cleft or mineshaft as evidence of their activity. In the dales below, the traces of water-powered mills, crushing gear or ironworks are scarcely more noticeable to the holiday-maker.

It was in this environment that traditional pencil making was centered. Although new processes, secret and otherwise, had been providing acceptable substitutes for pure Cumberland black lead, the new techniques tended to be complex and elaborate and raised the price of the best English drawing pencils. Thus artists, who tend to take a technical interest in their materials anyway, became especially interested at mid-century in learning all they could about pencil making employing native graphite. The editor of *The Illustrated Magazine of*

Art no doubt sensed this, and in 1854 an article entitled "Pencil Making at Keswick" appeared in that journal. The article demonstrates that, in spite of decades of short supplies of high-quality plumbago, some aspects of English pencil making had experienced few changes, even though machinery had been introduced in the early 1830s to take over some of the operations, done previously by hand.

In the Victorian manner, the magazine piece opens with a florid description of the cradle of the local industry:

> Situated in a slightly undulating valley, with the lake of glorious Derwentwater in the immediate vicinity, backed by Skiddaw, who rears his hoary peaks to an elevation of more than three thousand feet, and traversed by the river Greta, endeared to every lover of the English language by its literary associations, is the pretty straggling town of Keswick. Were this spot "unknown to fame," from the irresistible attractions which its neighbourhood presents to all lovers of the sublime and beautiful, there would be an interest felt in the spot by at least some sections of the community, as having furnished them with the means of embodying their own conceptions of taste and fancy by the pencil of the artist. . . .
>
> The pencil works of Messrs. Banks, Son, and Co., which we have to visit, are seated on the banks of the Greta, the waters of which furnish the motive power for all the machinery of the establishment.

Just as the railroad and its infrastructure may be seen to be unintrusive in George Inness's *The Lackawanna Valley,* so the waterwheel implied in this art magazine's word picture provides the natural and matter-of-fact connection between the bucolic evocation of the Cumberland valley and the largely unflourished tour through the factory that follows. Unromantic illustrations of men and boys working at the various stages of pencil making break up the text that describes technical and practical aspects of the industry, including the focus of the manufacturing process itself, which occurs at the benches where the lead is fitted into the grooves in the strips of cedar. But the scene is so striking that the author almost lapses into the style he had mostly left outside the factory before getting on with the technicalities:

> The men . . . are dressed in dark blue smocks,—this being the general costume of the place,—with loose

sleeves fitted tight at the wrist, and are sitting at very black shining tables. The men's hands, and the tools with which they are engaged, as well as most of the furniture of the apartment, look as if they had been fresh polished every morning by the servants, by the same processes by which they cleansed the grates and stoves; while their faces often exhibit tints and streaks of different colours. Each workman has a number of sticks of cedar, in which the grooves have been cut, and a number of slices of lead just as they appear after the sawing. He then takes one of the slices, and having seen that it is not too broad to enter the groove—for if this is the case he rubs it down to the proper dimensions on a rough stone which lies in front of him—he dips it in a pot of glue which is kept hot just beside him, and then presses it into the grooves. He then gives a scratch to the lead on a level with the surface of the wood, and breaks it off, so as to leave the groove properly filled. In the making of a single pencil, perhaps as many as three or four slice lengths are required; but however many there may be, each slice is fitted exactly endwise to the other, so as to leave no intervals.

The filled cedar rods were passed on to the "fastener-up," who performed the next operation of gluing another piece of cedar over the lead, clamping the assembly, and putting it aside to dry. Sometimes triple-length cedar rods were grooved, filled, and glued, with each dried and finished assembly yielding three pencils. The magazine article culminates in statistics on the factory's production of five or six million pencils a year. At one machine a man could round 600 to 800 dozen pencils a day, at another a boy could smooth and polish some 1,000 dozen a day, and at another machine as many as 200 pencils per minute were stamped: "Banks, Son, and Co., Manufacturers, Keswick, Cumberland" along with letters signifying the degree of hardness or softness of the lead.

While it is generally known that early pencils were not painted the way modern ones are, the art magazine's article on pencil making in Keswick clearly indicates that the practice, including the production of *yellow* pencils, which some pencil manufacturers would later claim to have introduced in the 1890s, was already known in 1854:

The fashion of varnishing pencils has come up very recently. It first began with inferior kinds, but it is now adopted with the best, and many sorts of pencils will

ROUNDING PENCILS

POLISHING PENCILS

CUTTING PENCILS TO LENGTH

GILDING PENCILS

Some steps in finishing pencils at Keswick, ca. 1854: note that these pencils were assembled in triple lengths to be cut to size after the polishing operation

indeed hardly sell without it. It brings out the colour of the wood, while it serves at the same time to prevent the pencil getting black and dirty during the cutting, and preserves them uniformly clean. . . .

Many pencils are now finished, but some have gilt letters instead of the mere impress on the wood. . . . The pencils that have gilt letters are usually coloured black, yellow, or blue, by which the fine tint of the cedar is altogether lost.

After describing how the pencils are wrapped for shipping, the author allows himself to imagine where the pencils might end up going—one "to the studio of an artist, another to the boudoir of a lady"—and to philosophize about the fact that it is part of a grander design that a pencil is created to be destroyed: "And we might conclude by moralising on the fact, that as it is by the wear and tear and destruction of the agent that its worth is developed, so it often is that men, in striving and labouring for society and the world, are themselves exhausted and consumed, and the elements of their physical constitution pass away, to mingle with, and to be absorbed into, the universe at large."

The plumbago from Borrowdale had certainly been absorbed into the universe at large over the three centuries since it had been discovered: being blown in dust from all the sawing and rubbing, being deposited on the furniture of pencil factories and on the hands and clothes of their workers, being carried in fabricated veins of lead in millions upon millions of wood-cased pencils made and exported around the world, being buried with the stubs of pencils no one wanted to hold on to, being laid down in notes in the margins of books like trail markers through forests of thought, being redeposited in thin lines of thoughts and images on countless sheets of paper, being twisted and crushed in the lines of crumpled manuscripts and sketches, being burned with the thoughts and images no one wanted, or no one wanted to remember or to build. So by the mid-1800s what had once been the world's purest source of plumbago was essentially worked out and had been diffused throughout the world in a three-centuries-long fit of black entropy.

The article on pencil making at Keswick reiterated the history of the social entropy that had also hastened the disorderly, if not at times chaotic, exploitation of the Cumberland plumbago:

When its commercial value was first ascertained, the proprietors found it very difficult to guard the mine from depredations; the practice of robbing it having become at length so common, that persons living in the neighbourhood were said to have made large fortunes by secreting and selling the mineral. About a century ago, a body of miners broke into the mine by main force, and held possession of it for so long a time as to succeed in abstracting from it an enormous quantity of lead, which they sold at so low a price, that the proprietor was induced to buy it up in order to restore the old rate of prices. Some years since the mine failed, and very little or anything has been obtained from it since, though there is Borrowdale lead still in existence. Messrs. Banks, Son, and Co., are part proprietors of the mine, their share at the last and final division of the produce being about five hundred pounds' weight of the lead.

So in 1854 pencil manufacturers such as Banks still had stockpiles of Borrowdale graphite with which to make their finest pencils, but that could only be done with pieces sufficiently large. Some pieces were only pea-sized when taken out of the mine, and they were used to make inferior pencils, after being "cut up, pounded down, and mixed together" along with a large quantity of imported graphite. Artists were finding that pencils were "very inferior in quality to what they once were," and that even those stamped "Warranted pure Cumberland lead" were found often to have "little or none of it in them."

Even at Banks's factory, whose ownership of the mines went back to the turn of the century, pieces of Borrowdale graphite that were too small to be fitted directly into wood cases were pulverized and, with the dust from cuttings, were being mixed with increasing quantities of inferior graphite, presumably to be formed into leads by being bound with gums or waxes or combined with clay in Conté-like processes or even by some still newer method like Brockedon's that compressed graphite dust into reconstituted blocks. One form of inferior pencil, made of two-thirds poor plumbago and one-third sulphuret of antimony, was called a plummet as late as the latter part of the nineteenth century. Very inferior pencils were made of a mixture of powdered graphite, paste, and the claylike fuller's earth.

In light of the many poor-quality pencils that must have been offered at the time, it should be little surprise that in

1854 an art magazine could expect its readers to be equally interested in a bucolic valley, its bustling industry, the exhaustion of one of the world's finest sources of graphite, and the philosophical implications of it all. A decade later another English art journal would also carry an article on pencil making in Cumberland, for the situation would not improve. But in 1854 it was only three years since the Great Exhibition. The workings of art and industry and, if not the universal then at least the international significance of it all were still the talk of London and the world.

Other countries were already holding or planning their own world's fairs, to be housed in their own crystal palaces. These temporary structures would evolve more over the next sixty years than would some of the smaller labor-intensive pencil factories, however, as would be seen in the television miniseries *The Murder of Mary Phagan*. While the men and boys in the Banks factory in Keswick were replaced by women and girls in the small National Pencil Company factory in Atlanta, where the young Phagan girl was murdered in 1913, the machinery depicted there was minimal and pencil-making operations were still largely manual.

Regardless of how they were made, no one in the early nineteenth century seems to have questioned the assumption that pencils made of pure Borrowdale graphite were the best anywhere in the world. However, even the best available product of craft or technology will have flaws and shortcomings, if not in the artifact itself, then in the process of manufacturing it, and so the artifact and the ways of making it can always be improved upon. In the case of Cumberland pencil, the increasing difficulty in obtaining large pieces of good Borrowdale graphite meant that the pencil lead was often a series of short pieces of graphite butted up against each other throughout most of the length of the pencil. This would work fine and the pencil would write or draw with all the wonderful qualities of the best pencils—until one used a penknife to sharpen a fresh point on a worn-down one. If the exposed lead was near the end of a piece of graphite, the sharpening process would leave a very short piece of unbroken lead encased in the wood, and when one tried to write or draw with this newly sharpened pencil, the point would most likely break off or fall out of the wooden shaft. Such frustrations were common even with early-twentieth-century pencils, when what was supposed to be a single continuous piece of lead somehow became broken into short pieces inside the wood. While this kind of hidden

fault in the lead would be prevented by more advanced manufacturing processes used in making quality pencils or by taking care not to break the brittle lead by bending or dropping
the pencil, most pencil users would prefer a more rugged
product. Thus pencils produced in industrialized countries
today contain relatively tough leads bonded to the wood. But
a recent visitor to China complained that he could not sharpen
a Chinese pencil because the lead kept falling out.

Clearly the best remedy to the problem of discontinuous
pencil leads was to make the lead in one tough continuous
piece. The early mixing of graphite sawdust and pulverized
graphite with waxes may have solved part of the problem, but
at the sacrifice of marking quality. The Conté process also
made it possible for leads to be formed in one piece, but some
artists believed that in marking quality they were inferior to
the best pure Cumberland graphite. Brockedon's technique of
subjecting powdered graphite to great pressure produced a
high-quality pencil, but at a high price. And so one had to
choose among the disadvantages of discontinuous lead, arguably less quality, and high price.

The best of anything can be as elusive as the best pencil, as
illustrated by the difficulty encountered when the "best" iron
was sought for the construction of the Britannia Bridge. According to the account of Edwin Clark, the resident engineer:

> It was essential to use the finest possible quality of iron
> for so important a work, and the contracts accordingly
> were entered into, specifying "the best plates." On find-
> ing, in some instances, these plates very inferior, and on
> making inquiries on the subject, it was discovered that
> among the ironmasters the word "best" means, literally,
> the ordinary quality of plates, or common iron; the term
> "best-best" being applied to plates of a higher price, and
> even "best-best-best" being sometimes applied to signify
> what, in our ignorance, we imagined to be expressed by
> "best." A higher price for good iron was accordingly de-
> manded and paid.

Getting better iron or better bridges or better pencils usually requires more than having command of "a very peculiar
dictionary of technical terms," which Clark found necessary
for ordering the proper iron for the Britannia Bridge, for there
are always limits as to how long a string of "bests" can be and
still produce a noticeably better product at a price someone is

willing to pay. As with pencils made in Keswick in the mid-nineteenth century, pencil manufacturers have continued to make their products in a range of qualities, and one must know the "peculiar dictionary" of the pencil industry to know what means what.

11/ From Cottage Industry to Bleistiftindustrie

*P*encil makers in Germany in 1800 had neither the high-quality graphite available in England nor the knowledge or inclination to exploit the new clay-and-graphite process for making leads that France had pioneered. Political and cultural traditions in Germany had worked against the international development and growth of business, and pencil making had by and large remained a cottage industry employing methods passed down from master to apprentice. But as the eighteenth century gave way to the nineteenth, old ways also began to give way to new. There was to be a lessening of restrictions imposed by the trade guilds on craft practices, and out of the ranks of the traditional craftsmen came the new breed of *Fabrikant,* or manufacturer and factory owner. But it would be well into the nineteenth century before this enterprising new kind of pencil maker would overcome the difficulties created by a long period of neglect of technological development.

The Staedtler family business, for example, which was started in 1662 by Friedrich Staedtler, *Bleistiftmacher,* was successively run by Johann Adolf, Johann Wilhelm, and Michael, each of whom was a *Meister,* but Friedrich's great-great-grandson, Paulus Staedtler, called himself a *Fabrikant* even before he passed his examination for the title of *Meister* in 1815. He was able to advance his position despite the traditions of the craft and guild system no doubt in part because of his own ambition and personality, but he was also able to do so because he happened to be running the pencil business when political, social, and technological factors made change possible.

With the incorporation of the Free Imperial Town of Nuremberg into the new Kingdom of Bavaria, the town's Trade Inspection Board was dissolved in 1806. While the liberalizing influence encouraged new pencil makers like Johann Froescheis, who that same year bought an old factory and started what today is the Lyra Bleistift-Fabrik, the pencil-making business generally was not in a position to capitalize on its new freedom because of the unavailability of good English graphite and the German adherence to old lead-making processes, which could no longer produce pencils that could compete in the free marketplace. Not only were French pencils made with clay of much better quality than the German sulphur-based ones, but with the Conté process it had become "possible to make pencils whose writing properties were in no way inferior to the so-called English pencils." Even if clay-and-graphite pencils were not as good as the best English ones, those were becoming scarce indeed, and thus standards of excellence were to change.

The very existence of the German industry became threatened by foreign goods, but the individual pencil-making concerns appear not to have taken any steps to help themselves. Thus in 1816 the Bavarian government built a royal pencil factory in Obernzell, near Passau, in the German coal district, in order to experiment with making pencils in the French manner. While this endeavor was not ultimately a commercial success, it did spur on the private manufacturers around Nuremberg to take another look at how they made pencils. Paulus Staedtler represented the local manufacturers in carrying out tests with clay-and-graphite leads, and his efforts were so successful that he soon adopted the new method for his own factory.

Contemporary with the relaxing of regulations and the emergence of the new lead-making process there was the introduction of horse and steam power into pencil making, which thus shared in the Industrial Revolution. The increased efficiency, productivity, and distribution made possible by the newer ways not only gave impetus to businesses to expand but also encouraged individuals to strike out on their own. Johann Sebastian Staedtler, in keeping with the old guild custom, had begun working in his father's factory in 1825, but in 1835 he applied for a license to open up his own pencil-making business "in all towns and cities of the Kingdom of Bavaria." It became Johann Staedtler's intention to establish a factory that would incorporate mills to grind graphite, kilns to bake leads, and machines to cut, slot, and shape wood for pencil cases, which at the time were being made out of imported Florida cedar as well as domestic alder and lime. Staedtler's was an innovative and ambitious undertaking in the inaugural year of the first German railway, running between Nuremberg and Fürth, and only a year before the introduction of the first stationary steam engine in the area, but he was to succeed admirably. In addition to making black-lead pencils out of graphite and clay, the company also manufactured fine colored leads, using cinnabar and other natural pigments for the crayons, as colored pencils were called.

While sixty-three different types of pencils would be shown by the firm at the 1840 Nuremberg Industrial Exhibition, J. S. Staedtler, as the firm is still known today, began by concentrating on the manufacture of red-ocher crayons, which the family business had long specialized in and which young Johann Staedtler had vastly improved upon while still with his father's concern. According to a contemporary report, the new red crayons had "the great advantage over crayons manufactured from ordinary ruddle in that they can be sharpened more easily, always have the same hardness, and retain their color unchanged." Other pencil manufacturers soon used J. S. Staedtler as their own supplier of red crayons, and the business expanded to about one hundred employees when the factory was passed on to the three eldest Staedtler sons in 1855. One of the brothers, following his father's example, eventually left to start his own pencil-making business, Wolfgang Staedtler & Company. By the 1870s, J. S. Staedtler was producing about two million pencils annually, but both surviving Staedtler firms were beginning to face difficulties in an increasingly competitive world pencil market, and the J. S. Staedtler Company was sold by a grandson of its founder. From 1880 J. S. Staedtler would be owned by the Kreutzer family, which in 1912 would buy out the failing Wolfgang Staedtler firm to gain total control over the Staedtler name in pencil making.

Such a family saga is not unique. Kaspar Faber, who had hung a shingle outside his cottage in Stein in 1761, also made a go of his small local business, but since it was not in the merchant's interest for retail customers to know where their favorite pencils were made or by whom, Faber, like Staedtler, could not put his name or address on his products. Rather, he could only distinguish his different grades of pencils from each other and from those of other makers by means of a nonsense mark or device, such as a harp, star, moon, or crossed hammers. But it was through such trademarks that the Fabers, Staedtlers, Froescheises, and others were to establish a growing demand for their products, and the lyre registered in 1868 by Johann Froescheis's son, Georg Andreas, is claimed to be the oldest pencil trademark still in use.

When Anton Wilhelm Faber took over his father's pencil-making business in 1784, he carried on pencil making in the traditional way. Even into the nineteenth century, Faber continued to make pencils by smelting "Spanish lead" and sawing it into pieces to be fit individually into wood cases. The firm of A. W. Faber was not prospering when the proprietor's son, Georg Leonhard, took it over in 1810. Competition in Nuremberg and its environs had increased a great deal, and the local demand for pencils was not nearly as great as the potential supply. Only better or cheaper pencils could compete. Better pencils were difficult to produce without obtaining English graphite or adopting the newer French process, and cheaper pencils were driving out the good, to the general detriment of the German industry. While Georg Faber did make some improvements in the management of A. W. Faber and its manufacturing processes, the business suffered in an oppressive and xenophobic environment.

When Georg Faber died in 1839, his oldest son, John Lothar, took over. From his youth he had studied all aspects of his father's pencil business, and had gone to Paris in 1836 to broaden his education. In Paris, Lothar Faber observed the manner in which Paris firms maintained close business relations with markets throughout France and abroad, and he saw the advantages of a world market for pencils. He also visited London and learned more about trade before returning to Nuremberg to take over the factory, whose work force by then had dropped to about twenty.

Among the changes that the young Faber felt were essential to the growth of his business was the introduction of more and better products, which would also necessarily sell for higher prices. Adopting the French process of baking ceramic leads from a graphite-and-clay mixture enabled Faber to introduce a new line of smooth-drawing pencils in different degrees of hardness and softness. Because of their predictable uniformity and their variety, these "polygrade" pencils were attractive to artists and engineers, but there was not a very large outlet for the high-priced pencils in Nuremberg alone. Thus Faber personally traveled throughout Germany and all over France, England, Italy, Austria, Russia, Belgium, Holland, and Switzerland to establish outlets for his pencils. The new foreign markets were to grow in importance as Faber continued to introduce new and improved pencils over the years, and foreign demand continued to grow not only for A. W. Faber products but also by association for pencils of other German manufacturers. For such contributions to the worldwide growth of Germany's industry, the innovative marketer would eventually be granted a patent of nobility and become Baron von Faber.

Faber pencils were so widely known by midcentury that "Faber" came to be used by some as a generic term for pencil. However, their quality appears to have suffered from an increasing decline in the availability of good graphite. One observer, writing in 1861, found it hard to believe that good lead pencils were impossible to come by, but "having given fair trials to most of the well-known makers," he concluded that it was "next to impossible." His complaint was "common to architects and draughtsmen" of the time, he asserted, and he went on to elaborate:

> Since the Exhibition year, 1851, I have been able to find no pencils equal to what Faber's were: then they were perfection; now they are no better than the others. It is too bad that manufacturers who have made really good articles, to take medals at the Exhibitions, should take advantage of their well-known names to offer to the public an inferior article, at the same price, making believe it is equal to what deservedly took the medals.
>
> Any pencil-maker who would produce pencils capable of making a fine, firm, black line, the same throughout, and insure all having the same letter on them being alike, would certainly confer a benefit on himself and the public. . . . As it is, one cuts two pencils of the same letter, and the chances are, one is hard, the other soft; and often the same pencil is hard at first and soft at the end.

The Conté process was apparently more widely known than mastered in the wake of the Great Exhibition, and the resulting international competition made an advantage once held hard to retain. A. W. Faber's pencils might never have been able to build on their early lead in the industry and overcome some of the ground they had lost had a new discovery in the Orient not given them an inestimable advantage. The one aspect of A. W. Faber's pencils that was to make even the inconvenience of a poorly glued case tolerable was that the better ones would contain graphite from a new mine that was widely acknowledged to yield the best-quality marking material to be discovered in three centuries. The story of how Faber cornered the market on this new graphite began far from Germany, with a French merchant residing in Asiatic Siberia who had heard about discoveries of gold being made in California.

On a business trip through the mountainous region of eastern Siberia beginning in 1846, Jean Pierre Alibert looked for signs of gold along the sandy beds of several rivers flowing into the Arctic Ocean. He found no gold, but in one of the mountain gorges near Irkutsk, he came upon some pieces of pure graphite. Since they were smooth, round, and highly polished, Alibert reasoned that they must have been carried a great distance by the stream, and so he began systematically to follow the stream and its tributaries to their sources. In 1847 he located the source of the graphite itself, about 270 miles west of his first discovery, in a branch of the Saian mountain range, at the top of Mount Batougol, near the Chinese border.

Supplies had to be brought in from hundreds of miles away, and they could be taken to the mountaintop only by reindeer, but Alibert was not discouraged. He started a farm at the base of the mountain to raise what food he could, and he gradually peopled a colony with workmen. The three hundred tons of graphite that Alibert removed from the mine in the first seven years was no better than what would once have been considered waste at Borrowdale. Eventually, however, a rich and unbroken deposit was uncovered, and it yielded some pieces of pure graphite weighing as much as eighty pounds. The Russian government encouraged the exploitation of the mine, and Alibert submitted samples of the richest material to the Academy of Sciences in St. Petersburg, which judged it to be of the same quality as the famous Cumberland graphite. The Imperial Academy of Fine Arts also examined Alibert's graphite and reported that they found it "to be of excellent quality for drawing-pencils of every kind, and that it not only by far surpasses that used at present in the manufacture of all other lead-pencils, but is equal, nay, superior even, to that formerly obtained from the now exhausted mine of Borrowdale, the pencils made from which enjoy such a high reputation throughout Europe."

Alibert was awarded a royal silver medal, and Mount Batougol was renamed Mount Alibert. He also traveled to England, where he assured himself that the Borrowdale mine was essentially worked out, and he asked English pencil makers to examine his new find of graphite. They agreed with the Russian Academy that the Siberian graphite was "in no way inferior to the Cumberland lead," and Alibert concluded that he had discovered something in Siberia as good as California gold. He received the Cross of the Legion of Honor from the French Emperor, and the Society for the Encouragement of Arts and Sciences, which did not believe prospects to be good for producing artificial graphite, awarded Alibert a gold medal. Alibert deposited in various museums specimens of his graphite, and their beauty and value earned him further honors, from Spain, Denmark, Prussia, Sweden, and Norway, as well as Rome and elsewhere.

Since he believed that A. W. Faber was the largest pencil manufactory then in existence, "and that it deposited the highest quantity of high-class goods throughout the civilized world," Alibert approached the firm with an offer to give it exclusive rights to purchase his Siberian graphite for the manufacture of lead pencils. An agreement was reached in 1856, and it was sanctioned by the Russian government, which controlled the mineral rights. But having the finest graphite is not the same as having the finest pencils, for it took "five years more of incessant labor and study before [Faber] had successfully mastered the difficulties of the new material." It was not simply a matter of sawing the graphite into slips and gluing them in cedar cases. Although that would have made fine pencils, their hardness and blackness could not have been controlled beyond the range in which the graphite occurred. The real value of the Siberian graphite lay in exploiting its purity by properly grinding it, mixing it, and baking it in order to produce pencils in a spectrum of uniform and reproducible hardnesses, and it was developing the proper manufacturing and production processes that took five years. But such an investment would help Faber maintain its international market.

Even though Faber pencils had become world travelers, Faber pencil workers did not stray very far from Stein or Nuremberg. A world market meant that not only quantity but also variety had to be added to production, and the growing business had to increase its workforce. It was clearly desirable to have long-term and loyal workers, for then there would be less need for training and defection with trade secrets would be minimized.

A savings bank was instituted, and after their deposits reached a certain minimum workers earned 5 percent interest, but, being intended for future needs, the deposits could not be withdrawn except in an emergency. The factory maintained a fund to pay the workers when they became ill. Employee housing was constructed and land sold and money lent to those workers who wished to own their house. A school, church, lending library, open garden plot, outdoor gymnasium, and other amenities were provided. There was also a nursery, which during working hours cared for the young children whose mothers were "unwilling, or cannot afford, to give up working in the factory."

Some of the things Faber gave funds for, such as the village church, were for the larger community of Stein, but most of the benefits were for the smaller community of the A. W. Faber worker family. The Fabers reported participating in sports and festivities with their employees, and the Faber family dwellings and factory buildings became intermingled. Lothar Faber is almost deified in one of the firm's own histories:

> He himself dwells near them, and truly in their midst. The gardens and parks surrounding his house and those of his brother [Johann] enclose the factory buildings on three sides, while the river Rednitz runs between them and the village itself. The slight eminence on which the dwelling-house is built makes its turreted roof a prominent object on the northern bank of the little stream, while the pointed Gothic spire of a bright, cheerful church at the southern end of the village . . . throws over the entire neighborhood a halo of peace, quiet, and plenty.

Whether A. W. Faber's omnipresence in nineteenth-century Stein was seen this way by the workers, or whether they found the presence oppressive, is not recorded in the company histories. However, there seems little doubt that pencil making did benefit Stein, whose population would triple by the end of the century from about eight hundred when Lothar Faber took over the business. But long before then, to honor the fact that A. W. Faber had been making pencils for one hundred years, a celebration was held on September 16, 1861.

The festivities began early in the morning and continued

throughout the afternoon, with games and prizes and dancing around the maypole, to be interrupted only by the arrival of a handwritten congratulatory letter from the King, which noted that Faber's "well-earned reputation . . . both at home and abroad" honored the industries of Bavaria. The King also expressed approval of Faber's attention to his workers' "moral and economic well-being" and wished the firm continued prosperity. The letter was signed: "Your affectionate King, Max." After reading it, Faber called for three cheers for King Max.

Then he expressed his indebtedness to the artists who had used his pencils and furthered their reputation, read a poem based on his motto, "Truth, Integrity, Industry," and unveiled an "allegorical picture . . . representing the activity of the mercantile enterprise of the manufactory and containing an appropriate allusion to the jubilee festival." The artist reviewed "the history of the lead pencil and its application to art," thanking Faber for his contributions and calling for three cheers for the pencil factory. Among the presentations made to Faber was the announcement of a trust set up to pay in perpetuity a choir of boys, "who ever after will keep his birthday by the singing of hymns at break of day beneath his window, while he lives, and over his grave after his death."

The centennial celebration was such a success that another holiday was planned to celebrate the twenty-fifth anniversary of Lothar Faber's proprietorship. Although August 19, 1864, was the actual date of his silver jubilee, September 19 was chosen as the holiday. On that morning Faber was presented with a painting depicting a graphite mine, cedar trees being felled, and other suitable pencil motifs, all surrounding a poem and the signatures of the presenters. In the afternoon a procession from the church was led by "a herald who, instead of the usual staff, held an enormous pencil in his left hand, while the trappings of his horse were tastefully ornamented with designs composed of all sorts of pencils," and followed by a line of vehicles and floats:

> The first car was representative of graphite mining, and bore miners in German and Chinese costumes, the latter being intended as an allusion to the graphite mine in Siberia which furnishes its output to the A. W. Faber manufactory. The second car showed the process of washing the graphite, and the preparation of the lead. The third represented the manipulation of the wood; the fourth the glueing of the pencils; the fifth the planing and

finishing of the pencils; and the sixth the polishing, binding and lithographing process; the seventh carried a ship decorated with the American star-spangled banner and the German, British and French flags. The vessel was manned with white and black sailors, and laden with Florida cedar wood. As eighth car a specimen of the delivery van was chosen, in which a workman stood with a truck-basket, showing the manner in which the pencils formerly were delivered. The ninth car was laden with flowers, fruit and vegetables, and was suggestive of the transformation of the waste land and impoverished fields of former days into the large and charming von Faber estate. These cars were followed by four workmen, who supported on their shoulders a pencil eight feet long and of corresponding thickness, blue polished, pointed at one end and fitted with a white tip at the other.

This was not the end of the procession, for still to come were an enormous writing slate and a model of the church, but the cars representing the stages in the pencil-making process were the center of attention. And those riding in the cars did not hold artificial poses, for the passengers, the "sawyers, planers, groovers, washers, gluers, finishers, markers, printers, and the girls engaged in polishing and tying, were all busily employed at their work tables and machines." There was also a working steam engine. When the remarkable procession arrived at the park, the surprised proprietor was presented with speeches, diplomas, and cheers. Then the procession passed slowly by Faber and his family, each car having "an orator, who explained its meaning in well-chosen verse." Faber then spoke to the workers, giving them "a short explanation of the principles upon which he had built his establishment." He was proud, on looking back on twenty-five years, that "in spite of the periods of stagnation and depression that had occurred during the time, and in consequence of which most great manufactories in Germany had been compelled to dismiss large numbers of their hands, the workpeople of his factory had been fully engaged, without interruption, at their full rates of pay." He promised them that this would continue.

Other significant events in the history of the Faber family were also the occasion for reflection, though perhaps not always so public. In 1877, when his son Wilhelm was married and assumed some of the management of the firm, Lothar von Faber presented him with an album, on whose first page was a letter that read in part: "I dedicate this album to you in commemoration of an important epoch in your life, namely, your marriage and your entrance upon a career of independent activity for the business of the firm of A. W. Faber. . . . The management of the business of the firm, whether its wares are sold in monarchies or republics, is carried out on the monarchial system, which alone I consider right."

The mature Lothar Faber's emphasis seemed to have been on the business and management as opposed to the engineering and production aspects of pencil making. If the firm of A. W. Faber and its pencils achieved an international reputation under his leadership, it was not necessarily because of his attention to technological advances in pencil making. The procession of vehicles that celebrated Faber's twenty-five years as proprietor did not emphasize the latest pencil-making machinery but rather carried an army of specialized workers. Pencil making in Germany into the 1870s adhered to old methods and old designs. Pencils were still made individually, or perhaps in triple lengths, in very much the same way in which a slip of Borrowdale graphite was glued into a single piece of grooved wood, to be covered by a thinner piece of wood. Furthermore, German pencils formed in this way were said to be only tied together with string while the glue dried, thus resulting in the weak joint that often separated and frustrated pencil users.

This is not to say that the Faber plants were not mechanized, for in the 1830s and 1840s the Germans did introduce some innovations into their pencil industry, including a reliance upon machinery to cut and groove the wooden slats into which the leads would be inserted. Pencil leads of graphite-and-clay mixtures came to be extruded in long strips from presses, albeit in square shapes to fit the square grooves. Silver and gold foil came to be used for stamping top-of-the-line pencils, and the A. W. Faber Company was producing pencils of a hexagonal shape in the early 1840s. But any mechanization or modernization of processes was in service to making pencils pretty much the way they were made in the cottage of Kaspar Faber, one pencil at a time. Even though machines had multiplied outputs sufficiently to supply an enormous export trade, the availability of inexpensive German labor provided no incentive for full mechanization.

It was not until Faber's centennial year of 1861, fifteen years after Alibert's discovery of the raw material, that pencils of Siberian graphite were placed on the market, and they would

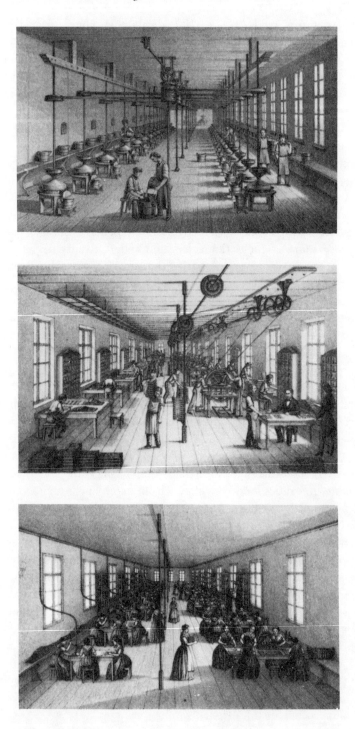

Some views inside a German pencil factory operating in the mid-nineteenth century

not be available in America until 1865. Not only could A. W. Faber henceforth make excellent pencils out of the pure Siberian graphite; the firm could also combine the pulverized material with fine Bavarian clay to make an unsurpassed line of the finest polygrade artist's pencils whose hardness was uniform and reproducible. This also made it possible to extend the range of the standard assortment of drawing pencils that Faber had been making since the late 1830s. At first these were made in seven more or less uniformly spaced degrees and labeled, in order of decreasing blackness and increasing hardness: BB, B, HB, F, H, HH, HHH. And while thirteen different degrees of Wolff's "purified black lead pencils" could be bought in London by mid-century, the availability of Siberian graphite made it possible to extend the range, in both hardness and softness, to sixteen degrees. They were shown at the London Exhibition of 1862 and were hailed as representing the only progress in pencil lead since cakes of Brockedon's compressed graphite were displayed at the Great Exhibition of 1851.

Where the letter designations originated remains a matter of some uncertainty, but the actual grading of pencils to indicate the darkness of their mark apparently started in France with the ability to control the hardness of the lead by the proportions of clay and graphite employed. Conté used the numbers 1, 2, etc., to indicate *decreasing* hardness, which is the opposite of their meaning today. Letters may first have been introduced in the early nineteenth century by the London pencil maker Brookman, B standing for "black" and H for "hard," with the number of repeated B's or H's indicating increasing blackness or hardness. (The divergent interests of artists in the degree of blackness and of draftsmen in the degree of hardness of a pencil may be one rationale behind the seemingly asymmetric B and H designations.) As pencil users showed increasing preferences for the degrees of lead around B and H, the designation HB may have been introduced for a "hard and black" pencil between a B and an H, and F, possibly for "firm" or "fine point," for a pencil falling between HB and H. Both German and French writers credit English words for the letters for pencil grades.

In America, where the Thoreaus used numbers to grade some of their pencils, they also employed the letters S and H, in a somewhat more consistent use of the language. By the end of the century, the Dixon firm would be advertising pencils for artists and draftsmen in eleven grades, distinctively ranging from VVS (very, very soft) through MB (medium black) to VVVH (very, very, very hard). While pencil grading systems would appear to become somewhat unified in the twentieth century, in fact the grading itself has never been standardized and exactly what a designation like HH or 2H means still varies from manufacturer to manufacturer.

Even though the system of grading had been adopted from the English, the system of making leads had been invented by the French, and the best new source of graphite had been discovered by a Frenchman in Siberia, it was a German's marketing sense that eventually established the German pencil as the norm. With the initiatives led by Lothar Faber, Nuremberg became the great center of the world pencil trade, with as many as twenty-six factories employing over 5,000 persons and turning out 250 million pencils a year well before the end of the century, with pencils made from Siberian graphite the standard to be emulated.

Since A. W. Faber had gained exclusive use of the output of the Alibert mine, it did not hesitate to remind its customers that it alone had made "the appellation 'Siberian Graphite' a household word amongst artists, engineers, designers and draughtsmen generally." The company's 1897 catalogue, for example, featured its top-of-the-line "Siberian Lead Pencils," each of which was made, reminiscent of the best of English pencils, "of one single piece of graphite," and listing some of the "most eminent artists of Europe," including Eugène Viollet-le-Duc and Gustave Doré, among those who had "borne testimony to the excellence of these pencils." The 1897 American price list also carried, in large type, a message purportedly from the firm's namesake, who had long been dead:

> I call special attention to my registered trade-marks, which consist of the name and letters "A. W. Faber," or, for some of the low-priced qualities, the initials "A. W. F."
>
> Please observe that without exception the label on each dozen of my lead pencils bears the name of A. W. Faber and a fac-simile of the signature . . . as well as the words "MANUFACTORY ESTABLISHED 1761."

This last reminder was repeated at the top and bottom of every page of the price list, which was also an extravagantly printed catalogue with colored plates of the company's pencils. Evidently such extreme measures were necessitated because, then as now, imitation was the sincerest form of flattery and the quickest way to make a mark in the highly competitive world marketplace.

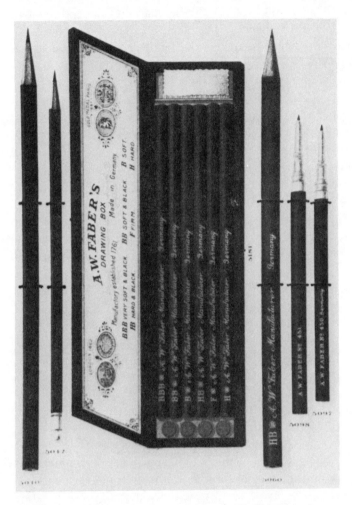

A page from a late-nineteenth-century A. W. Faber catalogue, showing "hexagon drawing" pencils, a box of "English" drawing pencils with drawing pins and rubber eraser included, and mechanical "artists' pencils"

The value of an established name made it especially important to Lothar Faber that a male heir carry on the German business. There seemed a reasonable hope of this happening when Lothar's son, Wilhelm, married Bertha, the daughter of his uncle, Eberhard Faber, the New York pencil baron. Two sons were born of the marriage, but both died before reaching five years of age, and Wilhelm continued in the business without enthusiasm until his own sudden death in 1893. When the patriarch, Lothar von Faber, who had led the business for over half a century, himself died in 1896, his wife took it over until her granddaughter, Ottilie, married Count Alexander zu Castell Rüdenhausen. The count obtained a royal decree that permitted him to change the family name to Faber-Castell, and this line carries on the pencil-making tradition today.

Before tragedy struck, there might have seemed a surfeit of Fabers. In 1876 Johann Faber, a younger brother of Lothar Faber and the partner who had managed the engineering and production aspects of the A. W. Faber firm in Stein, had retired from the family business. Since he had worked in the manufacturing department, Johann knew all the secrets of making an excellent pencil, and he began his own factory at Nuremberg in 1878. Leaving behind an A. W. Faber factory with equipment as much as forty years old, he outfitted his new factory with the latest equipment and soon expanded outside Germany, setting up houses in London and Paris. But even with the best of pencils to offer, he had "many prejudices to overcome when introducing his goods into the market":

> At that time there existed a number of pseudo-Fabers who made inferior pencils in imitation of the well known "Faber" brand, fraudulently stamping them with the name of "Faber" but coupled with different initials to those which appeared upon the genuine articles. All this was calculated to make the public distrustful of another (although genuine) "FABER" mark.

Among the problems that Johann Faber encountered was the publication by his brother Lothar of announcements to the effect that any "Faber" pencils without the initials "A.W." were "spurious imitations." This led the brothers to the law court, and in 1883 a verdict was handed down in favor of Johann, compelling the A. W. Faber firm to acknowledge the legitimacy of Johann Faber pencils. While this legal resolution of a family squabble may have caused A. W. Faber to word its claim about imitators more cleverly, it did not make it any easier for Johann Faber to export his pencils from Germany. So the new company in an old market had to send forth its representatives. Johann's sons Carl and Ernst soon took over the company, and they traveled extensively throughout Europe to set up outlets for their products. By the time of the World's Columbian Exposition, held in Chicago in 1893, Johann Faber pencils were being sold worldwide.

Johann Faber had obtained supplies of Siberian graphite from elsewhere than the Alibert mine, which the new company claimed had "in all probability fallen in and become destroyed." The "great purity" of the new Siberian graphite was supported by the analysis of the chief chemist to the Bavarian industrial museum, displayed with fists pointing to the high carbon content:

☞ **Carbon**	**94.5** ☜
Kaolin	3.1
Silicic acid	1.6
Oxide of iron	0.4
Chalk and magnesia	0.2
Foreign matter	0.2
	100.0

All pencil makers felt it was important to be able to claim (or at least to suggest) that their top-of-the-line pencils were made of the best graphite. While Johann Faber could advertise the chemical analysis of his new raw material, which enabled the company, as the pencil maker personified, "to complete the assortment of his pencils," other pencil makers used other means to emphasize the quality of their best products.

While a rose by any other name may still smell sweet, a pencil with the wrong name may not sell very well. Manufacturers have known this for some time, and thus they have stamped their pencils with names and claims that are designed to evoke associations with quality. Empty claims of pencils containing Cumberland or Siberian lead may have been pure lying, but uses of color and foil, and the association by gilt, have been pure genius.

Faber pencils had been creating severe competition for the L. & C. Hardtmuth Company, which had factories in Vienna and Budweis, when Franz von Hardtmuth, the founder's grandson, got the idea of making a top-quality pencil that would sell for three times the price of any other pencil in the world. After the necessary research and development, the new pencil was ready to be produced. According to one story, it had already been decided that its colors would be those of the Austro-Hungarian flag, and since the graphite was black, the pencil had to be painted golden yellow, but its suggestion of the Oriental source of the finest graphite also made yellow a brilliant choice. Hardtmuth then needed a distinctive name that would connote quality and value, and hence the pencil was called Koh-I-Noor. It was ready to market in 1890 and became a tremendous success, especially after it was exhibited at the Columbian Exposition in 1893.

Even decades after the pencil was introduced, the American Koh-I-Noor Pencil Company was proud of its name and did not hesitate to state to its potential customers, who in all likelihood did not see the Great Exhibition or the juxtaposition of carbon in the form of black plumbago and "the great diamond of history," that the name Koh-I-Noor "seemed entirely appropriate for the great pencil of history—hence its application." Whether the Koh-I-Noor pencil was just "great" or was really considered "great-great" or even better, the company did realize that since it spent more to make a better pencil, it had to charge more than the price of a common pencil, and so it also would later explain: "In goods of such superior quality there is economy."

Pencils with the Koh-I-Noor name are still sold today, but one must know the "peculiar dictionary" of the trade (or check the prices) to know if one wants the "Koh-I-Noor Writing Pencil," the "Koh-I-Noor Quality Pencil," or the "Koh-I-Noor Deluxe Writing Pencil." While they all may write without their lead breaking, one will no doubt write better than the others. And knowing this about pencils, we can appreciate why the engineers building the innovative Britannia Bridge wanted the "best-best-best" iron, even if they did not know the ironmaster's name for such iron.

The emergence of Siberian graphite as the standard to come up to, coupled with the success of the yellow-finished Koh-I-Noor, later advertised as "the original yellow pencil," apparently led manufacturers to use such names as Mongol and Mikado in order to associate their pencils with the Orient and thus with the source of the then finest graphite. For this same reason, painting pencils yellow, though it really had been done as early as the middle of the nineteenth century in Keswick, most likely to cover the imperfect wood used in some of the pencils, became a sign of quality in the last decade of the century.

It was actually customary at the time to finish in dark colors, such as black, red, maroon, or purple, pencils that were not left in natural or varnished cedar. The best pencils, which were made with the best wood, had been "natural polished." In an 1866 description of pencil making in Cumberland, the author described the varnishing of pencils as unnecessary and the "most unpleasant part" of their manufacture: "It may make a pencil look well, but it certainly cannot improve its contents, and this is proved by the fact that the *best* pencils are never coloured with varnish." But the tremendous success of the Koh-I-Noor changed that.

Today about three out of four pencils made are yellow, regardless of their quality. And there is a story about yellow pencils that, like many a pencil story, is of obscure origins but has had frequent retellings. Supposedly a pencil manufacturer once took a certain number of identical pencils and finished half of them in yellow and half in green for a certain office. The office distributed the pencils to its employees, who began complaining about the inferiority of the green pencils—their points broke more easily, they were harder to sharpen, and they wrote less smoothly than the yellow pencils. Apparently, by the middle of the present century, when this experiment was conducted, yellow had become so firmly established as a sign of "pencilness" in the minds of pencil users, though they may not consciously have known of its Asian allusion to quality graphite or of its association with a pencil named after a legendary diamond, that a finish of any other color was assumed to indicate an inferior pencil. Whether or not the judgment of pencil behavior is psychosomatic, yellow has become well established as the preferred color for a writing pencil, as it has for school buses and highway signs. It makes them all highly visible, whether on a busy desk or highway.

But long before Siberian graphite and yellow pencils were the norm, and long before the brothers Lothar and Johann Faber feuded, there was a plethora of pencil makers competing for business, and still young America was a prime new market. By midcentury the center of the pencil business in America had moved from the Boston area to New York City and its environs, where wholesale dealers took a manufacturer's product and distributed it, saving the manufacturer the trouble of selling his own product from door to door. At about the same time the German pencil industry had sought to gain a foothold in the new and growing market. In 1843 A. W. Faber had appointed J. G. R. Lilliendahl of New York City as its exclusive agent in the United States. This was the first attempt by a German pencil manufacturer to establish a permanent presence in America, and it signaled an era of fierce competition.

12/ Mechanization
in America

Although some Boston stationers and hardware dealers were selling American-made pencils as well as English ones in the 1820s, that is not to say that even the few small-scale Massachusetts pencil makers of the time could always find an outlet for their products, as Joseph Dixon was to learn. Even toward the end of the century, when the company he founded was one of the most successful marketers of pencils in America, its promotional literature could still declare:

> Strange to say, there is prejudice in the American mind, at times at least, against American products. . . . Dixon pencils at the outset had all the inborn prejudices of stubborn Americans to contend with, and it was only little by little that this prejudice was overcome, and that Americans were brought to know that the home-made or domestic product was not only quite as good, but in many particulars greatly superior to the foreign or imported article. To the credit of the Joseph Dixon Crucible Company be it said that they have always sold their goods as American goods and never have they trucked to the prejudices of their customers by putting on an imported stamp. To-day their advertising goes forth as "An American industry, American materials, American capital, American brains, American labor, and American machinery."

By the 1890s, the Dixon Company had grown so large and successful that its letterhead could ignore the foreign competition and so proclaim: "Established 1827. The oldest house in the trade. The largest concern of the kind in the world." Such superlatives have abounded in the pencil industry, and to be able to understand how they can arise and how they can appear to contradict one another and yet each contain a semblance of truth it is necessary to understand how the American pencil industry did go from rags to riches in the course of the nineteenth century. It is a story of men and machines, usually in that order.

Joseph Dixon was born in 1799 in Marblehead, Massachusetts, the son of a shipowner whose vessels sailed back and forth between New England and the Orient, where one of their ports of call was Ceylon. Since graphite existed in abundance in Ceylon and since it was heavy and compact, the ships used it for ballast and dumped it in the bay when they reached America. As a boy who had never seen a pencil, so the story has been told, Dixon learned from his friend Francis Peabody that graphite was baked with clay to make a good pencil lead, and he carried out some crude experiments. While Dixon may have conveyed to others the idea of a baked lead, they would also have had to do some experimenting. Nevertheless, soon being without money to continue his own research and development, Dixon is said to have gotten a job at a kiln to earn some money and learn more about baking ceramics.

When he was twenty-three years old Dixon married Hannah Martin of Marblehead, the daughter of a cabinetmaker, Ebenezer Martin, and the young couple's small cottage also became Dixon's laboratory. He continued to experiment with graphite and clay, and he developed some hand-cranked machines to extrude pencil leads and cut and groove cedar slats. But the pencils he produced, perhaps because he did not refine his raw materials sufficiently, were evidently not acceptable to the local merchants. A dozen Dixon pencils dating from about 1830 were found to have gritty leads that were not laid evenly in the wood case, which itself was only roughly finished. Even the label, lithographed by the multifaceted Dixon himself, was flawed, for it contained a typographical error, the *a* being omitted from Salem, the Massachusetts town where Dixon had set up his successful crucible works.

Dixon had begun importing graphite from Ceylon, which meant that ship captains ordered their ballast unloaded at the dock rather than dumped in the bay, and so he had an inexpensive and plentiful raw material ideally suited for making crucibles, so named because they once were marked with a

Joseph Dixon, early American manufacturer of crucibles and other graphite products

cross. Crucibles were used as receptacles in which to melt metal before casting it, and since the molten mass would not fuse with crucibles made of graphite, they had a distinct advantage over plain clay ones. Dixon evidently made excellent crucibles that did not crack under the repeated thermal abuse they were subjected to, lasting up to eighty firings, and so foundries did not prove to be a sufficiently lucrative market for replacements. Thus Dixon looked for other uses for the abundant graphite, and he identified such products as stove polish and pencils. The stove polish proved to be a commercial success, but pencil making did not.

Dixon continued to tinker. He worked with another American inventor, his contemporary Isaac Babbitt, to develop a material that would not disintegrate under the heat generated by friction. The result was an antifriction alloy known as babbitt metal, which came to be widely used in bearings for machinery. Dixon also worked with early cameras, devising a mirror that enabled a photographer to see a correctly oriented image in a viewfinder, and from his familiarity with photography and lithography he developed a photolithography process to foil counterfeiters.

When the Mexican-American War broke out in 1846, there was a sudden demand for graphite crucibles to be used in making iron. To meet the demand, Dixon opened a new factory in 1847 in Jersey City, across the Hudson River from New York. Since pencils were made at one end of the crucible factory, Dixon's may also be said to have been the first pencil factory in the New York metropolitan area. Coincidentally, Dixon's works were opened when the world pencil-making business began to expand. However, there is little doubt that it was not pencils that kept the business going, for after the first year of operation in New Jersey, Dixon reportedly had made a $60,000 profit on crucibles but suffered a $5,000 loss on pencils.

Since steel was an emerging industry, Dixon renewed his experiments with graphite crucibles, and he patented some innovative uses of them in the 1850s. Crucible steel was the kind of high-grade steel that was to be used for the suspension cables of the Brooklyn Bridge, and for a while it could only be made using graphite crucibles. That these were then the dominant product of his business was reflected in the name that Dixon gave it when failing health forced him to reorganize in 1867: Joseph Dixon Crucible Company. While Dixon had been making lumber pencils of "solid black lead, one-half inch square, four inches long," since at least the early 1840s, it was only when the Germans began to set up factories in America that the Dixon Company began to make quality pencils in earnest.

Orestes Cleveland, Joseph Dixon's son-in-law and the president of the company since 1858, began preparing to make quality American pencils in the mid-1860s. By 1872 a local reporter touring the Dixon works was allowed to see the fruits of American ingenuity:

> By a private door we enter another room; here is where the plans for making pencils by machinery have been

worked out under Mr. Cleveland's own eye. In this room are three turning lathes, a planer, a portable forge, vises and tools without end. We cross the yard from these buildings and enter a new brick building put up expressly for the pencil department. The basement is devoted to staining the wood and artificial drying. Every piece of the cedar undergoes a careful examination before going to the machines. On the main floor the wood passes through a machine which planes and grooves them ready for the leads. They are landed in another room where the leads are laid in, and the two parts of the pencil glued together. Passing into another room, they go through a "shaping" machine, and fall into baskets that carry them up on the next floor; here these unfinished pencils are piled into a hopper, and hardly stop a moment until they are varnished, dried, polished, the ends cut smooth and even, and the gold stamp put on, and then they are finished, no hand labor being used until they are put up in packages of a dozen each, these packages put into boxes of six dozen each; and these boxes wrapped, labeled and packed into wooden boxes containing from five to fifty gross each, for shipping. The operations in this department are original and peculiar. They are all patented, even the method of putting them up for sale.

We have used them about two months, and find them [*sic*] wear longer without sharpening than any pencil we have ever used. They are smooth, black, and pleasant, and are a credit to American mechanical skill.

The "new pencils" were ready for sale early in 1873, and in announcing them the Dixon Company recalled its founder's early difficulties with the American prejudice that "nothing is good that is home-made." However, in the half century since Dixon's early attempts to sell his pencils and crucibles a new prejudice had developed, which Dixon aimed to exploit: "Certain Germans are trying to make pencils here, calling them American pencils, but we are the ONLY AMERICANS who have undertaken the production of fine pencils, and our success is greater than we had dared to hope for." The success was attributed to "purely American principles," which meant "every manipulation by machinery instead of by hand labor; producing perfection and absolute uniformity."

The Joseph Dixon Crucible Company knew that it had to protect itself against the imitators that were sure to follow, for it had had to sue many a manufacturer and dealer who had sold imitation Dixon Stove Polish by emphasizing the name Dixon, including: James S. Dixon, Dixon & Co., W. & J. Dixon & Co., George M. Dixon, J. Dixon & Co., J. C. Dixon, and Charles S. Dixon. To protect its pencils from such exploitation, Dixon planned to identify each one with a "skeleton crucible" and to use a new system of grading. These unique markings were registered as trademarks, as were the words "American Graphite."

Among the first machines developed to make the new American pencils were cedar-working designs, and in 1866 Dixon was granted patent No. 54,511 for a wood-planing machine for shaping pencils. Such machines could process enough wood for 132 pencils per minute, but even so the growing demand produced by the Civil War for pencils would be difficult to keep up with. The Jersey City company has been described as the "birthplace of the world's first mass-produced pencils," principally because of the mechanization that its founder and his successor promoted. And the machines were not seen only as aids to human hands, for the shaping machines were covered with hoods and connected to ducts through which all the shavings and dust were sucked by a vacuum down to the engine room, where they were used for fuel. Dixon's factory, like all pencil factories, was permeated with the pleasant aroma of cedar, but in Jersey City the caloric value of the shavings was also appreciated.

The demand for pencils seems to have been growing at an unprecedented rate at the time, and in the early 1870s it was estimated that over 20 million pencils were being consumed in the United States each year. The favorite style of writing pencil at the time appears to have been the black round No. 2, and since the lowest retail price of a pencil was five cents, pencil making was a million-dollar business. Furthermore, because tariffs of as much as thirty to fifty cents per gross were imposed on foreign pencils in the wake of the Civil War, only the highest grades of foreign pencils were generally imported. Hence the American pencil companies had exclusive control of the market for inexpensive pencils, and inexpensive pencils were even more likely to be used profligately. One contemporary report observed about the waste of pencils that "only three quarters of each pencil is really used, and the remainder . . . thrown away. In effect the people of the country waste no less than $250,000 worth of pencils by throwing them away before they are used." It was in this environment that the Joseph Dixon Crucible Company was making pencils in the 1870s, but there was so much competition that the business could not be neglected.

In 1873 Dixon purchased the American Graphite Company of Ticonderoga, New York, and that location would eventually give its name to a famous line of yellow pencils. When Elbert Hubbard, who was given to hyperbole, published his preachment, *Joseph Dixon, One of the World-Makers,* in 1912, he could claim that "the Dixon Company are the largest consumers of graphite in the world. They are also the largest consumers of cedar." But while Dixon took pencil making beyond a New England cottage industry, the company's diversification made it possible for others to claim that they brought about the beginnings of the modern pencil factory, principally because the Dixon Company did not sell pencils on a significant commercial scale until the 1870s, even though writing implements were eventually to be the firm's most important products.

For all his apparent mechanical genius, Dixon's son-in-law appears to have preferred politics to business, and in 1880 the firm was put under the receivership of Edward F. C. Young, an experienced bank president. Young brought the business back to health, and it continued to prosper in the twentieth century under his son-in-law, George T. Smith. The company operated in Jersey City until the mid-1980s, when its buildings were bought by a developer who is preserving the landmark by converting the complex into apartments and a "self-contained neighborhood setting" known as Dixon Mills, with a commanding view of and ready access to lower Manhattan. Dixon Crucible is now owned by Dixon Ticonderoga, Inc., a holding company named after its most recognizable product, the yellow-and-green pencil; the corporation's headquarters are in Vero Beach, Florida, with manufacturing facilities in Versailles, Missouri, and elsewhere.

Because crucibles were the first business of the early Dixon Company, with lumber pencils being made only at one end of the factory, the distinction of having built the first lead-pencil manufactory in America has also been attributed to a great-grandson of Kaspar Faber. John Eberhard Faber was born in Stein in 1822, when his father, Georg Leonhard Faber, was running the third-generation pencil business, which was to be taken over by Lothar, who in turn brought his brother Johann into partnership in 1840. Their father did not expect his

An aerial view of the Joseph Dixon Crucible Company works in Jersey City, with the Statue of Liberty in the background

youngest son to go into pencil making and hoped he would become a lawyer. However, while he did study law, young Eberhard "became absorbed in the study of ancient literature and ancient history, and placed Virgil on a much higher plane than Justinian." So instead of becoming a lawyer in Bavaria, the young scholar came to America and began to represent the German House of Faber in 1849. By 1851 he had taken over as sole agent for A. W. Faber products in America and had established a branch of the German firm at 133 William Street in New York City. In addition to pencils, he sold on commission German and English stationery products, and he later acquired large tracts of forest land on Cedar Key, an island off the Gulf coast of Florida, from which he would be able to ship cedar-wood slats ready to be made into pencils. Indeed, securing cedar tracts may in fact have been the principal purpose in Eberhard Faber's coming to America, for the wood was by far the best in the world for making pencils, and, when Eberhard migrated, A. W. Faber had already had an agent in New York since 1843.

Since the cost of importing finished products was already great, and since the Civil War had caused tariff, freight, and marine insurance rates to rise, the Fabers wondered if they could manufacture some pencils at a more reasonable cost in America. Although New York was closer to the supply of Florida cedar, it was farther from that of Bohemian clay and Austrian graphite. So it was decided that leads made in Stein could be shipped to New York and assembled in the Florida cedar, with machinery being used to offset the considerable difference in labor costs between New York and Nuremberg.

Although war was also making it more difficult to obtain cedar from the Confederate territories, where it grew, the Fabers went ahead with their plans, and with German financial assistance Eberhard opened his factory in 1861—the centennial year of A. W. Faber—by the East River at the foot of Forty-second Street, on the present site of the United Nations Building. According to Johann Faber's history, before very long "the ingenuity of the Americans, combined with German ability and experience" in pencil making, resulted in "a number of entirely new machines," without which the large-scale industry that pencil manufacturing has become would not have been possible.

It was the best of times and the worst of times to start up a pencil factory in America, for the war apparently created a great demand for pencils—presumably because the soldiers used them to write letters home. Although at least one chronicler insists that Union soldiers used pencils rarely, "except during periods of active campaigning," and Confederate soldiers "paid three dollars for a bottle of ink or made their own from pokeberries, rather than write home with a pencil," evidently many soldiers did use them. But whether it was the demand for pencils or the scarcity of cedar, in 1863 offers of ten dollars per gross of pencils were being refused in New York. However, Eberhard Faber seems to have had the advantages of possessing resources on both sides of the Atlantic: his factory made the grades of pencils on which excessive duty was charged, but the more expensive grades were still imported from Germany.

In 1872 the East River factory was destroyed by fire. Although Faber had been considering erecting a new factory on Staten Island, fire created an urgent need for a new plant, and three existing buildings were purchased across the river in the Greenpoint section of Brooklyn, near the intersection of West and Kent streets. Over the years several buildings were added to the original complex, and they are still identifiable as monuments to pencil making there by the Eberhard Faber trademark of a star in a diamond locked in the brickwork near the roof. One neglected brick-and-concrete-faced building, a major 1923 reinforced concrete structure, is undistinguished as architecture, but it is remarkable for being adorned with monumental yellow-tile pencils standing upright and properly sharpened between the factory's large sixth-floor windows. On each pedimentlike projection above the otherwise squat and boxlike building is, also in yellow tile, the star-in-a-diamond

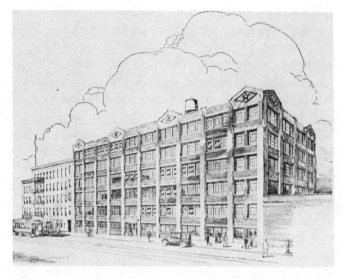

The 1923 reinforced-concrete addition to the Eberhard Faber factory in Brooklyn, with the company's trademark and yellow pencils decorating the facade

symbol that once appeared on the company's pencils and letterhead.

So explicit and exclusive a devotion to pencil manufacturing gave Eberhard Faber some claim to calling itself "the oldest pencil factory in America." But that claim, which ignored such pioneering pencil works as those of Munroe and Thoreau, was strictly true only if the qualifying "still in existence" was understood. The old Faber factory remained with the company headquarters in Greenpoint until the plant became obsolete, and the company decided to move its operations to Wilkes-Barre, Pennsylvania, in 1956.

The changing physical plant of Eberhard Faber was also accompanied by changes in identity and management over the years. When Eberhard died in 1879, the business was taken over by his sons, Lothar W. Faber and Eberhard Faber II. It is not clear when the American and German Fabers ceased to be one big happy family, but for some time in the nineteenth century the oldest pencil company in America was known as E. Faber Pencil Company, being so incorporated in 1898 with Lothar as president and Eberhard as vice president. This name was no doubt selected to be confused with the German firms of A. W. Faber and J. Faber, thus associating E. Faber's pencils with the well-known foreign products. The title of the American company was changed when it was reincorporated

as the Eberhard Faber Pencil Company in 1904, occasioned
by the outcome of numerous lawsuits between Eberhard and
A. W. Faber over "the right of title to the ownership of the
name 'Faber.' "

When Lothar Faber died in 1943, his brother, Eberhard II,
took over the company, with Lothar's son, Eberhard III, serving as vice president. Within two years both Eberhards had
died, however, and no Faber was available to head the company. In 1953 Eberhard III's widow, Julia T. Faber, began to
take an active role in managing the company, until the young
Eberhard IV could take over as president and chief executive
officer. The fourth-generation American Faber, like Henry
David Thoreau, had been in and out of the business since
graduating from college and, like his great-grandfather and
Thoreau, was more attracted to using pencils to write with
than to making pencils themselves. However, like his American predecessors, Eberhard IV was to oversee successfully the
business he became responsible for. In 1988 Eberhard Faber
was sold to the A. W. Faber-Castell Corporation, thus bringing together two old competitors.

But such were to be twentieth-century developments. Back
in the middle of the nineteenth century other factories were
being opened in the New York area. What began as an American office to represent a Bavarian firm started at Fürth
became a pencil business known by the partners' names,
Berolzheimer, Ilfelder, and Reckendorfer. But the simpler
and more American name of Eagle Pencil Company was soon
suggested and adopted. Daniel Berolzheimer's son, Henry,
took over the firm in 1861, and in 1877 the company bought
New York's first iron-frame building and began to emboss an
image of the American eagle on its products. Business prospered, and by the 1920s the firm was calling itself the "largest
pencil factory in America," because it had the largest jobbing
trade and sold an enormous quantity of inexpensive pencils.
One branch of the Berolzheimer family moved to California to
exploit incense cedar for pencil making, while the other remained to run Eagle. The family continued to control the
company well into the twentieth century, but in recent years
Eagle was renamed, after the family's Americanized name.
The Berol Company in America continued to be known as
Berol USA even after it was in fact owned by the Empire
Pencil Company of Shelbyville, Tennessee, which in turn was
owned by the Empire Berol Corporation, which was acquired
in 1988 by a New York investment group.

Other firms also began in the mid-nineteenth century. In 1861 an American druggist named John Faber, but of no relation to the German pencil Fabers, went into partnership with a restaurant owner named Siegortner and formed a pencil company. A suit was brought against the use of the Faber name, and when the suit was decided in favor of Eberhard Faber the upstart pencil company was sold to some Hoboken, New Jersey, importers. Edward Weissenborn, an "enterprising young man" who took charge of the new pencil factory, is credited with founding the American Lead Pencil Company. This company established the Venus trade name for its new drafting pencils in 1905 and began to open foreign offices and manufacturing facilities in the following decades. In 1956 the company's name was changed to Venus Pen & Pencil Corporation, and shortly thereafter it moved its executive offices from Hoboken to New York, at which time Charles of the Ritz purchased the company's stock. The company was sold to a group of private investors in 1966, and the following year Venus bought a pen company and became Venus-Esterbrook, Inc. The pen company's facilities in New Jersey were soon closed and the production equipment moved to Tennessee, England, and Mexico, but within a few years the latter two establishments were bought by Berol and a wood-slat plant in California was sold to the state. In 1973 what was left of Venus-Esterbrook was taken over by Faber-Castell, whose name was then changed to Faber-Castell Corporation.

In the nineteenth century, the pencil barons had many other things to worry about besides takeovers, whether friendly or hostile. As the demand for pencils was growing, the availability of at least one raw material was shrinking. The increasing scarceness of pencil wood during the Civil War has been related by Horace Hosmer, who then worked for pencil makers in the Boston area. In one of his reminiscences, Hosmer describes what led him to range about Thoreau's Maine woods looking for a substitute wood in the red cedar: "There was a great demand for pencils in 1862 and the stock of Florida Cedar in Northern hands was nearly exhausted, and the price was enormously high."

Not only was the proper wood itself important for pencil making; the appearance of the wood was also of prime concern, according to Hosmer:

A pencil maker offered $300, for a process of coloring Cedar black, as the foreign pencil makers did. I made many careful experiments and failed, but I dreamed out the process and succeeded not in getting the money because the man thought if I could do it he could, so I let the thing rest, but during the Civil War it was worth more than $3000 for coloring substitutes for Cedar which could not be obtained.

Hosmer's recollections also give some insight into how finishing work on pencils was subcontracted in the nineteenth century. Even though the center of manufacturing and trade had moved to New York, there was still work for experienced hands in Massachusetts. On one occasion, Hosmer related the economic difficulties he faced, emphasizing the size of his commitment:

In 1864 I took 10000 *Gross* of Faber's pencils to polish, letter and put on Rubber Heads. I had $3000 worth on hand and found that I could not do them at the contract price, because Shellac had risen from 18 cts to $1.25. Alcohol was [$]4.00 when I used to buy it for 55 cts in earlier years, and everything else in proportion. I sent my working girls home for the day, locked the shop and within three hours had an entirely new process which enabled me to earn $2300 the next year. I earned over $400 per month with the help of two girls. In 1867 I earned over $2000 in much less than a year on the same work.

Hosmer also finished pencils for the Eagle Pencil Company, and he reported that Faber once increased his pay from 20 to 80 cents per gross in order to keep him finishing their pencils. He economized in the new process that he developed behind closed doors by substituting glue for shellac and naphtha for alcohol, and "new mechanical contrivances enabled two girls to handle 120 Gross per day, *one process.*" Hosmer also reported that he lost his savings of five years when his bills were not paid by his trade in the South and that the war caused a Boston firm to lose payment for a million pencils they had shipped to Confederate states. Presumably Hosmer knew of this firm's loss because it was for them that he finished Faber and Eagle pencils—polishing, lettering, tying, and labeling them ready for market. The rubber heads that Hosmer put on pencils were, of course, erasers, but his operation was surely less mechanized and more humane than that depicted in the David Lynch film *Eraserhead*.

There had been known as early as 1770 "the very convenient method of wiping out writing made with a black-lead pencil, by means of gum elastic," or "Indian rubber." In his *Familiar Introduction to the Theory and Practice of Perspective,* Joseph Priestley reports having seen "a substance excellently adapted to the purpose of wiping from the paper the marks of a black-lead-pencil." He told where the very useful substance might be gotten: "It is sold by Mr. Nairne, Mathematical Instrument-Maker, opposite the Royal-Exchange. He sells a cubical piece, of about half an inch, for three shillings; and he says it will last several Years."

The well-known ability of a tree resin from the West Indies to rub out pencil marks is what caused it to be called "rubber." Almost a century later, "indiarubber" was so familiar and "old fashioned" as to have become a single, uncapitalized word, and in the meantime, since natural indiarubber did not erase all pencil marks completely, ersatz rubbers had been developed. But these were apparently also far from ideal, for as one commentator remarked in 1861: "As to that newfangled eraser, something between a hearthstone and a . . . knife-board, it would rub its way through anything." But whether good or poor, the pencil and eraser remained separate and distinct stationery items well into the latter part of the nineteenth century. The Faber firm for which Horace Hosmer finished the pencils claimed to have been the first to put on its pencils not only rubber erasers but also metal point protectors. However, while Eberhard Faber did patent in the early 1860s "a lead pencil with an angulated rubber-seal head . . . which serves as a seal, a preventer against rolling and as an eraser," a knoblike attachment that sounds like the wedge-shaped eraser tips still sold today, the first U.S. patent for attaching an eraser to a pencil was issued in 1858 to Hyman Lipman of Philadelphia. His invention consisted of a pencil with a groove at the tip, into which was "secured a piece of prepared rubber, glued in at one end."

In 1862 Joseph Reckendorfer had taken out a patent for an improvement on Lipman's patent, which he had bought for a reported $100,000, and he sued Faber for infringement. However, the Supreme Court eventually declared both patents invalid, because there was "no joint function performed by the pencil and the eraser; each performed the same function as before. The pencil was still a writing instrument and the eraser was still an eraser. Therefore the pencil and its eraser formed an unpatentable aggregation."

In 1872 the Eagle Pencil Company patented a pencil with an integral eraser inserted in one end of the cedar shaft. Other companies also made such pencils, which came to be known as "penny pencils," and they continued to be among the most inexpensive of styles. Even in the early 1940s they could be bought for less than a penny each. While pencils with attached or integral erasers remained in the minority throughout most of the nineteenth century, by the early decades of the twentieth about 90 percent of American pencils are thought to have come with attached erasers. Ironically, as erasers became more popular, there developed a concern for preventing the erasure of pencil writing and drawing that one wished to be permanent. *Scientific American* published numerous instructions on how to "fix pencil marks," including one recipe calling for washing them in skim milk.

A "penny pencil," with its eraser inserted directly into the wood case

Pencil catalogues from the turn of the century demonstrate that the rubber-tipped eraser was then still far from universally in use, and it was sometimes thought of as a gimmick to sell inferior pencils, which did, however, account for considerable volume. Better pencils made for drawing and drafting were not expected to have erasers, and a 1903 descriptive catalogue issued by Joseph Dixon indicates that attached erasers on school pencils were a matter of some debate. One section of the catalogue, headed "The Philosophy of Rubber," contains a short essay discussing the issue of "plain vs. rubber-tipped pencils." According to the Dixon company, which was making over 700 pencil styles, most of which appear to have been untipped: "Soon after the appearance of the Rubber Tipped Pencil, its use in schools became general; but within the last few years there has been a tendency among teachers and school directors to turn to the plain pencil, without the tip." The pencil company attributed this change mainly to three things: (1) "the Rubber Tip is the most expensive form of eraser," (2) "pupils soon soil the Rubber Tip, and it is then useless," and

(3) "pupils will do better work if there is no Rubber Tip on their pencils."

The essay goes on to explain that since "one of the duties of a teacher is to make pupils correct their errors," teachers should not want to encourage errors. Attached erasers, according to the pencil maker's logic, make it easier to correct errors and hence "it might almost be laid down as a general law, that the easier errors may be corrected, the more errors will be made." A fourth objection to attached erasers was also given —a medical reason. Apparently young pupils, "especially boys," not only put the rubber tips into their mouths but also were known to "swap" pencils and thus increase the chance of transmitting disease. But it was not only schoolboys who put their mouths to pencils, for stationery catalogues offered several styles of special pencils with the eraser replaced with a "mouthpiece, adapted to the use of persons who hold a pencil in their teeth." The mouthpiece, made of ivory or some other hard substance, prevented the "soiling of the lips or tongue with the coloring matter" of the pencil's finish. Why pencil chewing might not be an objection to all pencils is not made clear in the Dixon booklet, but a little later in the essay the unnumbered reason of competition finally does come up:

> The truth is, Rubber Tipped pencils have been used so extensively in schools for the reason that apparently they are cheap. We say "apparently" because in the long run, so-called "cheap" pencils are very expensive. No matter by whom manufactured, they are invariably made of the riff-raff of a factory. The leads in them are irregular in degrees of hardness, which renders them unfit for uniform work; they are apt to be brittle, so that they break easily; many of them are gritty, which makes them troublesome to write with, and wholly useless for Drawing; and the wood casing is generally hard, poor, cross-grained cedar, which frequently sharpens only with the greatest difficulty. The Rubber Tips on such pencils are little better than mere "catch pennies," being added simply as a means for selling them.

But as the next page in the catalogue shows, Dixon did make and sell rubber-tipped school pencils for those willing to pay the extra pennies. The company also sold separate erasers, including some encased in wood the way the ones on the penny pencils were and much as typewriter erasers are made today.

By the turn of the century American pencil companies seemed to offer about as many styles of pencils with erasers as without. But whatever the style of eraser on an American pencil, the process of attaching it was soon mechanized, and the "tipping machine" would serve the Smithsonian Institution as a metaphor for mass production generally.

In spite of the fact that a rubber-tipped pencil was carried in the 1864 procession honoring Lothar Faber's jubilee, late-nineteenth-century catalogues of German pencil companies showed many fewer pencils with rubber tips, and that was to be the general custom in Europe well into the twentieth century. Even today, a London stationer, for example, will have a wide selection of eraserless pencils and an almost equally wide selection of separate erasers, or "rubbers." Some European preferences were apparently changed after World War II, however, when Italian soldiers came by great quantities of eraser-tipped pencils and took them home.

But whether tipped or eraserless, European pencils are virtually all sold presharpened, and to this day the details of how a pencil is finished are significant. In Europe some of the best models have rounded ends formed by dipping them in paint. In America it is the ferrule, the means of connection between the pencil proper and the eraser, that gets the special treatment. While economy pencils will typically have a plain aluminum ferrule, better pencils have distinctively painted ones, historically of brass. A Velvet pencil has a royal-blue band, a Ticonderoga has alternating green and yellow bands, a Mongol has wide black bands at either end of its ferrule. Before the Cold War, the Eagle Pencil Company proudly advertised its Mikado as "The Yellow Pencil with the *Red Band," and a footnote identified the asterisk as signifying that the red band was registered with the U.S. Patent Office, thus claiming it as exclusively Eagle's. Even though renamed Mirado after Pearl Harbor and now made by Berol, this pencil still has the same characteristic red-banded ferrule.

No matter how appealing and distinctive a pencil and its associations and appurtenances may be, they are all incidental to the real point of a pencil, however, and what matters in the end is whether the pencil writes well. Similarly, no matter how convincing, the history, claims, and counterclaims of pencil manufacturers all have to be taken with a grain of salt, because they were, after all, very often trading on tradition and a name and competing in the marketplace.

This is nothing new. Just as pretenders to the Faber name

were deliberately misleading about the quality of the lead in their pencils, when there was any lead in them at all, so were English pencil traders before them. And *Caveat emptor!* was no doubt good advice when buying a pencil brush or even a *plumbum* in Rome. While there is no longer such a concern for counterfeit pencils as there may once have been, to this day the names of companies are confusing, and much of that confusion can be traced back to the intense competition of the nineteenth century that led to fierce battles for distinctive trademarks, as well as to deliberate confusion. Even family-related concerns separated by an ocean could not escape the problems of protecting their identities from each other. Finer and finer distinctions between pencils and their markings began to appear, and the Mongol pencil of Eberhard Faber was among the first products in the United States to have a trademark. Questions of proprietary rights led eventually to such awkward designations of pencil hardnesses as $2\frac{1}{2}$, $2\frac{4}{8}$ and $2\frac{5}{10}$, not to mention the decimal 2.5, as the arithmetical inclination to simplify fractions clashed with trademark protection laws.

When the pencil-making industry was brought to maturity and to worldwide competition, levels of mechanization and of research and development rose considerably. While an earlier Faber or a Thoreau could develop his own machinery and hold his own secrets of the business within his family, that no longer would be so as the nineteenth century gave way to the twentieth. Seemingly ever-present changes in the supplies of graphite and wood created the need to maintain scientific and engineering staffs and laboratories to create new lead formulas and explore the use of new woods, as well as oversee all the other details of designing and manufacturing ever-changing lines of pencils. Engineers had to lay out and develop new machines and processes to keep a factory competitive. Toward the end of the nineteenth century this was being done in America more so than in Europe, to the latter's disadvantage. With the expansion of the industry there was a constant need to deal with and answer the claims of competitors, and this was only possible by enlarging what once could have been a family business into one that required a range of machines and technical expertise beyond the ken of a few relatives.

13/ World Pencil War

As the nineteenth century drew to a close, the influence in America of European and especially German pencil companies began to diminish. One observer, writing in 1894, noted that in twenty years the cost of pencils had been reduced by 50 percent, at least in part because of the invention of machinery such as that used by Dixon in Jersey City. Furthermore, foreign pencils had been "gradually ousted," and America was believed to "export about as many lead pencils as we import." But readers of *Scientific American* were advised against trying to enter the booming business, because the few American factories were reported to "hang together like brothers, and the chances are that if we should put our spare money into a lead pencil factory, they would make it warm for us."

There were no doubt many and complex reasons for these developments, and they began with the shift from imported to domestic pencils, including a growing pride and confidence in American manufactures, as celebrated in such public displays as the Centennial Exhibition in Philadelphia. While formerly the transplanted European artists, engineers, and businessmen had brought with them their preferences for pencils made back home, now newer generations with fewer ties to Europe and fewer prejudices for things from the old country looked more to questions of quality, economy, and availability in making their stationery purchases.

Although Abraham Lincoln is said to have written his Gettysburg Address with a German pencil, the protective tariffs

imposed on foreign goods during his administration encouraged the development of the American pencil industry. By
1876 the duty on imported pencils was fifty cents per gross
plus 30 percent of the declared value. Such penalties on foreign goods, coupled with the growing demand for pencils in
America, made it a sound business decision to start a pencil
factory then, even with the relatively high costs of labor. But
if labor was expensive, manufacturers could keep other costs,
such as pilfering, down.

Even with protective tariffs and expanding markets, pencil
making was, as it had always been, a business of pennies. Just
as extraordinary measures had been taken to keep English
workers from taking home a little plumbago from the Borrowdale mine, so in America a century later pencil makers sometimes took extreme measures to reduce unnecessary losses. In
the 1870s, when the Joseph Dixon Crucible Company was
producing 80,000 pencils daily, which represented about one-
third of American consumption at the time, every pencil had
to be accounted for. A rigid system of discipline was in force,
and if one pencil was missing from a factory room, every
employee in that room was discharged if the pencil was not
found. According to one story, someone from the crucible
works one day went unauthorized into the pencil works and
helped himself to a single pencil, thinking that it would not be
missed. When the system of counting and checking revealed
the shortage, one of the pencil workers reported seeing the
crucible worker in the factory and the latter was called to
account for his actions. He confessed to taking the pencil and
returned it with apologies, but he was dismissed and never
reinstated. Such stories were no doubt behind the watchful
stares that any outside visitor was subjected to while inside the
factory.

But national pride, protective tariffs, and fearful employees
were not the only reason that the American pencil was able to
displace the European. A considerable number of some purely
engineering and technological factors tipped the balance. As
late as about 1869, English pencils, once the world's standard,
were still largely being made with leads cut from graphite,
whether straight from the Borrowdale mine or compressed in
Brockedon's process or mixed with clay and baked into sticks.
The wood case was almost universally still being formed the
way it was in centuries past, with a square groove cut with
a plane or saw, or perhaps with a simple hand-operated
machine, in a single pencil case at a time. When the groove

was filled with lead and planed smooth, a thinner piece of wood was glued on top, and the square pencils were shaped to roundness one at a time in a simple machine consisting of a pair of wheels in which the pencil was grabbed and pushed through revolving cutters.

Germany had introduced some more sophisticated machinery in the late 1830s, when it finally adopted the Conté process of making leads. Instead of using the original method of Conté, who had formed his leads by pressing the wet mixture of graphite and clay into square grooves cut into boards, the Germans began to extrude the leads directly through a die. The firm of A. W. Faber has claimed to be the first to extrude leads, but there is reason to believe that it was first done either in France or in England by Brockedon, who as early as 1819 was experimenting with drawing wire through holes in sapphires, rubies, and other gems. While extruded round leads were used for the earliest mechanical pencils, Faber was also perhaps extruding square leads to be put in the square grooves cut in wood. Square pencil leads were the worldwide norm for wood-cased pencils until the mid-1870s, and as late as the end of the century it could still be written that Faber pencils "can easily be recognized by their square lead," thus indicating their lag in modernization.

Round leads could not be encased in wood in the same manner that square leads had been for two centuries, a fact that Henry Thoreau seems to have mulled over. A square lead can be dropped into a square groove and covered with a flat piece of wood, but a round lead cannot be enclosed so asymmetrically. The groove to receive a round lead must be semicircular, and the covering piece of wood must also have a semicircular groove. Furthermore, the grooves must be of just the right depth and properly centered so that the mating pieces of wood just fit around the lead and form a square case. If the grooves are too shallow, the wood halves cannot mate; if the grooves are too deep, the lead can slip and work its way out of the wood case. The method of enclosing square leads, on the other hand, allowed for much greater latitude in cutting the single groove. If it was too narrow, the lead could be made thinner; if a little wide, adjustments could be made in the height of the assembly. By making a groove a little shallow and then planing down the lead, or by planing down the wood to meet a thin lead, a tight fit was always possible. And by not having to groove the covering piece of wood, a true square case could always be prepared before final shaping. The square lead in an old pencil was often visibly off center, but this was of little practical consequence when sharpening was essentially a whittling process. However, an off-center lead would be bent and broken by the rotary motion of the mechanical sharpener, which was introduced toward the end of the nineteenth century. Thus the development of round leads and rotary sharpeners were interdependent.

In America the development of a young pencil industry, without strong traditions in how things were always done, made it more natural to design machinery suited to making pencils more efficiently. While it has been claimed that even William Munroe in the early nineteenth century made pencils by grooving two slats each to half the thickness of a pencil lead, there would not have been an advantage to doing so if the lead was not round. As late as the 1880s, when "improved machinery" enabled "ten hands [to] make about four thousand lead pencils of the cheaper grade a day," the groove was still sometimes cut into only one of the pieces of wood.

It was not only the speed of the early machinery; what was also important was the fact that slabs of wood four pencils wide had four grooves cut simultaneously. Four leads were glued into the slab and a thinner, ungrooved slab was glued on top, so that four pencils were processed at a time. (Soon as many as six or more pencils were being made simultaneously.) The sandwich containing four leads was then inserted into a "molding machine," resulting in four pencils, selling for between eighty-five cents and two dollars a gross, and being "very good articles, writing smoothly and evenly." Such molding machines, with just a change of the blades, were capable of cutting out hexagonal as well as round pencils, and hence in the 1880s it was common to find many American pencils offered in both shapes. Dixon's 1891 catalogue offered identical styles of "fine office pencils" in both round and hexagon shape, with the latter priced more than a third higher. The same catalogue's cheap pencils were only about 20 percent more expensive in the hexagon shape. This suggests that it was not the shaping but the finishing that commanded the premium. The catalogue's best pencils, "Dixon's American Graphite Artists' Pencils," endorsed by "designers, drawing teachers, mechanical engineers, and artists generally," were offered only in the hexagonal shape, for at $9.37 per gross price was apparently no object in finishing them "in the natural color of cedar wood." By the turn of the century, some pencils were offered without a price distinction between

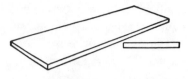

1. *Cedar pencil slat*—a little longer than a pencil, the width of six pencils, the thickness of a half pencil

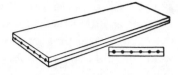

4. *Leaded*—the two grooved slats with leads between, glued together for shaping

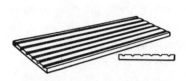

2. *Grooved for the leads*—with the depth of the grooves one half the thickness of the lead

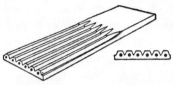

5. *Half-shaped, hexagon pencil*—formed from the leaded assembly under high-speed revolving cutters

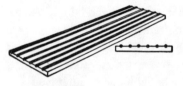

3. *The leads in the grooves*—ready for a second slat, similarly grooved, to be fitted over the leads

6. *Full-shaped, hexagon pencil*—the shaping process completed by repeating the cutting operation with the leaded assembly reversed

Woodworking and assembly steps in making a modern pencil, showing essentially the same process introduced with mechanization in the nineteenth century

shapes, and *fin de siècle* German pencil catalogues offered triangular and square models, but often still with square leads.

While the Germans had developed some pencil-making machinery in the late 1830s and 1840s, they apparently did not keep their plants modernized to the extent that advances were being made in America. By the late 1870s American technology had developed machines capable of such precision that a slat as wide as six pencils could be grooved to half the depth of the lead and mated with another to be shaped under revolving knives making nine thousand strokes per minute: "The machine separates and shapes probably fifty pencils while the foreign maker is shaping one, and requires nobody to complete its work."

This kind of mechanical capability was developed in America and not in Germany in the second half of the nineteenth century in part because of the high cost of labor in America and in part because the American pencil companies were setting up their new plants at a time when technology was more advanced and engineers could order and produce specialized machines capable of doing precision woodwork at high speeds. Once the Germans had mechanized to a certain level of technology, pencil barons such as Faber seem to have concentrated on social issues in order to promote worker welfare and loyalty. There was, however, an illusion of long-term worker security, for without maintaining a state-of-the-art physical plant, capable of making pencils as cheaply as anyone else, the security of the business itself was jeopardized.

Johann Faber, who left his brother's A. W. Faber plant to start his own firm in the 1870s, was in a position to set up a more modern plant. And his access to a new source of Siberian graphite, just as the Alibert mine was becoming worked out, made it possible for his pencils to gain a considerable reputation. In a Sherlock Holmes adventure set in 1895, Watson was informed that "Johann Faber is the most common maker's name." Holmes was right, as usual, for the relatively young firm then accounted for about 30 percent of the output of Bavaria's twenty-six pencil factories, twenty-three of which were located in Nuremberg. In all, almost ten thousand people were employed. But amidst all this apparent prosperity and fame, reports coming out of Germany were revealing that the industry was holding its position "with great difficulty." Johann Faber was being turned into a limited company, and its founder was lamenting the high customs duties imposed by the United States, which was by then producing "almost as many pencils as all the Bavarian factories put together." He was also reported to be concerned that the best cedar, which was almost exhausted, was still being exported to India, Mexico, Japan, and Australia "at extraordinarily low prices." Faber also complained about the duties charged by Italy, Russia, and France, and about the fact that French "schools and government offices, and even railway companies, are forbidden to buy German pencils."

According to a contemporary consular report, Faber also complained that the Americans had not replanted cleared cedar forests, and that only half as many pencils were being gotten from the poor-quality wood that was available, thus increasing the costs of pencils. Faber claimed that the American pencil industry was dumping its surplus of thousands of gross at a loss, thus further depressing world pencil prices. The English market especially had been swamped with cheap American pencils, and German companies were forced to compete at a loss. The consul concluded that "the position of the German pencil industry is not too brilliant."

At the turn of the century the situation was summed up by an item in *Scientific American* which noted that the Germans were "suffering severely from competition of American lead pencil makers" and that "the ingenious labor-saving machinery of American factories and their large scale of production, and specially cheaper prices at which they can supply themselves with cedar wood, are the chief causes for the failure of German makers to hold their own." In 1901 another consul reiterated that American success was "mostly due to the perfection of the machinery" and was also attributed to the control of the best cedar. A. W. Faber displayed its goods at the International Exposition in St. Louis in 1904, and described itself as having "1000 Workmen" and "steam and water power, together 300 H.P." But there was no mention of whether it had any new machinery.

With the coming of World War I, conditions worsened considerably for Germany while further opportunities arose for the American pencil industry. Great Britain was buying American copying pencils at the rate of a thousand gross per week from one manufacturer alone. While their use was not officially disclosed, it was conjectured that the pencils were being distributed to British and Allied officers for use in completing the vast paperwork associated with the war effort and in the field, where the nonerasable qualities of indelible pencils made them much more convenient than pen and ink. As the war continued, civilians in Britain had to pay higher prices for pencils. With the shortage of aniline dyes, a penny copying pencil cost four times as much, and "halfpenny cedars" were "almost unattainable." But in 1916 *The Times* of London reported some expected relief from an ally to the east: "The Japanese are producing excellent substitutes for German stationery materials, and in spite of the high freight to this country prices compare favourably with those obtained by the German manufacturers." By the end of the war England was the biggest importer of American pencils.

For Germany especially, the cost and availability of raw

materials were greatly affected during the war, and this drove up the prices of finished pencils. Clay, for example, had to be free of impurities such as silicon if it was to produce leads that did not scratch, and it could take as much as ninety-seven tons of water to wash the greater part of silicon out of three tons of clay. Yet during the war the Germans had to pay sixty times what it would normally have cost for inferior clay that was really only suitable for making drainpipes. Other ingredients were similarly affected, and it reportedly cost German pencil manufacturers thirty to fifty times as much as in prewar years for raw materials, while wages were ten to twelve times as high, making the total cost of production about fifteen to twenty times what it was before the war. Yet the prices of finished pencils were only about ten times higher. The German Pencil Makers Union of Nuremberg, which represented the industry, believed, however, that the prices of raw materials were beginning to drop in 1920.

But the Armistice did not make more cedar trees grow, and the influx of foreign orders caused prices for the American pencil wood to increase by as much as 50 percent. Among the larger users of cedar was Japan, which by this time had 117 pencil factories, with eighty in Tokyo alone, employing over two thousand workers. Almost 1.5 billion pencils were produced by Japan in the decade from 1910, and in 1918 alone almost 200 million were exported.

In addition to competition from American and Japanese pencils, the rise of prices in Germany meant that higher import duties had to be paid to bring German pencils into the United States. Pencil makers like Johann Froescheis and A. W. Faber sometimes protested such increases through appeals courts, but usually with little success. And when there were no trade barriers, there were more overt political barriers, as when the purchase of large quantities of German pencils by the London County Council for school use was protested. It was perhaps an understatement when *Scientific American* referred to the "displacement" of the pencil industry from Germany as "one of the consequences of the war."

Another consequence of the war was the seizure in 1917 by the Alien Property Board of the Newark, New Jersey, factory of A. W. Faber. The factory, including such assets as all the company's trademarks registered in the U.S. Patent Office, was sold to American interests. The firm was incorporated in New Jersey as A. W. Faber, Inc., but close relationships were established with the original company in Stein after the war.

After another disruption in relations during World War II, the company's name would be changed to A. W. Faber-Castell Pencil Company, Inc., with the German Faber-Castell firm acquiring some of the stock.

Long before World War I, Germany's strength in pencils was in the high-priced pencils intended primarily for artists, designers, architects, and engineers. While these pencils did not sell in the volume that cheaper pencils did, there was generally a higher profit margin for both the manufacturer and the retailer. Artists and engineers bought pencils on the basis of their quality and consistency, and once a favorite brand was settled on, it took a real technical improvement or effective salesmanship to get a pencil user to change. After all, it has been said that even blindfolded an architect or engineer could tell the difference between the 2H and the 3H of a familiar brand of pencil by the way it grabbed the paper.

A. W. Faber had introduced its polygrade pencils in 1837, providing a standard assortment ranging from BB to HHH. The user of such pencils was assured that the meaning of BB or any other polygrade designation would not change over the years. Thus the artist could confidently revise a sketch or an engineer a drawing with pencils marked the same but bought years apart. Faber's Siberian graphite established its polygrades as without peer, and they only began to be seriously challenged when the Alibert mine started to fail. After Johann Faber had founded his own firm in 1878, he offered his own polygrade pencils, some made from his independent source of Siberian graphite.

Faced with the prospect of being without the raw material necessary for its premier product, A. W. Faber asked its engineers to undertake a research and development program with the goal of making the highest-quality pencils out of graphite newly discovered in Europe and elsewhere. The resulting new process and its associated machinery were given a special name to serve as an exclusive selling point. Faber's new lead was processed in "Microlet mills," which were said to produce "a graphite composition superior to the Siberian graphite in both purity and quality." With the new lead a new line of drawing pencils was introduced in 1906, named Castell and painted green to distinguish them from the yellow Koh-I-Noors. While Faber would claim that the Castell was "the foremost pencil for engineers, technicians and draftsmen," the claim would often be challenged, as Faber-Castell was in fact challenging other distinctive pencils.

The Koh-I-Noor was being advertised as "the perfect pencil," "the world's best pencil," and the pencil with "a silken touch as light as a butterfly." *The Times* of London in 1906 reported receiving specimens of the "excellent and well-known" pencils, and an advertisement in *The New York Times* that same year told the pencil user that "in the Koh-I-Noor, made in Austria, you will find what you enjoy using most." While customers did not always remember the name of the pencil, they did remember its color and thus often asked for a "yellow pencil." The Koh-I-Noor's success attracted imitators, and the Hardtmuths warned that even "people familiar with the Koh-I-Noor [were] sometimes deceived by the colour of imitators," and the company asserted that "the colour and the exterior" of the pencil were all that could be imitated.

Koh-I-Noors began being imported into the United States shortly after they were exhibited at the 1893 Columbian Exposition, but the supply was cut off for four years during World War I. To prevent that from happening again, the Koh-I-Noor Pencil Company was incorporated in New Jersey in 1919, when supplies resumed. As long as the pencils were still made in Czechoslovakia, as part of Austria-Hungary came to be called after the war, a duty had to be paid on the finished pencils even though the cedar had originated in the United States. To avoid such a penalty, a factory was opened in Bloomsbury, New Jersey, in 1938. Then only uncased leads needed to be imported, with the assembly of the pencils in native American wood taking place in New Jersey.

No effort was spared in protecting and finishing the Koh-I-Noors: a Manx cat and her offspring were given the run of the storerooms, where otherwise mice would eat the uncased leads. After the leads were encased in wood, the pencils were given fourteen coats of golden-yellow lacquer, the ends of the pencils were sprayed with gold paint, lettering was applied in 16-carat gold leaf, and the grade designation was stamped on every other face of the hexagonal case, so that it could always be easily read. The finished pencils were inspected carefully, and one in ten was rejected—to be cut into shorter lengths for golf pencils and the like. Those Koh-I-Noors that did pass inspection were "packed in dozen lots in metal boxes that kept them straight even in humid areas where the porous cedar tends to absorb moisture and become warped."

High tariffs did not impede the sale of quality pencils in the United States in the nineteenth century because for some time there was little competition among these "fine goods," as pencils representing "the highest class of material grade and finish" were known. But as the young American pencil companies became more firmly established and experienced in making cheaper goods, they began to look toward competing more aggressively for the finer-goods trade. After all, the American machines would work no less effectively in shaping top-of-the-line pencils, if the company engineers could obtain and master the best raw materials. Since the Americans already had an advantage with respect to the supply of the finest cedar, it was only a matter of working clay and graphite to perfection.

By the late 1870s, for example, the Joseph Dixon Crucible Company was refining graphite to 99.96 percent purity. To achieve this, lump graphite was taken from the mine at Ticonderoga and pulverized under water, so that the particles floated to the top. Graphite to be used for pencils was further pulverized in Jersey City, and since graphite is an excellent lubricant, this was not a simple matter of using grindstones. When the graphite was sufficiently fine, it was "finer and softer than any flour." But, according to a contemporary reporter, Dixon's powdered graphite "does not cohere like flour; it can be taken up in the hand, just as water can, and is hardly retained more easily than water is; if one attempts to take a pinch of it between forefinger and thumb it is as evasive as quicksilver, and the only sensation is that the flesh is smoother than before."

To make pencil lead, the graphite had to be separated further according to fineness. To do this, the dust was mixed with water in a hopper and allowed to run slowly through a series of tubs:

> The coarsest and heaviest particles settle to the bottom of the first tub, the next coarsest and heaviest in the next, and so on, the movement of the water being made very gentle; on reaching the last tub, the powder, being twice as heavy as water and sinking in it if undisturbed, has so far settled that the water discharges from the top nearly clear. After the flow is stopped and the powder has been allowed to settle, the clear water is withdrawn by removing successively, beginning with the upper one, a number of plugs inserted in holes in the side of each tub, care being used not to agitate the contents so as to disturb the

deposited dust; this being done properly, the deposit is removed through the gates at the bottom of the tub. The separation is thus performed, by this ingenious process of "floating," more perfectly than it could be by any direct handling, *dry* treatment being wholly impracticable. For the finest pencils, the deposit from the last tub only is used, but for ordinary and cheap grades that from the two before the last will answer.

The clay was subjected to a similar floating process, and after being mixed with the graphite, it was ground for as long as twenty-four hours between flat stones, "thus securing the most perfect strength, uniformity and freeness from grit" for the best pencils. Next the lead dough was extruded through a die, the still-soft lead coiling beneath the orifice. (An unbroken coil 400 feet long was shown by Dixon at the Centennial Exhibition, and there was "no practical difficulty in making a coil long enough for an ocean cable or for Puck's promised girdle around the earth.") After the coil was cut into pencil lengths, which were then baked in a kiln, they were finally ready to be put into wooden cases: "For the cheapest pencil pine is used; for the common grades, an ordinary quality of red cedar; for all the standard grades, the Florida Keys cedar, which is soft and close-grained, and is so superior for the purpose that even the European pencil-makers are obliged to come to Florida for it." Dixon's fine goods were designated "American Graphite Polygrade" pencils, to remind the buyer with whose they were to be compared.

When the American Lead Pencil Company decided to introduce a line of drawing pencils, it named them after the Venus de Milo, which the company's president, Louis Reckford, associated with the Louvre and fine art generally. Venus pencils in seventeen degrees were first sold in 1905, and it was claimed that they contained "the first accurately graded black drawing leads ever produced in the United States." The Venus's distinctive color was to be dark green, but because of some defect in the paint, it cracked when it dried. However, "officers of the company liked the effect so much" that they adopted the crackled green finish as part of Venus's trademark. The pencil overcame prejudices against American-made drawing pencils, and by 1919 it was being advertised as "the largest selling quality pencil in the world." One of the circumstances that helped Venus achieve its popularity was its availability when

European pencils like Castells and Koh-I-Noors became scarce. Other American companies brought out lines of drawing pencils during the war.

Dixon's American Graphite artists' pencils had been graded with the company's idiosyncratic marks, ranging from VVVS to VVVH, to distinguish them from the European pencils, and their natural cedar had been associated with quality in the nineteenth century. But the increasing scarcity of fine wood, especially outside America, and the standard set by Koh-I-Noor and its competitors prompted Dixon around 1917 to introduce its Eldorado, finished in blue with gold lettering and graded according to the European system. A 1919 advertisement declared that "during the war, when most needed in the tasks of victory—Dixon's Eldorado, 'the master drawing pencil,' rendered a real National Service." While it was claimed to be a "real *American achievement,*" in place of Dixon's familiar "American Graphite" slogan on each pencil was "the master drawing pencil," to compare it with the Venus and the European standards of quality for such pencils. The Eldorado was intended principally for artists and engineers, who could request free full-length samples in their choice of degree, according to an advertisement in *Mechanical Engineering* in 1919. As competition increased after the war, Dixon looked to sell the Eldorado to a wider clientele, and an advertisement in *System, the Magazine of Business* in 1922 offered, for ten cents, special "system" sets of trial-length samples of Eldorados and other Dixon products.

Eberhard Faber declared that "Europe invented the pencil, but America perfected it." Faber called its yellow drawing pencil the Van Dyke, and advertised it as "good, to the last half inch," perhaps alluding to contemporary ads that showed stubs of Venus pencils whittled down to about three-quarters of an inch. According to advertisements in *Literary Digest* in 1920, discriminating pencil users would see the merit in "daily use of Van Dykes in HB grade in general business activity." A note on business stationery would get the reader a free sample, or two Van Dykes in one's choice of degree plus an eraser could be had for fifteen cents. The Eagle Pencil Company's drawing pencil was called Turquoise, which described its color, and it too competed for the quality-pencil business.

At the end of World War I the American pencil industry was optimistic. Although European manufacturers were once again producing and exporting, the Americans had confidence in their products and their future. They were sharpening their pencils and looking to further expansion of their markets. While business and marketing decisions would be important, the engineering of better pencils would also be increasingly necessary to back up any claims of quality and superiority.

14/ The Importance of Infrastructure

*T*he point of a pencil is its raison d'être; all else is infrastructure. But without the infrastructure the pencil point could not be held or sharpened or even used with any comfort or control or confidence. The point would be lost in the hand or lie broken on the desk. All technological artifacts require an infrastructure of some kind or other. The modern automobile would be ineffectual without a network of highways and service stations and parking spaces—along with their maintenance crews and mechanics and attendants. Airplanes would never get off the ground without airports and flight crews and air traffic controllers. Telephones need poles and wires and operators and switching devices and, today, long-distance carriers. Television requires producers and studios and actors and scripts.

But this is not to suggest that infrastructure precedes that which it serves. Henry Ford has been quoted as saying, "Cars must come before roads," and the use of black lead, in the form of plumbago from the Borrowdale mine, was used in its uncut native state before it was ever encased in wood to serve as a more sophisticated pencil that reached wide distribution and acceptance as a substitute for the pen. As black lead evolved from a local discovery and useful curiosity into a desirable commodity, however, the disadvantages of writing with a relatively scarce, brittle, and dirty substance came to be as annoying as were early excursions into the muddy, rutty countryside with a relatively expensive, unreliable, and uncomfortable automobile.

Infrastructure takes many forms, but it is always an underlying prerequisite for the effective functioning, dissemination,
and acceptance of inventions—once their novelty wears off.
And providing a proper infrastructure for a technological device or engineering structure can be as much a part of engineering as designing and making the central artifact. Indeed,
the very acts of manufacturing and construction require an
infrastructure, albeit transient or ephemeral or temporary, of
their own, and providing the tools or machines or molds or
falsework or centering required to make something can be the
greater part of the effort to produce it.

Just as the construction of the Britannia Bridge had attracted a crowd to the Menai Strait in the late 1840s, so the
completion of one of the largest cantilever bridges in America
attracted twenty thousand people to the upper reaches of San
Francisco Bay in 1927. There the closing span, which weighed
750 tons, was being hoisted into place on the Carquinez Straits
Bridge, and probably few if any of the onlookers were unaware
that it was not very many years earlier that just such an operation led to disaster in the construction of the Quebec Bridge.
Indeed, the hero of Willa Cather's 1912 novel, *Alexander's
Bridge,* is the chief engineer of the Quebec Bridge, and he is
killed in the fictional re-creation of the accident. But in 1927,
the weight of ten thousand men was lifted quickly and safely
into place, in part because the builders of the new bridge in
California had learned from the mistakes made at Quebec.

When he was asked about the Carquinez Straits Bridge, the
engineer David Steinman described one of the more difficult
problems associated with its construction: "Toredos—a worm
that eats wood." The bridge's first timber piles were destroyed
within a matter of weeks, and elaborate measures had to be
taken before "the destructive little shipworms could be van-
quished." The interviewer went on to ask Steinman about the
speed of building bridges in the 1920s, and he confirmed the
popular view that they were indeed built faster than bridges of
comparable size would have been in the old days. And while
answering the question, Steinman did something that illustrates the importance of infrastructure, as well as the fact that
infrastructure must sometimes be removed to ready an artifact
for use or actually be consumed in the act of using the artifact.
Yet sufficient infrastructure must also remain as long as the
artifact is to be useful. What Steinman did was he "pulled out
a pocketknife and began sharpening a pencil."

Not only was Steinman's pocketknife, the descendant of the

penknife used to sharpen quill pens, part of the infrastructure of the pencil; the wood that he was whittling away is also to this day the infrastructure without which the pencil lead would be little more than a dirty piece of brittle graphite, perhaps wrapped in paper or string, but certainly not nearly the popular implement that it is in the hands of the engineer, writer, or artist.

The artist Saul Steinberg, whose work frequently appears in *The New Yorker,* has been described by others as a "draftsman," and he has described himself as "a writer who draws." Though he studied architecture in Milan's Politecnico, he never practiced the profession, believing that "the study of architecture is a marvelous training for anything but architecture. The frightening thought that what you draw may become a building makes for reasoned lines." Freed from reasoned lines, Steinberg could draw buildings that did not need to be built, and he could draw drawings of drawings. And what Steinberg as artist has often drawn on, in both a figurative and literal sense, are artifacts of technology. Automobiles, skyscrapers, bridges, and railroad stations have all appeared in his drawings, and he has actually physically drawn on other artifacts, including chairs, bathtubs, boxes, paper bags, and pieces of wood that he has carved into the shapes of pencils.

Steinberg's pencils are not meant to be functional, of course, and so they do not need to have the reasoned lines of a carefully sharpened point or the comfortable shaft shaped like a rounded hexagonal or a smooth circular cylinder. Furthermore, because they are not meant to be used as pencils, the wood out of which Steinberg makes his pencils need only be capable of being whittled and painted. If Steinberg's pencils looked the way he wanted them to look on his "Tables" of the 1970s, then they were successful. The weight, stiffness, strength, color, and virtually all other physical properties of the wood were immaterial to the visual effect he was trying to achieve. The wood of Steinberg's objets d'art had to be satisfying as a medium to be worked in by the artist, and it had to give him the results he wanted to achieve, but it was an end in itself and did not have to work as a piece of infrastructure for a piece of lead.

The pencil is the subject of an English riddle that has come down through oral tradition: "I am taken from a mine, and shut up in a wooden case, from which I am never released, and yet I am used by almost everybody." But the real riddle about pencils has always been, in America and elsewhere: How are they made? Specifically: How is the lead gotten into the wood? Is a hole drilled into a stick, to be stuffed like a sausage? Is molten lead poured in or is a long piece of brittle lead delicately threaded into the hole? Is the wood case constructed around the lead, much as a crate is built around a piece of machinery for shipping? While we have seen that the manner of encasing black lead in wood has actually changed over the centuries, the modern techniques have made it as difficult to separate the lead from the wood as it is to tell the dancer from the dance.

No matter how it has been made, the wooden case enclosing a pencil lead has always been a kind of enigmatic falsework wrapped riddlelike around a mystery. You have to know the trick to figure it out. Ever since graphite was first encased in wood, it has been the wooden structure that is formed first, the cedar grooved to receive the long pencil lead, a virtual obelisk that is easily broken. After the leaded slat of the modern process is covered with a mating slat of wood, the assembly is not ready for use until some of the wooden centering is cut away, to be discarded and forgotten, leaving the short pencil point to make a daring bridge between pencil and paper—a metaphorical bridge that can carry from mind to paper the lines of a daring real bridge, which can cause jaws to drop, or the words of a daring new philosophy, which can cause eyebrows to arch. But black lead is not necessarily so easily matched with wood as an answer is to a riddle.

As in Steinberg's pencils, the wood in a real pencil must have good whittling qualities, of course, but in addition it must be suitable as a piece of infrastructure for the lead that it supports. And the wood in a real pencil plays a very curious role, as described by Charles Nichols, who as director of engineering for a pencil manufacturer could not eschew reasoned lines or the reasoned qualities he demanded of a suitable pencil wood. Unlike the artist Steinberg, who could make drawings of rocks defying the law of gravity, the engineer Nichols could not make pencils with woods that did not meet the requirements of a suitable infrastructure, including some unique aesthetic ones:

> Lead pencils are designed for a purpose diametrically opposed to the purposes for which most wood products are produced. In short, [pencils] are designed for destruction, and yet the unused portion of a pencil must remain

sturdy and pleasing to the eye. Since wood is the material representing the major portion of a pencil, it must be selected for characteristics which lend themselves to strength, freedom from warpage, and smooth cutting qualities.

The wood case of a real pencil is what makes the pencil work, just as the suspension cables make a great bridge work. Neither the wood of a pencil case nor the steel of a bridge cable is the real point of the object, yet these are the dominant elements that give the artifact its psychological and visual characteristics and make it appear "sturdy and pleasing to the eye." But no matter how satisfying an artifact might be from the point of view of appearances, it must ultimately function properly. The bridge must hold its own in the wind, must not rust in the rain, and must not sag in its old age. The pencil, as Nichols points out, must not be weak, must not warp, and must not splinter or split under the cutting edge of the pocketknife or pencil sharpener. No matter how fine the quality of the lead inside, if the wooden infrastructure of a pencil is inferior the point may be lost, as a steel bridge could be lost if founded on worm-eaten wooden piles.

Imagine what would happen, for example, if the pencil wood had no strength at all. Then the pencil might snap not at the point but at the fingers where the pencil as a cantilever beam met them. A weak wood could be used to make a stronger pencil by making it of a larger diameter, like the fat pencils designed to match the gross motor skills of a child's hand, but such pencils would be uncomfortably thick in an adult's fingers. And imagine what would happen, for example, if the wood of a pencil warped. Not only would the pencil take on the unattractive appearance of an inferior piece of lumber, but as it warped it would separate or, if it did not separate, it would bend the lead and secretly break it into little pieces that would fall out of the pencil when we tried to sharpen it. Imagine what would happen, for example, if the wood splintered and split each time we tried to sharpen an otherwise fine pencil. Then not only would the pencil be ugly but the cone of support that the wood gives to the lead would be jeopardized and we would find our pencil leads breaking very easily.

Thus the successful development of the pencil as we know it depended not only on finding the right kind of natural graphite or the right kind of clay and the right kind of mixing and processing techniques, but also on finding the right kind of wood in which to encase the right kind of lead. Had the wood lacked any one of the right qualities, had it been weak or had it warped easily or had it split on sharpening, then it would not have provided a suitable means for encasing the lead. A strong, straight wood case that proved impossible to sharpen would make an impossible pencil. A straight and easily sharpened case that bent like a piece of balsa wood could not displace the metalpoint stylus. Any wood that did not possess sufficiently desirable properties for a pencil case would no more suit the purpose than a metal that did not have simultaneously the strength, stiffness, and toughness to hold up the roadway of a bridge without breaking, stretching, or cracking too easily. Since the modern pencil owes so much of its success to the existence of a suitable wood, it is not surprising that it was woodworkers and cabinetmakers who made the first successful wood-cased pencils. They would naturally have known the qualities of different woods and thus have been able to identify the best woods to use.

Before the first modern pencil was made, red cedar was imported from Virginia and Florida by English makers of clothes chests. Thus when the idea arose of enclosing sticks of Borrowdale graphite in wood, the properties of red cedar were already known to be ideal. Those who had worked with it were the same craftsmen who were looking for a pencil wood that would not warp or splinter, and so red cedar is believed to have been used in pencils as early as the seventeenth century. While other furniture woods, such as deal and fir, were also used from time to time, it was red cedar that proved to be far superior to all others for pencils.

But there are no inexhaustible sources of naturally occurring trees. The use of their wood not only in the falsework for stone structures but also in the permanent work of timber structures, not to mention the burning of wood in producing heat for making iron and warmth for making homes, consumed timber fast in many industrializing countries. Forests were cleared for their land as well as for their wood, and timber became so scarce in England in the eighteenth century that alternatives to it for making iron and bridges actually led to the development of new iron- and bridge-making techniques.

It was not only eighteenth- and nineteenth-century hard technologists who became concerned about the supply of trees for the continuation of the Industrial Revolution or the pencil-making industry. In 1924 Melvil Dewey, the inventor of the decimal system of library classification and a proponent of simplified spelling, lamented the fact that "one seventh of all English writing is made up of unnecessary letters; therefore one tree of every seven made into pulp wood is wasted." And while Henry Adams did not worry about trees explicitly, he did about waste and, pondering the meaning of history, wrote of becoming frustrated with some of the exhibits in Chicago in 1893: "Historical exhibits were common, but they never went far enough; none were thoroughly worked out. One of the best was that of the Cunard steamers, but still a student hungry for results found himself obliged to waste a pencil and several sheets of paper trying to calculate exactly when, according to the given increase of power, tonnage, and speed, the growth of the ocean steamer would reach its limits. His figures brought him, he thought, to the year 1927."

Not only did "historical exhibits," which seem to have traced the development of artifacts from their humble proportions and projected them into their leviathan and titanic futures, apparently allow no limits, but the framers of the exhibits for 1893 apparently did not ask questions about the limits to growth that so troubled Adams then and, apparently forgotten even to the great historian, also troubled Galileo a quarter of a millennium earlier. In 1927 it would be almost three hundred years since Galileo, thinking like an engineer, had tried to calculate the limits of great structures like ships and worried about the waste of material in wooden beams.

But it was not the size of the pencil, which after all was just about right for its purpose, that was growing at the time of the Columbian Exposition. Rather it was the size of the industry, making pencils by the hundreds of millions and demanding the trees with which to do it. Only one-fifth of a typical tree might be suitable for use, with more ending up as sawdust than as pencil casing. With the growth of the pencil and other wood-consuming industries throughout the nineteenth century, the future supply of red cedar became as questionable as that of Borrowdale graphite had earlier in the century. This should not have taken anyone by surprise, however, for the Swedish naturalist Peter Kalm had predicted as early as 1750 the approaching end of the supply of American wood. The German pencil baron Lothar von Faber took steps to ensure his own supply by planting in 1860 four hundred acres in Bavaria with seeds from the red cedar. But since the trees grew so slowly, it would be the turn of the century before experiments showed that the wood from the Bavarian stand was totally unsuited for the manufacture of pencils.

As late as 1890 a pencil manufacturer could say that all the cedar he needed could be obtained from fallen trees that if not used would just decompose in Florida. Indeed, the best pencil wood came from the shells of rotting trees that had fallen over from old age. Cedar was so plentiful in Florida, Georgia, Alabama, and Tennessee that farmers built barns and fences out of it. But the expanding pencil industry was using cedar at such an accelerating rate that fallen trees could no longer supply all the needs. No other industry was so dependent upon a single species of wood as the American pencil industry was on red cedar, and yet, according to a contemporary report,

> the supply is gradually disappearing, and it is necessary every year to go further and further back into the virgin forests. Cedar cruisers know every region of the country where they can get any stock. Old cuttings have all been gone over repeatedly. Old stumps have been dug out. Even old log houses have been taken down. Large quantities of old cedar planks from barns are being bought, and fence rails have been picked over. The common practice is for the pencil manufacturers to put up a fine new woven wire fence for the farmer who has a fence with enough cedar to make it worth while, and the farmer who has a picket fence of cedar can get the best wire fence money can buy.

Whereas a manufacturer could boast in 1890 that "the average pencil in everyday use costs about one-quarter cent to make," and that he was "content with one hundred percent profit" in selling to dealers who in turn sold the pencil for five cents, in about twenty years there was "in the ordinary lead pencil three-fourths of a cent's worth of cedar." In the highest-grade drawing pencils, wood constituted as much as 40 percent of the final cost to the manufacturer. In 1911 *The New York Times,* which was unsympathetic to the "lamenting" pencil industry, stated in an editorial that the public was "paying, and paying high," for the manufacturers' lack of foresight. In addition to buying old barns and fences, the editorial suggested "the planting of anywhere from two to a hundred little cedar trees whenever [the pencil makers] cut down a big one, or even whenever they want to cut down a big one and cannot find one to cut down."

While the editors wanted the pencil makers to solve their own problems, the government had already become involved.

By 1910 the question of the wood supply for future pencil manufacturing had become so acute that the U.S. Forest Service had undertaken an investigation of the possibility of using woods other than the red cedar (known by botanists as *Juniperus virginiana*) and the closely related southern red juniper (*Juniperus barbadensis*). Among pencil manufacturers this latter tree was considered essentially the same as red cedar, even being called red cedar, and both were known as pencil cedars. According to the chief of the Office of Wood Utilization of the Forest Service, writing in *American Lumberman* in 1912, the office's study was prompted not only by the growing scarcity of red cedar but also by "the fact that there were many woods on the National Forests, now little used, whose physical and mechanical properties seemed to fit them for pencil making." The desirable properties were described by the forester as follows:

> A good pencil wood should be of an even texture, that is, the summer wood of approximately the same hardness as the spring wood; it should have an even straight grain; it should be soft and slightly brittle; of a dark red color; rather light in weight; nonresinous and slightly aromatic. The wood which contains all these properties in the highest degree is the red cedar, and for many years it has been the exclusive pencil wood. . . .
>
> The raw material for a pencil is known as a pencil slat and is $7\frac{1}{4} \times 2\frac{1}{2} \times \frac{1}{4}$ inches. These slats are manufactured in the southern states, where the tree grows principally, and are shipped in bundles or crates to the manufacturers. The manufacturers formerly required all slats to be $2\frac{1}{2}$ inches wide (a dimension which makes six half-pencils), but owing to the scarcity of the material now, they are glad to take a large amount of the slats in narrower widths, in some cases wide enough to make but two half-pencils instead of six. The slats are separated into three grades, the first grade being dark red and entirely clear, and used for the highest grade of pencils. The second grade admits of few defects, and the third grade contains much of the white sapwood of the tree. The second and third grades are used to manufacture the cheaper class of pencils, pen holders, etc.

Not only was wood graded in this way, but for the best pencils the slats were further separated according to their hardness, so that the two halves of a pencil could be made of woods as nearly identical as possible and thus provide the most consistent sharpening qualities. The forester's article, which noted that the high cost of red cedar had stopped its use in everything but pencils and clothes chests, also went on to describe the pencil-making process, which in 1912 was little changed from that developed in the latter half of the nineteenth century. (Woodworking still dominates the process. While the slats are now as wide as three inches, they are still first grooved, then glued sandwichlike around now as many as eight or nine precisely spaced parallel leads. The individual pencils are then cut from the sandwiches and finished according to their quality.)

In 1912, it was estimated, over one billion pencils, about half the world's production, were made of American cedar, with fully 750 million pencils turned out in the United States alone. This amounted to about eight pencils per capita. The population of red cedar trees continued to diminish, becoming practically nonexistent by 1920 in Tennessee, where some of the finest pencil wood was at one time grown. Pencil manufacturers were still buying up old fence posts and rails, railroad ties, and log cabins made of red cedar. This supply source was also limited, of course, and eventually substitute woods had to be employed for making pencils. A dozen different American woods were tested for their ability to replace red cedar. While none was found to be the equal of it, three trees were believed to be "excellent substitutes." These were the Rocky Mountain red cedar, the alligator juniper, and the western juniper. The report noted that all three "grow very scatteringly, however, and their exploitation would be costly." But, if not the best, then at least good substitutes for pencil cedar were also to be found in some of the more available trees—namely, the Port Orford cedar, the big tree, the redwood, and the incense cedar. This last *(Libocedrus decurrens),* growing principally in southern Oregon and northern California, would ultimately become the wood of choice to make pencils, but its acceptance was slow in coming.

While it had the strength and feel of red cedar, incense cedar lacked two qualities that had come to be associated with fine pencils: the substitute wood had neither the proper color nor the proper odor. While the color and odor of the wood had no bearing on the physical function of the pencil, these cosmetic factors impeded the sales of pencils made of the new wood. Pencils made of the white and relatively odorless incense cedar, which was a misnomer as far as the pencil industry was concerned, came to be accepted only after the wood was dyed and perfumed to simulate red cedar. Incense cedar to this day is dyed to lend a uniform color to the wood, and it is also impregnated with wax to act as a lubricant during the pencil-making process. The waxed wood also makes for easier and better sharpening.

By the mid-1920s pencils in France were being made of basswood and alder, properly treated after being dried. While pencils from these woods are not comparable in quality to those of red cedar, the economics of the situation were striking. At the time the substitute woods could be obtained for approximately $16 per ton, after treatment, while American cedar was costing $115 or more per ton. For similar reasons, English pencil manufacturers were developing cedar forests on the slopes of Mount Kilimanjaro in Africa, and this Kenya cedar, known locally as *mutarawka*, was available in France for about half the cost of the American variety. In the early part of this century, when it became known that there was an untouched stand of red cedar on Little St. Simons Island, off the coast of Georgia, the Hudson Lumber Company, which was associated with the Eagle Pencil Company, bought the island. Little St. Simons had been an old Indian retreat, and the vast quantity of oyster shells that had been thrown on a midden leached lime into the sandy ground, and this provided excellent soil for cedar growth. However, being also exposed to cold ocean winds, the trees grew gnarled and crooked, and the timber was expensive to move to the mainland. Since this cedar proved impractical to use, the island eventually became a private retreat of the California Berolzheimer family, whose Hudson Lumber Company today harvests wood for virtually all American pencil companies and for more than a hundred foreign companies.

In the early part of this century some European manufacturers were using Russian alder, Siberian redwood, and English lime for their pencil cases. However, these woods were somewhat hard and had an uneven texture; thus they also had to be treated, resulting in a good but far from perfect pencil. According to one critic: "The treatment gives the desired softness, although it has been noted in sharpening some pencils made from these woods on a machine that the spring wood has a decided tendency to roughen, showing the treatment is not yet perfect. Furthermore, the sapwood retains its white color, which is to a certain extent undesirable since it soils quickly and gives the pencil an unclean appearance. The wood also has not the cedar odor."

In the meantime, since they were having so much difficulty finding a wood that was a readily available and suitable substitute for red cedar, manufacturers looked for other means of encasing the pencil lead. Research and development efforts in the late nineteenth century had come up with the paper-wrapped lead pencil, pioneered by the Blaisdell Pencil Company of Philadelphia, as if returning to the string-wrapped graphite stick of centuries earlier, and this re-innovation worked well technically and showed great promise. Manufacturers thus went to considerable expense in investing in the development and installation of machines to make the woodless pencil, but the product failed for unanticipated psychological reasons. "The pencil-using public preferred something to whittle on," and the paper pencil never did come into widespread use, except where it covers thick and colored leads that would suffer a lot of breakage and waste when sharpened with a knife or mechanical sharpener.

By 1942 almost a billion and a half pencils were being produced in the United States annually, enough for over ten pencils for every man, woman, and child in the country, and virtually all of them were wood-cased. The manufacturing process had been developed to a high degree of sophistication, with the woodworking machinery operating to perhaps the highest tolerance of any woodworking equipment, in part because of the importance of conserving the casing material that is so critical to pencil manufacturing. It is for this reason that the triangular-shaped pencil, as comfortable to hold and as attractive as it may be, is not made in great quantities. Cutting triangular pencils out of a practicable sandwich of lead and wood is extremely wasteful of wood.

When in the early part of this century Dixon wanted to improve its pencil-making operations, the company focused on the woodworking aspects of the process to increase efficiency. By this time, machines were generally made in Germany, but a Boston manufacturer of woodworking machinery, the S. A. Woods Machine Company, devised a way of increasing the speed of the cutters by a factor of three. Also, German machinery separated hexagonal pencils by cutting them apart at their corners, but Woods proposed cutting them apart at the flats, thus getting less sawdust and an extra pencil out of each pair of slats. In all, production was increased about five-fold with the new machine. The hexagonal pencil is now gen-

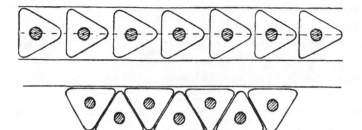

Two ways of shaping triangular pencils from leaded slats, showing the practical way to be wasteful of wood

erally preferred over the round by manufacturers because nine
hexagonal pencils can be gotten out of the same wood it takes
to make eight round ones. The pencil user seems to prefer
hexagons also, buying about eleven of them for every one
round pencil.

While machines might be efficient for making pencils, during World War II rotary pencil sharpeners were outlawed in
Britain because they wasted so much scarce lead and wood,
and pencils had to be sharpened in the more conservative
manner—with knives. But the importance of economy in the

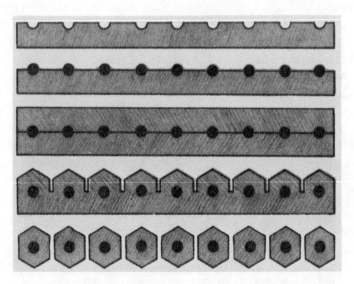

End views of steps in forming hexagonal pencils with a minimal
waste of wood

use of wood did not begin with the war or even with the pencil industry, as a biography of Marc Isambard Brunel, the great engineer of the first tunnel under the river Thames and the father of Isambard Kingdom Brunel, makes clear. The older Brunel was born in France in 1769, served as architect and chief engineer for New York City in the 1790s, and moved to England in 1799. There, one of his first and most innovative contributions was to design new sawmills and introduce efficient sawing techniques into the industry. He developed a mechanized system for making more than 100,000 wooden pulley blocks annually for the British Admiralty, utilizing machines designed by him and produced by Henry Maudslay, but the system earned a royalty that was disappointingly small when compared with the £17,000 that were saved each year by the government.

Brunel also developed the use of the circular saw, which made it possible for wood to be removed in usable pieces rather than in the wasted chips of previous techniques used for rabbeting and grooving. Machines with revolving knife blades were also used to cut away wood in shavings rather than chips or sawdust, and the shavings began to be used to make hat and pill boxes, thus conserving paper and reducing Britain's dependence on imports for those items. Thus it seems natural that machines that were descendants of the kinds developed by Brunel came to be used in the wood-intensive pencil industry.

Today woodworking in the pencil industry is believed to be as sophisticated as that anywhere. When Charles Nichols described the machining of grooves in the pencil slats, he paid special attention to the quality of workmanship:

> The tolerances in this phase of the operation are of necessity extremely close, since the accuracy of successive operations and the quality of the ultimate product depend entirely on the care and precision used in the machining operations. The center distances between the individual leads are maintained within tolerances of plus or minus 0.0005 in. and the finished thickness is maintained within plus or minus 0.001 in. The diameters of the lead grooves are also maintained with tolerances of the same order.
>
> The face of the slat on the grooved side must be absolutely flat and smooth in order to guarantee a first-class glue joint, and it will at once become evident that the

center distances between the leads must be maintained in order that two slats may be properly aligned face to face and so that the pencils which are finally shaped may have the leads concentrically disposed within their cross-sectional area.

While the language might be a bit technical and stuffy-sounding, the intention is clear—to make a good pencil. For that was the purpose for all of the pencil industry's woodworking machinery—to make a good case for a good piece of lead. While Thoreau could have used his ruined wooden pencil case as a penholder, the pencil industry in the mid-twentieth century, relatively speaking, did not have as many trees to spare, and it did not want to have to create unnecessary sawdust or reject pencils that did not have their leads centered.

Since the supply of red cedar had been virtually exhausted before reforestation was practiced, and since the supply of western incense cedar was not unlimited and it would take almost two centuries to replace it by reforestation, finding other wood substitutes continued to be an ongoing problem for pencil manufacturers. After World War II talk of "pencils of an absolutely new kind" began to be reported. One company executive who apparently did not want to be named, but who was quoted in an article dealing primarily with the Eagle Pencil Company, teased the reporter: "Wait till you hear about the pencil of tomorrow and how it may be made. Plastics, maybe, one piece, the whole thing extruded from a tube!"

In the 1950s, when oil was three dollars a barrel and "plastic" was the buzzword, at least one pencil manufacturer did more than talk; its owner dreamed of making the plastic pencil. The Empire Pencil Company spent a reported twenty-five years and an undisclosed amount of money developing a process whereby, rather than being the end product of as many as 125 separate manufacturing processes, a pencil could be extruded from globs of molten plastic, powdered graphite, and wood flour. In the early 1970s the company introduced the result of its efforts, a new pencil-making technique known as the Epcon process. Since the average two- to four-hundred-year-old tree yields hardly enough wood to make 200,000 pencils, the developer of the process could estimate that "many, many thousands of cedar trees will be spared the woodsman's ax each year as a direct result of this revolution in the making of pencils." The product was called the "first *new* pencil in 200 years." Starting with colored pencils and then introducing standard writing pencils, by mid-1976 Empire had manufactured a half billion "Epcons" in a tightly guarded plant at the end of Pencil Street in Shelbyville, Tennessee, which is "Pencil City" to its civic boosters. (Pencils have been made there and in nearby Lewisburg since the early 1900s, when red cedar was still abundant in the area. Today California incense cedar is shipped to the factories.)

In Connecticut the Berol Corporation, which the Eagle Pencil Company had become, also began making plastic pencils. Berol called its product the "first *all*-plastic pencil" because, unlike the Epcon, which had to be painted in the conventional way, Berol's pencil was the result of a "triple coextrusion" process that included the finish. Production of the new pencil occupied only 10 percent of the floor space required for wood-pencil production, and the cost of the plastic materials amounted to only four-tenths of a cent per pencil, which was half that of the incense cedar that would have been needed. Berol's pencils, which were being produced under license from a Japanese manufacturer, were being turned out at the rate of fifty feet, or eighty-five pencils, per minute in the mid-1970s.

Whether the plastic pencils will save trees or will suffer the fate of other unorthodox pencils will be decided in the marketplace. One thing is clear, and that is that the Epcon, like the end result of any quest for reproduction in another medium, is not exactly the same as that which its maker set out to copy. While the Epcon has very many admirable qualities, such as an extremely smooth-writing lead that is perfectly centered, it also has different qualities. Like that of old German pencils, the Epcon point becomes soft in a flame, and the Epcon is noticeably more flexible than a wood-cased pencil. While the manufacturer suggests that this may lead to improved writing comfort, that may in fact depend upon what a writer expects a pencil to feel like in the hand, for comfort may be in the mind and not in the hand of the beholder. And whether the clean break that one gets when snapping an Epcon in two is preferable to the splintering of real wood when one is angry is again a matter of psychology and not of technology.

Such psychological factors in the acceptance of the products of technology are often underestimated, if not completely ignored, by those who concern themselves with the effects of technology on society. Indeed, just as a bridge cannot be built without economic and political support, so a consumer product cannot be marketed successfully if it violates the aesthetic and psychological sensibilities of its intended users. Things that might work perfectly well from a purely technical point of view can be total failures from a political or a business perspective. The Ford Edsel is a famous case, and the introduction of "New Coke" is a more recent example of the inability of engineers, whether of pencils or automobiles or soft drinks, to impose their supposedly totally technical view on society. Engineers neither can nor really want to do such a thing, and whether they are looking for a new pencil wood or a new soft-drink formula, they do not want to ignore or overlook any factor—physical, chemical, or psychological—that might jeopardize the success of their design. Engineers, no less than anyone else, do not want to fail, but they or their marketing partners may so misjudge the performance of their innovations, or get so excited about the quality of the lead in their new pencil, that they forget the importance of its wooden infrastructure.

15/ Beyond Perspective

When Konrad Gesner wanted to describe in 1565 the then new invention that has evolved into the modern pencil, he used an illustration and a paucity of words. Indeed, while it has become a cliché that a picture is worth a thousand words, neither Gesner's illustration alone nor even the illustration supplemented by his brief verbal description was sufficient to convey unambiguously how one would *make* such a newfangled gadget. Imagine trying to make a pencil in 1565 having no more information than Gesner's words: "The stylus . . . is made . . . from a sort of lead (which I have heard some call English antimony), shaved to a point and inserted in a wooden handle."

What sort of lead? What exactly is "English antimony"? How large a piece is used? To how sharp a point is it shaved? How is the handle made? Of what kind of wood is the handle made? How thick is the wood? How far into the handle is the lead inserted? How is the lead held in the handle?

Without knowing the answers to these questions one might not easily succeed in making a pencil that worked as well as the one Gesner found so remarkable. The wrong kind or size of lead would not make a good mark or would break too readily or would tear the paper or would not take a point. The wrong kind of wood might give a pencil that would split too easily under the pressure of writing and drawing or might crack or warp and leave one, if one were an inveterate naturalist like Gesner, deep in a ravine or up a creek without a pencil. Nevertheless, Gesner's brief description together with the illustration would give one the basic idea of a pencil, and knowing that he at least had found an efficacious combination of "English antimony" and wood would give one the confidence that finding such a combination was not an impossible quest. One could at least proceed by trial and error, learning from the ways in which unsuccessful combinations failed to measure up to the standard of Gesner's pencil.

A modern rendering of Konrad Gesner's pencil

While the word "stylus," even without an illustration, should convey the overall shape and proportions of the instrument, the verbal description alone gives little indication of the relative size and thickness of the lead and wood, and without knowing those proportions, there is no telling what one might come up with. To the contemporary of Gesner, the word "stylus" itself should actually have invoked an image or mental picture of metal styluses in use at the time, even though these were generally of much more slender proportions than the new device containing the unfamiliar material. But imagine that there was no picture at all, neither an actual picture accompanying the words nor a mental picture evoked by the words. What if the pencil Gesner wanted to describe had not been descended from a familiar styluslike artifact?

Gesner's words as words alone are insufficient to evoke an unambiguous picture. Should the lead be inserted crosswise in the wood, much like an ax head in a wooden ax handle? Or should the lead be bound to the wooden handle much like an arrowhead or a spearhead might be bound with leather to its shaft? And what should be the shape of the lead? Should it be like an ax head or an arrowhead or a needle or a stick? Should the end shaved to a point be inserted into the wooden handle? Who is to say what the verbal description might conjure up in the mind of someone who had never seen a pencil? Just imagine what a twenty-five-word description of a suspension bridge might evoke in the mind of the uninitiated.

Today, of course, the pencil has become such a familiar object that it is hard to imagine that anyone would really need a definition, whether verbal or pictorial. Nevertheless, a standard college dictionary must define "pencil" just as it must define the articles "a," "an," and "the." While these entries are seldom if ever consulted by native speakers, the language student may need them to clarify distinctions that might be virtually impossible for a native speaker to articulate. Thus the definition of "pencil" in a desk dictionary, while no more descriptive in words than Gesner's, would be enough to evoke in the foreigner's mind another language's equivalent and, most importantly, a picture, for this dictionary definition is not illustrated:

> **pen'cil** (pĕn'sĭl; -s'l), *n.* [OF. *pincel,* fr. L. *penicillum, penicillus,* dim. of *penis* tail.] . . . **3.** A slender cylinder or strip of black lead, colored chalk, etc., usually incased in wood, for writing or drawing.

That Gesner felt it was necessary to include a picture of a pencil in his book while a modern dictionary sees no need to illustrate its definition of "pencil" tells us how common the object has become. But to recognize that an illustration is no longer necessary is not to say that a picture of a pencil is no longer part of its definition. The commonness of the pencil is such that a few suggestive words in *Webster's* are sufficient to evoke a mental picture of a pencil in every reader's mind. While the object in that picture may be a round or hexagonal pencil, may have an eraser or not, and may be yellow or colorless, it will convey the essence of pencilness.

There is hardly an artifact of engineering or technology that can be separated from its physical appearance, and thus it is not surprising that engineers and technologists think and create in terms of pictures. It is for this reason that the naïve view of engineering as "applied science" is simply not valid. It is actually the theories and equations of science that are applied to the object of an engineer's imagination—once there is a picture of what is to be theorized about or analyzed with equations—and so science is really used as theoretical engineering. Heavenly explanations of our origins may posit: "In the beginning was the word." But earthly explanations of the origins of our artifacts must start: "In the beginning was the picture." Science is really thinking "on second thought," and science is applied "after the artifact," when the object has been pictured first in the mind of the engineer.

Eugene Ferguson, a historian of technology who has written eloquently on this idea, has labeled the notion that artifacts must be derived from the words, equations, or theories of science as "a bit of modern folklore." He goes on to explain what some might term "right-brain activity":

> Many objects of daily use have clearly been influenced by science, but their form and function, their dimensions and appearance, were determined by technologists— craftsmen, designers, inventors, and engineers—using nonscientific modes of thought. Carving knives, comfortable chairs, lighting fixtures, and motorcycles are as they are because over the years their designers and makers have established shape, style, and texture.
>
> Many features and qualities of the objects that a technologist thinks about cannot be reduced to unambiguous verbal descriptions; they are dealt with in his mind by a visual, nonverbal process.

Ferguson's forceful and convincing case for the importance of pictures in the development of technological artifacts is not often articulated even by technologists themselves. But that is not surprising if indeed they tend to create and think not in words but in pictures. Nevertheless, there have been some clear and authoritative expressions of the picture-first view. David Pye, theorizing on the nature of design, demolishes the notion that form follows function, not only refuting the "folklore" that science precedes technology, but also writing: "If there had been no inventions there would be no theory of mechanics. Invention came first."

This is no less true in the history of bridges than it is in the history of the pencil. And even when the theories of engineering science have been developed to explain how artifacts work, the creation of new artifacts still begins with pictures rather than with words or equations, which are nothing but the sentences of science. And to manufacture a complex new artifact takes engineering drawing, which serves as the microscope of the engineer, magnifying the details for the worker to see.

Obviously, one could not use a pencil to make a sketch of a design for the first pencil. But the first lead pencil probably did not need a physical sketch to be conceived, for sticking a piece of black lead into a tubelike holder most likely was in direct or near-direct nonverbal imitation of the way a piece of metallic lead was stuck in a reed or quill or the way a tuft of animal hairs was stuck into the hollow handle of a pencil brush. Indeed, it was the marvelous new object itself that served as a model for Konrad Gesner's 1565 illustration of it. But Gesner was not trying to show an engineer how to make a pencil; he was showing naturalists a grand new and clearly distinct species of artifact that had evolved out of instruments less fit for writing and drawing without ink.

As the pencil that we know today evolved in slow stages from that early prototype, there probably never existed nor was there ever a need for any elaborate drawings to indicate to the early craftsmen what to modify. There may have been rough sketches made when a master wished to show an assistant how to shape a piece of wood or perhaps marks made on the wood itself, but these would have been considered of little value, if they even survived the making process, once they had served their purpose.

The earliest of black-lead pencils, like the one illustrated in Gesner's book, were no doubt round because that was the natural and comfortable shape in which brushes had long been made, and thus it would have been the shape that immediately came to mind and the shape to emulate. The crafter of Gesner's pencil may not even have thought to make it in any other shape. Early craftsmen producing pencils would naturally imitate the brushes, but as woodworkers came to make more and more pencils, it was no doubt easier and faster for them to fashion square cases. The lead was square because that was the most logical and efficient way to slice the block of graphite, and making square wooden casings would also be the most logical and efficient way to saw them from larger pieces of wood. Producing a round pencil would have required an additional finishing step.

But square pencils are uncomfortable to use, and this may have led the early craftsmen to create the octagonal pencil. The eight-sided shaft, which would require only shaving off the four corners of the square, would be much more comfortable to use than a square one and yet quicker to make than a round pencil. With the advent of machinery and mass production, the shape of pencils came to be a question more of efficient use of material and machines than of a workman's time. But machinery can shape a pencil in a variety of ways, round at least as easily as polygonal. The shape of a pencil then becomes again a matter of decisions beyond those of craftability. The hexagonal model is an efficient compromise between the uncomfortable square and the more comfortable circular cross section, and the shapes and other qualities of many of the most familiar objects that we use have evolved out of similar compromises between economy of manufacture and suitability of use.

Some of today's better pencils have been made in a "hexa-round" shape, in which the six edges of the pencil are rounded to remove, at least in part, the objections of pencil users like John Steinbeck, who wrote for six hours each day: "Pencils must be round. A hexagonal pencil cuts my fingers after a long day." Artists generally also prefer round pencils, not only because they are comfortable but also because they can be turned and twisted more easily while drawing, thus giving the artist more control over the line. But because a visual artist tends to hold a pencil lightly and more like a brush, the question of its digging into the artist's fingers is often moot. What look like carpenter's pencils have long been found convenient for representing in single strokes individual bricks, stones, and the like, but when a round sketching pencil with a "big flat lead" was advertised the copy noted that it was "easily and comfortably held." Thomas Wolfe, who wrote by bearing down hard on blunt soft pencils, wore a groove in his finger, and for someone who wrote as much as he there is no shape that could have prevented that entirely. Nor is there likely to be any size or shape of pencil that does not produce aching fingers or an aching hand after a hard day's write.

Engineers and draftsmen also are likely to twirl their pencils while drawing a long line so that the point wears evenly and the line is as uniform as possible. But unlike artists, who tend to keep their pencils and brushes in an old paint can when not in use, engineers tend to lay their pencils down on their drawing boards, which are usually tilted off the horizontal. Thus the hexagonal pencil has the advantage of not rolling down the slope. If the hexagon is good, then the triangle would be even better, and triangular pencils have been made over the years, with manufacturers arguing that the triangle is the shape that most naturally and comfortably fits the three fingers that grasp the pencil. (That this makes sense can be seen by bringing the thumb and first two fingers together and looking at the triangle formed among the tips.) The 1897 Sears, Roebuck catalogue offered both Dixon and Faber triangular pencils, which it claimed had these advantages: "This shape prevents the fingers from becoming cramped while writing and also the possibility of their rolling from the desk."

While most writers do not worry about their fingers rolling from the desk, the copywriter for Sears, then proud to call itself the "cheapest supply house on earth," apparently did. But it was not a faulty pronoun reference that kept the triangular pencil from rolling over all others in the marketplace. The triangular pencil was simply wasteful of wood, and thus it could not compete with shapes "cheaper" to manufacture and, therefore, able to be sold at a lower price. The triangular pencils in the Sears catalogue sold for thirty-eight to forty cents per dozen. Faber's "Hexagon Gilt," at forty-nine cents per dozen, was the most expensive pencil offered, and only Faber's "Bank" pencil, with a slightly tapered or conical shape, and the yellow-polished "Black Monarch" drawing pencil, pictured in a package labeled "F. Faber," sold for as much as forty cents per dozen. On the other end of the price range, Dixon's "American Graphite" pencils, round in shape and plain cedar in finish, sold for three cents per dozen. Thus at least twelve round pencils could be had for the price of a single triangular pencil.

However, aside from price and aside from what was to be the shape of the future pencil, imagine what might have been the case if, rather than black lead being found in Cumberland in the sixteenth century, graphite had been first discovered in New Hampshire in the late nineteenth. Imagine that the new find came to the attention of someone living in the age of steam locomotives who knew about chemistry, ceramics, wood, metal, and rubber. In other words, imagine that the modern pencil had not evolved from the brush but had come in a flash of genius to some inventor, some engineer. But if nothing like the pencil that this designer imagined then existed, how would he communicate his ideas to those from whom he wished to obtain a patent, to those from whom he hoped to raise capital to produce his invention, and to those whom he hoped would actually manufacture his graphite pencil?

While the engineer might first make some sketches, with a quill pen and ink perhaps, and a prototype of the pencil, he would eventually also make mechanical drawings of the object to define unambiguously and precisely what its shape and dimensions would be. On such drawings he might specify what the diameter and length of the lead would be and how much tolerance there would be allowed in the size and straightness of each piece of lead. He might specify how large the groove in each of the two halves of wood should be, including the allowable limits on the size and eccentricity of the groove. He might specify the dimensions of the hexagonal sides of the pencil, if he had reasoned that they should be hexagonal, including the final degree of smoothness and symmetry that was to be achieved. He might specify the details of the eraser and the ferrule, including the exact means of their attachment to each other and to the end of the pencil. The drawings might also specify how the pencil was to be sharpened, if it was to be, how it was to be painted, if it was to be, and how it was to be stamped with the company's name and other information, if there was to be any. In short, the drawings would be as complete and unambiguous as possible a means of describing size, shape, and quality of the yet to be manufactured object that would be a revolutionary writing and drawing instrument.

While this example is hypothetical, it is not unrepresentative of the problems faced by nineteenth-century engineers in communicating increasingly complex and revolutionary concepts. With the maturation of the Industrial Revolution, as its new machines and structures became more and more massive, expensive, and difficult to describe, careful mechanical and structural drawings, based on prior calculations as well as on experience, became essential to the making of prototypes of new devices. The engineer, practicing independent of craftsmen, was less likely to have his own machine shop, and thus he had to convey to some machine shop in two-dimensional drawings exactly what it was he wanted to be made in three dimensions.

Vitruvius stressed the importance of drawing for Roman architects and engineers, and he also recognized that it was the ability to picture the yet unrealized that set the architect-engineer apart from the layman:

> In fact, all kinds of men, and not merely architects, can recognize a good piece of work, but between laymen and the latter there is this difference, that the layman cannot tell what it is to be like without seeing it finished, whereas the architect, as soon as he has formed a conception, and before he begins the work, has a definite idea of the beauty, the convenience, and the propriety that will distinguish it.

But in Roman times conceiving and drawing often meant merely setting out the plan and proportions of buildings, fortifications, and the like. The experience of craftsmen would determine such things as the thickness of walls and columns. Ancient drawings of machines and engines of war look to us much like the attempts of children to draw three-dimensional assemblages of objects, and thus the drawings were mainly suggestive. What they represented could hardly be visualized, let alone constructed, by anyone without some direct experience with the crafts involved in making the device.

Perspective drawings appeared in the fifteenth century, and these were so faithful to reality that they were readily understood by craftsman and scholar alike. Thus such drawings as those in Leonardo's notebooks and the illustrations in Agricola's treatise on mining, which include "exploded" views of how the various parts of machines fit together, made the transfer of involved technology much more achievable. And the mass production of illustrations made possible by the printing press further advanced the rate of diffusion of inventions.

When engineering science caught up with engineering practice in the nineteenth century, there was a need not only to draw pictures of machines and structures but also to give detailed descriptions of their individual parts. This became especially important if conceptually new parts were to be made in quantity beforehand and were expected to fit together interchangeably, as they were in structures like the Crystal Palace. The development of analytical theories of beams and other structural elements enabled calculation to displace experiment, and thus obviated the sizing of parts by trial and error. Design calculations determined the size and shape of the parts, and calculations determined how large or small a girder could be if it was to fit easily between two columns without pulling or pushing them so far off the vertical that the next girder might not fit correctly and the desired visual effect of a long nave of orderly girders and columns might be lost.

The means that engineers employ for achieving on paper the proper two-dimensional description of a three-dimensional object is orthographic projection, and its advantage over perspective drawing can be seen by looking at a sharpened hexagonal pencil. When the pencil is being used to write or draw, it is naturally seen in perspective by its user. And exactly how the pencil appears from this perspective depends, of course, on exactly how it is being held in the field of vision. When I write with a pencil I tend to twirl it now and then to keep the point feeling fresh, but when I stop to look at exactly how I am holding the pencil, I sometimes see only two faces of its hexagonal shaft. How many faces do I not see? Could my pencil be square as easily as hexagonal? If I twist the pencil a bit I can see three faces at a time, but that by itself could mean the pencil was octagonal as easily as hexagonal. Of course, no matter how I hold the hexagonal pencil in the writing position, I could never see any more than three faces at one time, and so I might never conclude absolutely from a single view of it that the pencil's shape was indeed a regular hexagon. Since a single perspective drawing must necessarily be of the pencil in a single orientation, such a drawing would not be sufficient to convey the exact shape of the pencil. So how could I convey that unambiguously?

The problem is at the same time stated and solved in the illustration on the cover of Douglas Hofstadter's *Gödel, Escher, Bach.* In the illustration a cube of wood is carved in such a clever way that the block takes the shape of the letter *G, E,* or *B* depending on which face is viewed directly. If only a single face is viewed, we might naturally conclude that the entire block is in the shape of the letter we see. If two faces are viewed simultaneously in perspective, we might conclude that the block was carved with a pair of letters, and which pair would depend upon which two faces we see. But in a perspective drawing showing all three faces simultaneously, we see all three letters simultaneously. This is the case on *Gödel, Escher, Bach,* and when from three light sources three shadows are cast by the block, these shadows show the projections of the three perpendicular faces. Thus from these three shadows we might reconstruct the block more accurately and more confidently than from any single perspective drawing. Such shadow views taken together and properly arranged on a single flat surface form what is known as an orthographic projection of the block, and when they are drawn in a standard arrangement of two or three or more, they can give the trained eye all the necessary information to picture and make the object in three dimensions. A perspective drawing is unnecessary.

Picking up the hexagonal pencil again provides another good example of the advantage of orthographic projection. When we look down along the shaft of the pencil directly at the eraser end, all we see is a circular eraser within a circular brass ferrule. When we look directly at the pointed end, all we see is a circle of lead centered in a hexagon of wood. When we look directly at the face of the hexagon stamped with the brand name and hardness of the pencil, we see only three faces of the hexagonal shaft but we also see clearly that the pencil is sharpened to a conical point and that the eraser is about a quarter inch long and the ferrule about a half inch long. When we twist the pencil through ninety degrees about the axis of its lead, so that the brand name is just out of sight, we see only two faces of the hexagonal shaft. Whether drawn full size or to a specified scale, when taken together all these flat views contain more than enough dimensional information to construct a real pencil, assuming that we have the right materials and know how to process and join them, even though we have never seen or held an actual pencil.

But conversely, even if an artist has seen and held an artifact we cannot at all be assured that the artist's depiction of it is sufficient to make the artifact. This was embarrassingly illustrated in a 1981 issue of the construction weekly *Engineering News-Record.* The colorful cover was almost entirely filled with a close-up drawing of the sharpened end of a pencil, something the art director and editors must have thought was a natural way to announce the theme of the magazine's annual review of the "Top 500 Design Firms." Unfortunately, the drawing of the pencil, ostensibly a common yellow hexagonal one, was as confusing as one of Escher's endless staircases or a blivit, the impossible object that looks two-pronged and square on one end and three-pronged and round on the other. The magazine cover prompted readers to write and "*ENR* got almost as much mail as it does when a decimal point is misplaced." What the artist had drawn showed a "hexagonal pencil" with some confusing aspects, ones that are often repeated by careless cartoonists: (1) the scalloped border between the sharpened end and the unsharpened shaft of the pencil curved the wrong way, (2) the flat and slanted faces of the hexagonal shaft were drawn with equal width, and (3) the border between the lead point and the sharpened wood cone was wavy. While the last characteristic might result from an eccentric piece of lead sharpened in a wobbly sharpener, the first two could hardly be produced by a drunken machinist working on

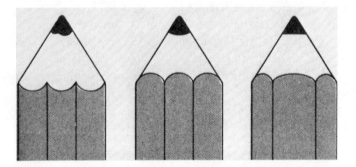

An "impossible" pencil and successive corrections by a reader and an editor of *Engineering News-Record*

a rubber lathe. Trying to make a real pencil look like the *ENR* drawing would be frustrating at best, and to avoid such impossible dreams being sent to carpenters and machinists, engineering drawing has evolved standard practices like orthographic projection.

Although orthographic projection was used in Albrecht Dürer's 1525 book on the geometry of drawing, and theoretical foundations were laid down in Gaspard Monge's 1795 book on descriptive geometry, the techniques and conventions derived from the work of these pioneers did not become universally employed in what has come to be known as mechanical or engineering drawing until the nineteenth century, when it became virtually indispensable for conveying information to machine shops and iron foundries. Up until about the middle of the nineteenth century, engineering drawing was learned in the long tradition of architectural drawing, and many early machines, such as large steam engines, were designed with iron structural elements cast in the forms of columns of the classical orders. Functional brackets were adorned with the classical motifs that students had learned in drawing classes, which consisted largely of copying increasingly complex architectural drawings, and one can only speculate on how much of what came to be known as Victorian architecture and structure was influenced by this practice.

Through the middle of the nineteenth century architectural drawing was learned by most, though not all, draftsmen by tracing. Hence technique was learned at the expense of theory. On the other hand, mastering orthographic projection in particular required not only becoming proficient in using the pencil but also understanding the conventional arrangement of and the standard relationship of the orthogonal views to one another. The mid-nineteenth-century state of the art of engineering drawing was recorded succinctly in the preface to one of the earliest textbooks on the subject, *An Elementary Treatise on Orthographic Projection, Being a New Method of Teaching the Science of Mechanical Engineering Drawing*, by William Binns. According to Binns, his successful course, giving "the ABC . . . of representing all kinds of engineering structures," was designed in 1846 for the students in the College for Civil Engineers in Putney, and, contrasting its novelty with older courses, he wrote:

> . . . the usual mode of teaching . . . is from the "flat"—
> that is, from copy—the practice being to lay before each

student of the class a drawing of some part or parts of a structure which he is requested to copy. This being done, another drawing, probably more elaborate, is laid before him; and the same course is pursued until he becomes tolerably expert with his instruments and brushes, and eventually is enabled to make a very creditable or even highly finished drawing from *copy*. If, however, at the end of one or two years' practice the copyist is asked to make an end elevation, side elevation and longitudinal section of his black-lead pencil, or a transverse section of the box containing his instruments, the chances are that he can neither do the one or the other.

Binns's terminology is standard to this day. The elevations he speaks of are essentially what the eraser, point, and shaft views of the pencil are. A longitudinal section of the pencil would be a drawing of the pencil's innards exposed by a straight saw cut all the way from the point through a diameter of the eraser, thus slicing the lead and wood in half lengthwise. And a transverse section would be the innards exposed by a saw cut anywhere along the length of the pencil but parallel to the flat end of the eraser. A transverse section very close to the point would show only a circle of lead; a section through the conically sharpened wood would show a circle of wood centered about the lead; a section anywhere in the main body of the pencil would show a hexagon centered about the lead; a section through the ferrule would show a thin ring of metal surrounding either the wood and the lead it contains or an eraser, depending on where the section was located; and a section through the exposed part of the eraser would show a solid circle of eraser.

Binns actually asks the student to draw the pencil on his table. Posed first in the most elementary of contexts, the problem is straightforward: "To find the end elevation and plan of a black-lead pencil." But later on in the same chapter Binns discusses the conceptually more difficult problem of showing in a drawing the inside as well as the outside of an object. His description not only elucidates the concept of a section drawing but also by the choice of example provides further insight into the variety of products and concerns of nineteenth-century pencil makers:

The object of a section is to show the internal configuration or arrangement and combination of parts of which anything is composed. As a familiar illustration of the application, let us suppose that a manufacturer of black-lead pencils has to make a number of those articles to order. Now, since there are pencils of various forms and lengths, some with lead throughout, and others with lead extending scarcely more than half the length of the wooden part, it will be necessary in giving the order to explain these things, as well as any peculiarity in the shape of the wooden part, which, for example, we will suppose to be of an elliptical form. The most convenient mode of describing this little object would be by the aid of a drawing, showing an end view and longitudinal section . . . of that class of pencils whose locomotive propensities over the drawing board and on to the floor are interfered with by its flat or elliptical form.

Except for the fact that Binns's British arrangement of the two views of the pencil would be reversed in an American drawing made today, his illustration is clearly sufficient to give all the necessary information about the size and shape of the lead and wood to make what we might recognize as something like a carpenter's pencil or an artist's sketching pencil.

A projection of or a section through the place where the sharpened cone of wood meets the unsharpened hexagon of a common pencil is less easily described, by Binns's or any other drawing method, because the characterization goes beyond normal visual description. Here the geometry will depend to a great extent on how regular the hexagon is and how well centered the pencil was in the sharpener. By inspecting in turn the six sides of a pencil where the paint meets the sharpened wood, we can see variations in the pattern that will be great or small depending upon the quality of the pencil and the accuracy of the sharpener, but the scallops will always point in the same direction, as the *ENR* artist learned too late. While a section of the pencil at this location is thus less a matter of defining the pencil than of providing information about its manufacture and sharpening, the problem of describing that section in the ideal case of a cone intersecting a hexagonal cylinder is the kind of problem that until very recently all engineering students had to wrestle with and solve with pencil on paper in a course in descriptive geometry. But now the general trend toward theoretical studies and the use of computer graphics threatens to make engineering drawing itself a lost art.

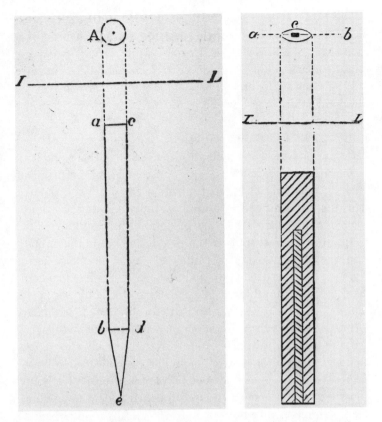

Left: William Binns's elevation and plan of a black-lead pencil, from his book on orthographic projection
Right: Binns's end view and longitudinal section of a pencil designed not to roll easily off the drawing board

Not only did the use of orthographic projection become standard engineering drawing practice with the maturation of the Industrial Revolution, but the manufacture and use of engineering drawing instruments also soon became more or less codified. Drawing instruments have their origins in antiquity. The Egyptians used looped string to scribe true circles, and the Romans used compasses made of bronze and rulers of wood and ivory. By the second century permanent drawings were being made by using a reed pen to draw in ink over the lines scratched on animal skin or papyrus.

Pens made from the quills of birds replaced reed pens by the seventh century, and during medieval times paper making was introduced in the West. With the use of paper widespread during the Renaissance, the method of marking on it with silverpoint was developed. In this method the paper was prepared by applying a wash of finely ground pumice suspended in a very dilute solution of glue and flour paste. When a silverpoint was drawn across such paper, a line varying from pale gray to black could be produced by controlling the pressure on the point. Leonardo used such a method for many of his mechanical drawings, which he then went over with a pencil brush or quill pen and ink.

By the sixteenth century making drafting instruments had become a trade throughout Europe, and by the eighteenth London became renowned for "mathematical" instrument making generally. But instruments were useless without a drawing medium, and according to Hubert Gautier de Nîmes, whose 1716 treatise was the first on bridge building, "the graphite stick, with a file to sharpen it, belongs in the military engineer's kit." George Washington's set of drawing instruments bears the date 1749, and includes a divider, two compasses with removable legs, a compass leg for holding a pencil lead, a compass leg with a pen, and a ruling pen—essentially the same equipment as in the set that students learned to draw with two centuries later. And Washington's kit almost certainly contained some extra graphite or pencil leads. In Washington's day, in a manner not unlike the use of silverpoint, drawings would first be laid out in pencil and, when all lines were complete, inked over with mechanically guided pens to produce a final draft. For his fieldwork, Washington the surveyor had a red-morocco pocket case holding a folded scale, dividers, and a three-inch-long pencil.

In engineering drawing, also called mechanical drawing and even in the past instrumental drawing, because of its use of tools like Washington's, the weight or thickness of a line is significant. Heavy solid lines are used for the visible outlines of objects, dashed or broken lines are used for indicating hidden parts of objects, and light lines are used for lines giving dimensions. In the mechanical drawing of a pencil viewed lengthwise, for example, the outline of the pencil that we would see on the desk before us would be drawn in heavy solid lines, and a pair of less heavy, dashed lines would extend the length of the pencil to indicate the outline of the lead inside. If we wished to show the dimensions of an actual pencil on the drawing, which might not be actual size, we would do so with light lines that would come close to but not touch the edges of the drawing of the pencil itself. The inking pens in Washington's mechanical drawing set were adjustable to make different thicknesses of line, and thus they were well suited to giving the proper weight of line in a final drawing. The draftsman needed only to concentrate on guiding his pen properly and on not cutting the paper.

The use of graphite was much more difficult in Washington's day. In order to achieve different thicknesses and different weights of lines, different sharpnesses of pencil point and different pressures on the pencil had to be used. Thus pencils made by the Conté process were a great improvement; since they came in a variety of grades, marks of different degrees of darkness could be had by changing pencils instead of changing the pressure on a single piece of lead. Certainly by the middle of the nineteenth century well-supplied draftsmen everywhere could change pencils as easily as they changed pens to vary the width and weight of lines.

With the ongoing publication of textbooks on technical drawing, the correct use of the pencil that had evolved among draftsmen, architects, and engineers was also put in writing for students to learn in schools rather than by apprenticeship. While the bulk of these textbooks concentrated, like Binns, on such topics as the principles of orthographic projection, virtually all of the books soon began to include a description of the drawing materials of the engineer. Special attention was often given to the choice, preparation, and use of the pencil.

Although A. W. Faber introduced in 1873 a "refillable" mechanical drafting pencil, which had the advantages of not having to have its wood case whittled away every time more lead was needed and of not becoming shorter as the lead was used up, architects and engineers have never been fully converted from the classic wooden pencil. To this day the high-quality, eraserless wood-cased pencil, usually called a drawing or drafting pencil, is most often the instrument that is discussed in textbooks. A student could still read in the latter part of the twentieth century: "A satisfactory drawing pencil must be straight and well made, with a uniform quality of lead exactly centered in clear, straight-grained wood."

Many textbooks describe in some detail the grades and designations of drawing pencils. Although the National Bureau of Standards tried to standardize pencil grades, they have always varied somewhat, along with their designations, depending upon the manufacturer, when the pencils were made, and the source of the graphite and clay. One analyst found the graphitic carbon content, for example, to vary from about 30 to about 65 percent in a variety of different pencils bearing the same grade designation. Nevertheless, pencils are sold, bought, and used according to their grades, and generally the hardest pencil available has been designated 10H or 9H and the softest 7B, and sometimes even 8B or 9B, depending on which manufacturer's pencils were being taken as the standard. Thus a full range of twenty-one pencil grades might read, from hardest to softest: 10H, 9H, 8H, 7H, 6H, 5H, 4H, 3H, 2H, H, F, HB, B, 2B, 3B, 4B, 5B, 6B, 7B, 8B, 9B. (The equivalent writing pencils in terms of these grades are now roughly as follows: No. 1 = B, No. 2 = HB, No. 2½ = F, No. 3 = H, and No. 4 = 2H.)

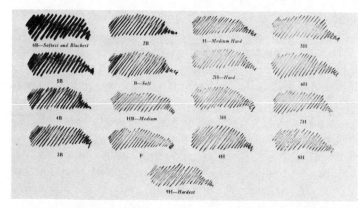

Marks made by seventeen different grades, or degrees, of Koh-I-Noor pencils

No matter what the designation of a pencil, the individual particles of lead it will leave on a writing surface will all be equally black, but their size and number will vary according to the pencil's hardness or softness and the nature of the writing surface. Since paper is a mass of layered fibers, the paper actually acts like a file in shearing off and catching in its small recesses some of the pencil lead. "How the paper eats up pencil line," was how John Steinbeck saw it. But the darkness of the mark made by a particular pencil on a particular kind of paper actually depends on the density of particles of lead left by the pencil stroke. Since rougher paper will have more of a file action than smooth paper on a pencil point, to make a dark mark on a relatively smooth piece of paper takes a relatively soft pencil.

Engineers generally do not use pencils softer than an H on their mechanical drawings but will use softer grades for tracing and sketching. Architects and artists tend to prefer the softer pencils, especially in combination with textured papers. Pencil manufacturers know this, of course, and Derwent pencils, for example, are packaged in sets according to their intended use: Designer Set (4H to 6B), Draughtsman Set (9H to B), Sketching Set (H to 9B).

While the hardest pencils will write on metal and stone, the choice of pencil for use on paper generally is governed by the purpose of the line to be drawn. If the line is being used for graphical computations, where accuracy is of the utmost importance, then a very hard pencil is preferred because it will take an extremely sharp point that will not dull quickly. For light lines on engineering drawings, pencils in the 4H range might be used. The bulk of technical drawing might be done with pencils nearer an H in grade, with mechanical drawings done in 3H and 2H. Drawings from which blueprints are to be made might be done in 2H and H pencils. The pencils designated 2B and softer are generally unsuited for engineering drawing because they require constant sharpening and tend to smear.

The kind of point put on a drafting pencil depends in part on the kind of work being done and in part on the draftsman's style. While all work can be done with the familiar conical point, which is easily formed on a sandpaper pad by twirling the pencil as it is sharpened, there are distinct advantages to using a variety of other points. A bevel or elliptical point is easily formed by sanding one side of the lead without turning it on the sandpaper pad. This is useful in drawing circles, and it will stay sharp longer than the conical point. A wedge-shaped point is formed by sanding the lead, without twirling it, on two opposite sides, and this point has the advantage of staying sharp when drawing long lines. Furthermore, the flat side of the point will sit flat on the T square and other guiding edges and thus will keep itself aligned. Some textbooks even suggest sharpening both ends of a drawing pencil, which traditionally will not have an attached eraser, and putting a different style of point on each. It is the sharpness and hardness of the pencil point and not the pressure applied to it that is supposed to determine the weight of the line.

While using a sandpaper pad to sharpen drawing pencils enabled the engineer to form his lead into just the point he wanted for the work at hand, the method was also very messy and potentially disastrous. A great deal of dust was always being produced and it could ruin any drawing over which it might spill or be blown. For this reason the drawing textbooks

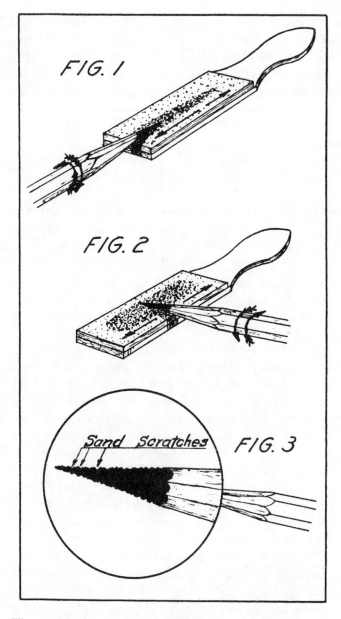

The use of the sandpaper pad to point a pencil, showing the cause of notchlike scratches that can weaken the lead also cautioned students to keep their sandpaper pads in envelopes and not to sharpen their pencils over their work. Engineers were thus incidentally taught, along with their graphical ABCs, to anticipate accidents and to take steps to avoid them. Executing a correct and clean drawing the first time was good practice for executing safe and sure designs generally.

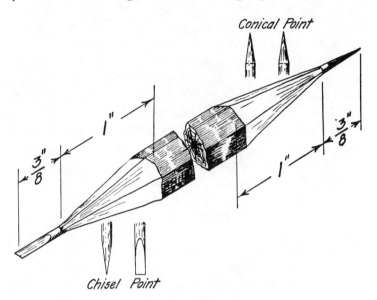

Two kinds of points put on drafting pencils

In spite of the messy means of sharpening pencils, the properly sharpened pencil was the foundation of the engineering drawings of exceptional style and beauty that began to be produced in the eighteenth century. Color washes over pencil and ink began to be used on drawings, with the different colors indicating different materials and functions of the structural and mechanical parts. The practice reached its peak in the colored engineering drawings of the nineteenth century. But before a drawing could be inked and colored, it was laid out entirely in pencil, and the proper use of the pencil meant not only that the right lines were drawn in the right places but also that the paper was not indented by the pencil point. Pencils were used for first drafts because the inevitable errors in layout could easily be corrected with an eraser. But if the pencil was wielded by too heavy a hand, it would leave not only a line but also an indentation. The line would thus be difficult to erase if need be, and even if its mark could be removed, the line would provide an uneven surface for future pencil and especially ink lines. Thus the draftsman attempting to make a dark line with too hard a pencil could jeopardize hours or days of work.

Color continued to be used in the early years of the twentieth century, but with decreasing frequency as copying methods that could not reproduce color became more widespread. Blueprints were available from the 1870s, and the practice of using different colors in drawings had virtually ceased by 1914. By 1925 some drawings were being finished in pencil only, and within a decade this was common practice. The use of pencil for a final drawing meant that the lines had to be made heavy and yet hard pencils had to be employed to achieve sharp lines and avoid smudging, so that quality reproductions and blueprints could be made directly from the pencil drawings. But drawing instruments had long been made of light construction and soft materials, because little pressure was to be put on them for drawing either pen or pencil lines. The breakage of old and the anticipated shortage of newer instruments made in Germany apparently created a new market for heavy-duty drafting equipment. One American manufacturer argued for the new style of drafting by noting that building a first-class battleship required "three large freight-car loads (180,000 pounds) of drawings," and, under the pressures of war, drawings made the old way would have been "obsolete before they could have been inked in." While this company advocated the "new standards for draftsmanship," employing the heavy-handed use of the pencil rather than the pen, it also hedged its bet by including inking pens in the drafting sets it marketed.

One of the main concerns of a good engineering drawing was the quality of prints that could be made from it. India ink had the advantage of blackness, but the pencil had speed in its favor. However, both ink and graphite drawings were subject to smearing, especially when exposed to water and perspiration, and lines became fainter and the drawings dirtier as they were repeatedly revised and used. With the advent of waterproof Mylar polyester film as a substitute for paper and cloth, and the development of pencils suited to its surface, it became possible to produce drawings that could be washed. New drafting pencils were developed containing plastic that bonded to the Mylar drafting film. While this new process had clear advantages for making long-lasting original drawings and clean prints, it showed no immediate advantages to the draftsperson. The new pencils had an unfamiliar grading system; they did not produce very dense-looking lines; their marks were difficult to erase; they had a "crayony" feel; and their points were apt to break. To forestall total rejection of the new drafting medium, the Keuffel & Esser Company, which had introduced Herculene drafting film and worked with Staedtler to come up with the compatible Duralar pencil, took out advertisements around 1960 giving recommended practices to "smooth the path to acceptance."

The end of the pencil drawing, whether on paper, cloth, or film, has been prophesied by advocates of computer-based systems, but it is unlikely that drawing by hand will ever totally disappear, even though fewer and fewer engineering students take formal drafting courses. Until recently drafting was taught to all engineering students, and the textbooks were not very different from those of the turn of the century. Students taking mechanical drawing courses in the mid- to late 1950s had to outfit themselves, as students had for decades, with 2H, 3H, and 4H pencils, sandpaper pads, erasers, triangles, a T square, a drafting board, and a beginner's drafting set, which could serve for a lifetime. They learned how to sharpen pencils and how to hold them and how not to dig them into the paper and how to ink over their lines. They never colored a drawing, and they were not taught that all drawings had to be inked. However, they did learn the meaning of orthographic projection and how to draw just about any mechanical concoction that would test the limits of graphical communication. They learned which lines should be solid, which dashed, which heavy, which light, and how to cross-hatch and letter. They learned how to determine on the drawing board what figure a cone and a hexagon would make at their juncture, and they also learned that you don't design something on paper that can't be made in the machine shop or in the foundry. It was in the tradition of an engineering apprenticeship. But even the finest-looking drawings are still not enough for successful engineering. According to an engineer's favorite saying, "any fool can tighten a nut with a pencil," but it takes an engineer's circumspection to make sure a machinist can find room to tighten the real nut properly with his wrench.

16/ The Point of It All

Since sharpening a pencil can be a distracting and bothersome interruption when one is working, it is important that a newly sharpened point not break too easily and that the pencil stay sharp for as long as possible. So in addition to producing pencils with leads in a range of hardnesses that will write smoothly without smudging and leave erasable marks of the appropriate uniform darkness, the pencil manufacturer must make leads able to take as fine a point as possible and yet not be so brittle and weak as to snap off too easily under the pressure even of a heavy hand.

It is not a simple task to find the proper combination of the right kinds of graphite and clay, the proper means of purifying and mixing them, the proper temperature and pressure at which to process them to give a lead smoothness, hardness, toughness, and strength, and the proper means to bond the lead to its properly fabricated wooden case so that the pencil will maintain a sharp point. Yet it is exactly this kind of problem, a problem involving the quest for competing qualities at a competitive price, that engineers face constantly, whether in making pencils or building bridges. One desirable quality is often gained at the expense of another, for how the properties of complicated materials will change with changing ingredients and methods of preparation, whether they be pencil leads or concrete, is not always easily predictable. While a new mixture might give a stronger material, it might also give a more brittle one that will be greatly weakened by a small crack. Making synthetic materials always means making real compromises.

Because compromise necessarily involves matters of judgment and because judgment is always a subjective matter, the business of finding the "best" combination of ingredients and the "best" way of processing them into a pencil or a bridge will necessarily produce a variety of pencils and bridges that their different designers will each consider "best." When Henry David Thoreau set out to make a high-quality pencil, he essentially undertook a program of research and development. His bookish research in Harvard's library was followed by experimental study among the tools and materials in the family pencil workshop, and this in turn was followed by his development of machines and processes for producing the best lead pencil then made in America. How Thoreau and his father determined exactly when to stop researching and start developing, and how they determined to stop developing and start producing pencils for the marketplace, involved decisions influenced by psychological and economic considerations as much as by scientific and technical ones.

Henry Thoreau's personality was such that he did not like to stay put, and thus he did not have the psychological constitution to keep improving on a pencil that he knew was already the best made in America. When he reached a certain stage in the development of his pencil, he made the judgment that it did not interest him to expend any further effort for marginal improvement. The idea of making fine pencils was for him ultimately to make money, not to make the perfect pencil. When the Thoreau pencil business began to make more money selling ground plumbago, the family did not hesitate to de-emphasize the manufacture of pencils. While theirs were still excellent pencils, there was also the growing competition from New York and foreign pencil manufacturers to consider. There can be much personal satisfaction in achieving a good new competitive product, but little in making a marginally better one. Thoreau's behavior is very much in keeping with the typical personality of a creative individual, whether writer or engineer.

On a less personal psychological level, research and development are motivated by a desire, whether felt by an individual or by a giant corporation, to provide something that can be claimed (and demonstrated) to be significantly better than its competition, and not incidentally because then a larger and more secure profit can be made. Since the competition is also likely to be carrying on its own research and development to improve its own product, lest it be left in an inferior position, one cannot go on indefinitely reading and experimenting and tinkering. There must come a time when the individual or the corporate body says, "Enough! This pencil is as good as we can make at this time. We must now manufacture and advertise it before the competition comes out with something that would cause us to go back to the drawing board."

Sometimes a research and development effort must be ended arbitrarily because the money to support it has simply run out. A fixed amount of funds or a fixed amount of time may have been allocated to the effort, and the engineers involved may have no choice but to pass on their findings to the production engineers, who can then begin making whatever improved pencil had been developed. For if the new product did not appear soon and begin to provide revenue for the company, there would be no income to fund research and development on still better pencils. And if such an effort were to cease, the company would soon be making inferior pencils and, if the marketplace had any sense, so making an inferior profit, if any at all.

A better pencil might be a less expensive one that writes just as well as any already on the market or a more expensive one that draws better than any other available. The claims made by J. Thoreau & Company in the nineteenth century for the superiority of their improved pencils are echoed in the claims made by later pencil manufacturers in their own trade catalogues and promotional literature. Their own words show that they were not unaware of the expectations of discriminating pencil users. One pencil maker stressed the composition of the lead itself and assured the buyer that, "sharpened to a needle point, it does not snap off under firm pressure." Another emphasized the importance of "a stronger lead-to-wood bond for maximum point strength." And still another manufacturer's 1940 catalogue offered the "smoothest pencil with the strongest point!"

It had been a chronic problem for the pencil industry that pencil points broke too easily, and often they broke first just inside the sharpened cone of wood, tearing it open under the slightest pressure. Understanding why and how this happened made it possible to correct the fault, by altering the manufacturing process; not incidentally, it provided the basis for an advertising claim. But sometimes pencil companies went too far in claims that could not be backed up, and they were forced to retract them. Although the Federal Trade Commission had detected no intent to mislead pencil buyers, in 1950 it persuaded the Eberhard Faber Company to stop claiming that laboratory tests proved that Mongols were 29 percent sharper.

As opposed to tests to "prove" advertising claims, real engineering research and development, which involves the exploration of new materials and processes for combining them, and which must be performed before any product testing can take place, is seldom the subject of discussion, oral or written, outside the laboratory itself. Even the technical details of the manufacturing process are still rarely expressed in print. Thus, when Charles Nichols, director of engineering for the Joseph Dixon Crucible Company, presented a paper at the annual meeting of the American Society of Mechanical Engineers in 1946, in which he described in some detail the woodworking operations involved in pencil making, he revealed nothing of a research and development nature and he certainly disclosed no trade secrets. However, the mere fact that a paper on the subject was read and published in *Mechanical Engineering* occasioned in a subsequent issue a comment from Sherwood Seeley, the director of the Research and Technical Division of Dixon. His communication, perhaps part of a conspiracy by these Dixon colleagues to shame the competition into sharing at least some information, read in part:

> To those who have unsuccessfully searched the literature for the details of pencil manufacture, it will be obvious that [Nichols's] paper has in it the element of pioneering. Since there is so little published on this subject, it is to be expected that the paper would be somewhat general and descriptive. However, it differs from others in that all steps of manufacture are faithfully reviewed in a technical manner and in that tolerance specifications are given. . . .
>
> Unless the good start made by the author is to die, other papers on the wood-cased pencil should be forthcoming from the industry.

The writer went on to suggest timely subjects, including the selection and preparation of wood, adhesion problems, shaping methods, materials-handling methods, and finishing and decorating pencils. Without the cooperation of those in the pencil industry, Seeley saw it as "inevitable that technical literature on wood-cased pencils will lie dormant." It did, for Nichols's paper was not followed by others "from the industry." This perhaps should not have surprised anyone, for three decades earlier Ainsworth Mitchell, a British research chemist who had been asked by a pencil manufacturer to look into the suitability of different kinds of graphite for pencil leads, wrote of his own frustration in finding that "the literature on the subject was exceedingly scanty, and that it was necessary to discover for myself the reasons underlying changes in the manufacturing processes in the evolution of the industry." But even Mitchell's papers gave away no secrets.

While pencil engineers no doubt read with interest what Nichols and Mitchell had to say, it could not have been news to the initiated. In a discussion appended to Mitchell's 1919 paper, for example, a Professor Hinchley, who "did not know of any maker now who produced a pencil that could compare with the old-fashioned Borrowdale graphite pencil," attempted to correct Mitchell on several historical points and spoke with authority about the pencil-making process:

> The process of drying to produce a first-class lead took about a month; it could not be done in less than a fortnight without risk of failure. If the drying was too rapid the inner portion became porous and resulted in pencils with rotten points. That had been a common fault of some of them to-day. The grinding operations that were carried out to-day were the finest that had ever been done. The grinding surfaces were trued up as accurately as in any engineering operation, and the graphite mixture of the finest pencils to-day was generally ground for at least eight days; the total period of manufacture was from four to six months, and necessitated fifty or sixty operations. That would give a notion of what was involved in making high-class pencils.

But if Hinchley spoke with authority, he gave neither references for his historical statements nor details for his technical ones, and he does not seem to have published any papers himself. As for the papers of Nichols and Mitchell, they appear to have been cited only by the rare scholar writing about the history of the pencil.

The paucity of technical papers describing manufacturing processes especially is due in part to the interest of engineers, as opposed to scientists, in getting on, like Thoreau, to other things rather than describing in words on paper what they had already done in actual graphite and wood. Furthermore, the nature of research and development also promotes the "loose lips sink ships" mentality. In 1960, for example, when a new

pencil lead was developed by one of the major pencil manufacturers, it was announced not at a technical meeting or in an
archival journal but in the advertising columns of *The New
York Times* to coincide with the launching of a marketing campaign. The story began under a picture of the full-page advertisement that was to run in trade publications, showing a close-
up of a pencil point sharpened to finger-pricking perfection
and emphasizing the strength of the new lead, but not being
precise about exactly how strong the lead was:

> The introduction of a new type of lead for pencils, de-
> scribed as all but unbreakable under normal writing con-
> ditions, is to be announced in Wilkes-Barre, Pa., today
> by Eberhard Faber, Inc.
>
> The announcement follows a year and a half of labora-
> tory secrecy, tight internal security and consumer testing
> on the order of [a] spy story.
>
> When the company was satisfied some months ago that
> the product was all it had hoped for, it began distributing
> the new Mongol pencil with the new lead trade-marked
> "Diamond Star." It looked like Eberhard Faber's familiar
> Mongol except for a tiny black dot.
>
> Advertising and other promotional efforts for the new
> Mongol were prepared under just as much security re-
> strictions as the pencil itself.

In part to maintain a competitive edge, the details of the
research and development activities of pencil companies are all
but undocumented. But there is also another reason that there
is little of a "scientific" or "technical" literature on the manufacture of pencils and other seemingly simple but in fact complicated products of technology that have developed out of the
crafts over a long period of time. Knowing the "science" of an
ingredient or a process is not the same as understanding how
that ingredient or process will affect the manufactured product. Mastering the chemical formulas for graphite and clay
and the thermodynamics of the kiln does not at all suffice to
make a better pencil lead.

Sherwood Seeley, writing on the uses of natural graphite in
the 1964 edition of the *Encyclopedia of Chemical Technology,*
mentions in passing that, while the quality of pencil lead depends not only on the quality of the graphite but also on the
quality of the clay used, there is no formula for identifying the
best clay. Suitable clays, he points out, are known commer-

cially as "pencil clays," and the best have come from in and around Bavaria. Yet even with "the ceramist's extensive knowledge, the only conclusive criterion of a clay for use in pencil leads is the result of tests made on pencil leads made with the clay."

If the ceramist, who may know a great deal about the chemical, thermal, and mechanical properties of clays, must ultimately resort to tests on actual pencil leads to determine which clay mixture and kiln temperature combination is the best for pencil lead, so must the engineer generally put any proposed design to the physical test. But if such be the case, what is the role of theoretical engineering science in the development of pencils and even less common artifacts?

Even though the ceramist may not be able to claim incontrovertibly without the confirmation of an experiment which pencil lead will behave better than another, he will be able to infer from his research and experience that certain desirable properties tend to correlate with certain constituents and conditions of manufacture. The ceramist will see a pencil lead as "a baked ceramic rod of clay-bonded graphite incased in wood," and he will be able to give reasons why: natural graphite is better than artificial; a wax-impregnated lead will not glaze over in use; the best kiln temperatures are between 800 and 1,000 degrees Centigrade; the degree of hardness depends on how much clay is used; increasing the clay content strengthens the lead.

The strength of pencil leads is an especially important property, as the advertising claims of pencil manufacturers attest. While the ceramist or materials scientist may understand how such factors as clay quality and content may influence the strength of "a baked ceramic rod," it remains for a special kind of engineering scientist, the mechanicist, to provide a mechanical explanation of the influence of such factors as writing angle and pressure, as well as a pencil point's shape, length, and sharpness, on its strength. Although the ability to predict empirically the effects of these and other factors has been gained by the experience of centuries, just as the ability to predict heavenly phenomena had been gained by observing the motions of the planets for millennia, a mechanical explanation of the strength of a pencil point is as desirable as a theory of planetary motion, if for no reason other than scientific curiosity. While predicting when and where a pencil point will break may not have the cosmological import of predicting when and where a comet will return, the pencil engineer, unlike the natural philosopher, can use his under-standing of mechanical phenomena to change the course of events.

As the solar system presumably followed the laws of nature before they were articulated in the equations of Newton's laws of motion, so a pencil point follows the laws of nature whether or not they be articulated in the equations of the theories of elasticity and strength of materials, which have the potential to explain the mechanical behavior of things from pencil points to nuclear reactor vessels. But, while equations can flow from pencil points and stronger pencil points in turn might follow from equations in a circular manner reminiscent of Escher's "Drawing Hands," there is no documented record of any theoretical work significantly influencing the fundamental development of the pencil. The first pencils certainly did not come from equations. Yet better pencils may.

Sometimes theory and practice can both advance by give-and-take from each other. In the case of bridges, in which a long history of actual structures built by stonemasons and car-penters carried the theory of structures across gaps of scientific knowledge into new frontiers of experience, the growing body of theory also helped design innovative and bolder bridges. But an artifact like the wood-cased pencil, manufactured in quantity rather than constructed in unique examples, thus risking in its failure perhaps only a family's fortune rather than a community's health and safety, is able to advance well into its maturity without the need of any equations explaining a priori its behavior or sharpening its point.

Yet the fact that the development of an artifact does not rest upon a theoretical foundation does not mean that a theoretical explanation of the artifact and its behavior cannot or should not be developed. The power of engineering science, which is essential for modern research and development, is in its ability to generalize and explain why existing things work, whether those things be steam boilers or pencil points, and thereby to predict how new and improved things should work, thus being the very source of new and improved designs. But theoretical answers and predictions do not come until interesting ques-tions are asked, and those questions are usually prompted by some failure or shortcoming of an already existing arti-fact.

In 1638, motivated by the inexplicable breakup of large ships and other disasters, Galileo sought to determine how large a timber beam had to be in order to support a certain weight, and he wondered what the ideal taper might be so that the beam was no stronger than it had to be. Galileo's beam was embedded in a wall and the weight was hung at the free end. Today we call this a cantilever beam, and even so elementary a problem by modern standards proved difficult for Galileo, who spent two days of his *Dialogues Concerning Two New Sciences* talking about it.

The problem of the strength of a pencil point is essentially the same as that of Galileo's cantilever beam, with the paper pressing up on the point instead of the weight hanging down from the end. And just as Galileo's beam had to be firmly fixed in the wall to support the weight, so the pencil lead has to be firmly fixed in the wood. No matter how strong a beam one has, adding enough weight to its end can cause it to be torn out of the wall if the masonry is not strong enough. Like a timber in crumbling masonry, a loose pencil lead in a weak wooden shaft spreads the wood apart and breaks off at the "pressure point," as one manufacturer called it.

As it happens, pencil leads are boiled in wax to impregnate them for smoother writing qualities. Every particle of graphite and clay gets coated with a lubricating film of wax, but this makes it difficult for the glue to adhere to the lead and bond it to the wood. Different pencil companies attacked this problem in different ways, but all had the same objective of creating a better bond between the wood and the lead in order to provide a stronger resistance to the writing force at its tip.

In 1933 the Eagle Pencil Company's research and development department achieved the objective of keeping the wood from splitting by both improving the bond and toughening the wood itself. First, by bathing the lead in sulphuric acid, the outer film of wax was burned off, and then by bathing the lead in calcium chloride, a sealing film of gypsum was deposited on the surface. In addition, the wood was impregnated with a resinous binder that locked the fibers into a tough sheath that could not be easily split by the pressure of the lead. The pencil industry's stock adhesive, hide glue, was then able to hold the lead and wood as never before. The combination of lead and wood treatment was termed by Eagle the "Chemi-Sealed" process. It is what enabled the company to claim a 34 percent increase in point strength for its Mikado pencils, and sales increased by 40 percent. Other manufacturers developed their own means of achieving a stronger pencil point, and the various processes go under such names as "Bonded," "Super Bonded," "Pressure Proofed," and "Woodclinched." These processes also prevented the lead from breaking up inside the wood if the pencil was dropped.

While pencil engineers working within the privacy of their home companies may have asked and indeed answered to their own satisfaction the question of exactly how and why pencil points break the way they do, the legacy of secrecy inherited from the craft tradition seems to have kept the researchers themselves from leaving so much as a marginal note about the subject in the archival literature of engineering science. Most of what is published appears in sales and advertising magazines, where successful campaigns are described.

But bonding lead to wood did not mean that pencil points would never again break, as every pencil user knows. Strengthening the pencil point where it joins the wooden shaft simply means that the location of least strength must now be somewhere else. No one appears to have publicly brought up the question of where and how bonded pencil points break until an independent engineer, Donald Cronquist, did so in a 1979 article in the *American Journal of Physics*. As do many engineering-scientific articles, Cronquist's opened with an observation and introduced an acronym:

> Some time ago I was clearing my desk top after completing an unusually long hand-written rough draft. I became mystified as I discovered a very large number of broken-off pencil points (BOPP's) lying between and behind the books and other reference materials on my desk. The BOPP's had apparently flown to these hiding places upon snapping off of my newly sharpened pencils. The mysterious thing about the collection of BOPP's was that they were almost all nearly identical in size and shape.

Since Cronquist could not find any explanation for the size and shape of a BOPP in the sharpening process or the nature of the lead itself, he looked for an answer in the physical shape of the pencil point and the forces exerted upon it by the paper. Thus, in the manner of an engineering scientist, Cronquist proceeded to look at the sharpened pencil point as a truncated cone sticking out of the wood casing. The question as posed is essentially the same one that Galileo asked when he wondered about the breaking strength of a cantilever beam sticking out of a wall.

Although Galileo did not completely solve the problem, the engineering-scientific theory of strength of materials that

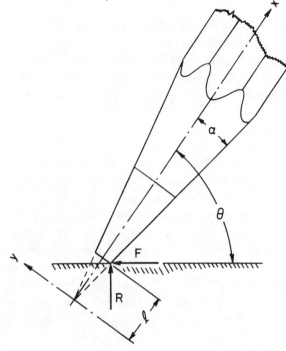

An engineering scientist's idealization of a pencil point and the forces exerted upon it during use

evolved from his great first efforts provides the mathematical basis for answering the question Cronquist posed. And while Cronquist's analysis employs equations that represent more realistically the forces in the pencil point, his method is still essentially the same as Galileo's, as any modern engineering scientist's would be. What Cronquist did was make assumptions about the way in which paper reacts to the push of a pencil point and thus exerts a force on it. He then assumed that this force acts on the truncated cone as if it were a cantilever beam being bent upward. Thus the conical pencil point is being stretched on the side facing the paper, and Cronquist proceeded to calculate the intensity of the stretching forces within the pencil-point beam at a representative distance from the tip of the point. Using mathematics no more complicated than elementary calculus, he then determined where the intensity of internal forces tending to open up cracks in the point overtakes the ability of the lead to resist that happening. Since this is where the point is most likely to break, assuming the lead is made uniformly strong and is free of nicks, Cronquist had in his equations a theoretical (and quantitative) prediction of what the size and shape of a broken-off pencil point should be, and he could compare his prediction with the experimental data points he found lying about his desk.

What Cronquist's equations predict is that the greatest tension in the pencil lead is at a location where the diameter of the lead is 50 percent more than the diameter at the tip. Since a real pencil will be sharpened not to a perfect needle point but to a somewhat flat point, the theoretical prediction is that when the pencil is pressed too hard against the paper, a piece will break off that is in the shape of the frustum of a cone, the diameters of whose base and top are in the ratio of three to two. According to Cronquist's result, for any particular pencil, the sharper the point, the more easily it will break and the smaller will be the BOPP. Cronquist's result thus also predicts what children seem to learn from experience: they will be less frustrated when they write with a blunter point. But no matter how sharp or dull the point, if it is pressed too hard it will break off, and all the BOPP's will be geometrically similar. This is the kind of generality that one can achieve with theoretical calculations. While they may be dispensable in the design of old craft items like pencils, where fractures are inconveniences but not disasters, calculations are essential for designing never before realized devices and structures like interplanetary probes and space stations. One does not want to have to analyze the broken-off parts of such complex and expensive artifacts; one wants to be able to predict when and how they might break so that one can beef them up before launching precisely to prevent such failures.

While Cronquist was satisfied that his analysis explained in general terms a mechanism for breaking a pencil point, he also recognized that he did not have a complete answer. For one, what he calculated did not take into account the different ways in which different people press a pencil onto a piece of paper. Nor did the theory allow for the fact that different pencil sharpeners might make pencil points with cones of different angles. Also, Cronquist's theory could not explain why the BOPP's scattered all over his desk had slightly slanted planes of fracture, rather than flat bases. Jearl Walker, who writes "The Amateur Scientist" column in *Scientific American,* reported on his own experiments confirming the general accuracy of Cronquist's predictions about BOPP's, as long as the pencil was held at a fixed angle, but Walker too could not explain why the point did not break straight across the lead.

It is the nature of engineering science, as of all science, that questions asked but imperfectly answered attract the attention of other engineering scientists. What, the reader of Cronquist's work will ask, did he miss? Did he jump to conclusions in his assumptions? Did he make a mistake in his math? Did he not manipulate his equations in just the right way? Did he not ask the right questions of his answer? The physical BOPP's, which anyone can replicate, are the hard evidence that there is still a puzzle to be solved, for it is the expectation of engineering scientists that the theoretical result can mirror the physical, if only the theory is polished, if only it is done accurately and carefully and completely enough.

Thus Cronquist's work attracted the attention of another engineering scientist, Stephen Cowin, who carried out a more detailed analysis that allowed for a more general force between the pencil and paper and for the effects of different cone angles produced by different pencil sharpeners. In his analysis, which appeared in the more mathematical *Journal of Applied Mechanics* in 1983, Cowin confirmed the accuracy of Cronquist's results, but he still did not explain why the fracture surface was slanted rather than being straight across the lead. Not only had its geometrical irregularity made it difficult to measure the diameter of the broken end of a BOPP and had even caused Walker to measure not the broken diameter of his data points but the more unambiguous shortest length along the side of a BOPP, the slanted fracture remained an unresolved issue and therefore a challenge to still further research work.

The reason that neither Cronquist nor Cowin could explain why the fracture plane of a broken pencil point slants is that, while they incorporated the conical geometry of the pencil point into their analyses, they did so only for the purpose of calculating distances and areas. The intensity of forces inside a pencil point, as inside all objects, depends not only on where the forces are calculated but also across what imagined surface they are calculated. As it turns out, in a tapered object like a pencil point, the maximum intensity of force will occur not across a plane perpendicular to the length of the pencil but across a plane perpendicular to the slanted edge of the pencil point closest to the paper. All other things being equal, which means essentially that the pencil lead is made of a uniformly good-quality graphite-and-clay mixture and that there are no nicks or cuts to disturb the geometrical uniformity, the pencil lead will begin to crack open across the plane of maximum intensity of force. Hence the characteristic slanted surface left at the end of a broken pencil.

Once engineering scientists begin to explain and understand how and why an object like a conical pencil point breaks, they gain confidence in their theoretical methods and feel that they can use them not only to analyze what they find all about them in nature and in technological artifacts but also to improve those artifacts and to design that which has not yet been made or built. In the case of pencil points, it is natural for the engineering scientist who has solved the puzzle of the size and shape of a BOPP to ask further, related questions. For example, what shape of pencil point most resists being broken, what shape is the strongest?

The lead in the carpenter's pencil illustrated by William Binns in his nineteenth-century book on orthographic projection is not round but rectangular in shape, and old (and even some not so old) pencil catalogues show that drafting, sketching, and "landscape" pencils have also been made with that kind of lead. And the pencil manufacturers were not unaware of its advantages:

> For technical and machine drawing the advantages of the lead pencils with broad leads are these: that they can be more readily pointed; that the points are stronger and finer than those obtained from round, square or hexagonal leads; and that, in consequence of the chisel-like form, the points are more durable and need less frequently be renewed. The pencils have this further advantage that, in line drawing, the lead can be applied directly to the ruler, with increased steadiness and precision of manipulation.

It is natural for the engineering scientist to ask if the theories of engineering science can predict if the wedge- or screwdriver-shaped point that would be easily formed on such a lead is indeed stronger than a conical point of the same thickness, and, if so, explain *why* this is so. The analysis predicts that it is and can explain why, but that is not to say that the analysis had any more to do with it than did Newton with the ways of the planets. The advantages for drawing straight guidelines with a flat lead or the heavier use expected of a carpenter's pencil might actually have caused the rectangular lead to be developed, which in turn may have dictated the flat, rectangular shape of that kind of pencil. Or, conversely, the flat shape of the pencil, which would have kept it from rolling off pitched roofs, may have suggested the shape of lead. The rectangular leads of drafting pencils, which would not have to be as strong as those intended to write on wood, did not have to be as thick as the lead in a carpenter's pencil, and hence they could still be enclosed in a more comfortably sized and shaped hexagonal or round case. Or, again, the hexagonal case would have been enough to keep it from rolling on the slight slant of a draftsman's table.

While Binns, in commenting that his drawing of the carpenter's pencil exhibits the "strength or thickness of the lead," makes it clear that thicker meant stronger to him, the wider range of hardnesses of drafting pencils meant that there was a considerable difference in strength in leads of the same diameter or thickness but of, say, degrees 6B and 6H. Since the harder the lead (the higher the proportion of clay to graphite), the stronger it is, and since excessive thickness is a liability in a hard drafting pencil intended for fine work, the diameter of the lead in drafting pencils can be made smaller as the hardness increases without diminishing strength. While it is an exercise in hindsight to confirm theoretically that this should be so, the fact that the theory is capable of explaining why it is reinforces engineering confidence in the theory, which may also be called upon to predict the strength of a new airplane wing.

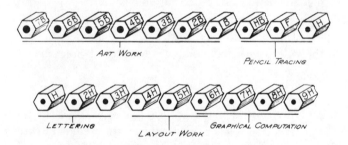

A full range of drawing and drafting pencils, showing how the lead grows thinner as it increases in hardness

The use of the Conté process allows pencil leads to be made in any shape, for it is just a matter of the shape of the die through which the graphite-and-clay mixture is extruded. (Hexagonal pencils with hexagonal leads were sold at the turn of the century, and today one can imagine heart-shaped pencils with heart-shaped leads for one's valentine.) It has been stated that Conté himself made round pencil leads, and he may also have made the different hardnesses of his leads with different diameters. Indeed, limitations of woodworking rather than of lead making may have prolonged the use of square leads and kept round leads from coming into common use until the later part of the nineteenth century. Even the existence of analysis capable of comparing the relative strengths of square and round leads would not have designed or developed the woodworking machinery necessary to form the cedar slats to contain the preferred shape.

Whether or not it is "useful" at the time, if the engineering scientists do publish their analysis, as Cronquist and Cowin did, then other engineering scientists can improve upon it to better solve the original problem or learn from it to solve other problems. Someone might ask, for example, "What is the very best shape for a pencil point to have?" and the question can be explored with logic and the infinite patience of mathematics rather than with the finite and frustrating hit-and-miss approach of trial and error. If logic and mathematics and the theory of pencil points predict a perfect shape for a point, then that can be tried. If the new pencil point is stronger and harder to break, then what began as a theoretical excursion could lead to a new practice in drawing or even to a new invention.

Engineering analysis can also evolve independent of the development of artifacts, and sometimes it can get carried away with itself by elaborating on what might have started as a simple and realistic problem. Thus pencil-point analysis could lead to theories of theoretical pencils with infinitely long or infinitely sharp points or to pencils with points shaped like barbs or bananas. While the theory might predict them to be best, who would want to try to sharpen a pencil in those ways, especially if there was an impatient boss waiting for a fresh pencil?

But that is not to say that such idealized theory and analysis are useless to engineering design and practice, for they are not. Even the seemingly fanciful exercise of predicting the size and shape of BOPP's is full of useful lessons for the student and practitioner of engineering. And not the least of these lessons is how analysis can seem to be so close to getting the right answer quantitatively and yet be so far from getting it qualitatively correct. While the slanted surface of a BOPP may not have affected its dimensions very much and thus may not have seemed to be important in explaining how the pencil point cracked and broke, it could be much more important for engineers using the same analytical methods in other problems to understand the slanted nature of the surface rather than the size of the broken-off piece. Thus the analysis can be a legitimate object of study in its own right, for without studying it as an analytical artifact, the very ways in which the theory itself can fail would remain unknown. And if the analytical tools of engineers are liable to failure, then what could one hope to see designed with those tools?

Theories developed in the abstract can also find unexpected applications. The magazine column in which Jearl Walker discussed breaking pencil points was headed: "Strange to relate, smokestacks and pencil points break in the same way." In it he described the familiar phenomenon that occurs whan a tall brick chimney is demolished. After one corner is knocked or blown out, the intact chimney begins to fall over like a tree, slowly at first, but then gaining speed until, suddenly, and for no apparent reason, the thing breaks in two in midair. The shape of the chimney will influence exactly where the break will occur, and predicting it is complicated by the fact that the structure accelerates as it falls. However, the problem is essentially the same as the pencil-point problem because in both cases it is calculating the intensity of the forces inside the object that is the goal of the analysis. When these forces reach a value greater than the ceramic pencil lead or the mortar between the chimney bricks can take, the object breaks. Another example of the same phenomenon is the fracture of bones away from the point of impact when a victim is struck by a car.

The reason accelerations present a further complication to a falling smokestack can be illustrated by balancing a pencil on its eraser end and then disturbing it. At first the pencil falls slowly, but then it gains speed until it strikes the desk top and bounces to a stop. If you had marked the spot where the eraser started, you would see that the pencil has not only fallen but also has moved away from its spot. Why? This behavior is predictable by equations derived from the principles of Newtonian mechanics, but it is the presence of accelerations that causes forces that can move pencils and break chimneys in ways that are counter-intuitive. Such counter-intuitive behavior was known even before Newton wrote his *Principia,* and in a book of mathematical recreations published in 1674 it was the subject of what might be called a parlor trick: "How to break a Staff which is laid upon two Glasses full of Water, without breaking the Glasses, or spilling the Water." The old book does not explain the karate-like phenomenon; it only describes how to reproduce it:

> First, place the Glasses which are full of Water upon two Jyont Stools, or such like, the one as high as the other from the grȯund, and distant one from the other by two or three foot, then place the ends of the Staff upon the edges of the two glasses, so that they be sharp: this done, with all the force you can, with another staff strike the Staff which is upon the two Glasses in the middle, and it will break without breaking the Glasses, or spilling the Water.

The phenomenon central to the parlor trick, in which the greatest intensity of cross-strain occurs away from the glasses, can be turned to practical advantage in the kitchen, for as the old book goes on to relate, "Kitchin-Boys often break Bones of Mutton upon their hand, or with a Napkin, without any hurt, in onely striking upon the middle of the Bone with a Knife."

While we can explain and predict such phenomena today, they are no less striking. But when strange mechanical phenomena become coupled with even stranger electrical, thermal, optical, and nuclear phenomena, as they do in such artifacts of high technology as nuclear reactors and solid-fuel rockets, the role of theory becomes increasingly important, for we do not want to be tricked by our own creations. Prediction of both desirable and undesirable phenomena by means of analytical theory is at the heart of modern engineering, but theory too can play tricks on us. How much better to discover those tricks in trying to explain some details of innocuous problems involving BOPP's or mutton chops than in failing to predict the strength of materials in massive engineering structures on which lives depend.

17/ Getting the Point, and Keeping It

When the office boy would hear Harold Ross wish that he were running a newspaper "out West" rather than dealing with the latest crisis at *The New Yorker,* the boy would look at the pencils on the editor's desk, for the young lad knew who had been responsible for sharpening them. That was among the first tasks he had to master:

> I went to Mr. Ross's office and took his pencil-box back to the office-boys' room. I inspected each one, as carefully as one might inspect the triggering apparatus of a hydrogen bomb, before I sharpened it, for my instructions from his secretary and from the office manager as well had indicated in a thorough briefing that Mr. Ross required his pencils to be a certain length, neither shorter nor longer than a fixed size, and without any teeth marks on them.
>
> . . . I was required to sharpen them into a fine point, but not *too* fine, for the lead was not supposed to break or crumble under the pressure of Mr. Ross's hand. It took me two weeks to master this detail, but I eventually succeeded and one afternoon when I was in Mr. Ross's office he said, "Son, you're a goddam good pencil-sharpener."

The desire for a perfect pencil point is no doubt as old as the pencil itself, but two independent though complementary technologies had to be developed before the ideal could be realized. First of all, the lead must be strong enough so that a slender cone of it can resist the concentrated stresses of ordinary writing and drawing. While the catalogues of nineteenth-century pencil makers pictured their products with perfectly sharp points, very little emphasis seems to have been placed on this aspect of the pencil. Rather, the uniformity of grading and the erasability of the mark were emphasized as signs of quality. The silence about sharp points also seems to have been due to the difficulty in achieving them, regardless of how strong the lead might have been. Thus the second technology that was required was that of making the point sharp.

For centuries, sharpening a pencil meant using a knife to whittle away the wood and shape the lead. Penknives had long been used to point quills, and they were naturally adopted to sharpen pencils. But using a penknife for sharpening either a pen or a pencil to a fine point was not a trivial task. In the seventeenth century, a schoolmaster's chief distraction seems to have been to whittle goose quills. While twelve-year-old students were expected to point their own pens, many seem not to have been able to succeed in doing so, and the master often sat at his desk working on quills while the pupils came up individually and recited their lessons before him. The fortunate schoolmaster had an assistant to cut the quills, and many students not only finally mastered the technique but also enjoyed practicing it later in life. According to one recollection: "The fashioning and refashioning of one's own [quill] was inevitably accompanied by care, pride, and pleasure in its use."

Pointing pencils seems not to have been so fondly remembered. Old pencils made with graphite dust held together with glue and other binders tended to be so brittle that special precautions had to be taken to sharpen them. According to one description: "If the point broke it was quite an undertaking to sharpen it again. First the wood had to be cut away and the graphite heated over a light to soften it. Then it was again drawn to a point with the fingers." After clay came to be used as a binder for the graphite, the ceramic lead could no longer be softened by heating, and so sharpening it had to be done differently.

Sharpening a clay-and-graphite pencil seems not to have been any more easily mastered than sharpening a quill. Even though red cedar had fine cutting qualities and presented no special obstacle to the penknife, it still took practice to make a neat cone of wood. While the way the wood was cut might be more a matter of aesthetics than function, using the knife for this was really tricky, for the point tended to break off before it could be finished, not only wasting lead but also requiring further whittling. Such frustrations led some to recommend that "it is better to restore the point by rubbing it on a piece of paper, and at the same time turning it round, than to attempt cutting it every time with the knife." Debates continued about accepted sharpening procedures, which did not all meet the Boy Scout's dictum of always cutting away from oneself:

> It is usually held that the correct way to sharpen a pencil is to hold the point against the right thumb, and cut away the surplus wood and lead by drawing the blade of the knife toward the thumb. This method is open to the objection that it is apt to soil the thumb and fingers. On the other hand, the more cleanly method of sharpening a pencil by cutting outward, that is, away from the body, is apt to result in too deep a cut and the consequent breaking of the pencil point.

Such an explicit description in 1904 of what everyone already knew was really introductory to the announcement of a new invention that avoided the shortcomings of the conventional method: a device that fit over the pencil and guided the knife to keep it from taking too deep a cut. Many other sharpening devices were patented in the late nineteenth and early twentieth centuries, each removing an objection to a previous invention. For example, the knife-guiding device still required a separate tool to be used. But a pencil-sharpening attachment patented about 1910 incorporated a blade that was self-guiding. Other inventions simply adapted existing tools to avoid deep cuts and dirty fingers. One inserted a small carpenter's plane upside down in a box so that the pencil could be pulled across the properly exposed blade and the shavings, which were "excellent for driving away moths," could be collected in the box for later use.

But such alternatives to the penknife were the exception. In an adventure first published in the early 1900s, Sherlock Holmes was asked to determine which of three students had sneaked into a tutor's room and copied down a question from a scholarship examination to be given the next day. The tutor, Hilton Soames, who had found "several shreds of a pencil which had been sharpened," as well as a broken point, concluded that "the rascal had copied the paper in a great hurry,

His First Pencil, a Norman Rockwell painting commissioned by the Joseph Dixon Crucible Company, showing the use of a penknife as a pencil sharpener

had broken his pencil, and had been compelled to put a fresh point to it." On inspecting the broken lead and the chips of wood left by the knife, Holmes concluded: "The pencil was not an ordinary one. It was above the usual size, with a soft lead, the outer colour was dark blue, the maker's name was printed in silver lettering, and the piece remaining is only about an inch and a half long. Look for such a pencil, Mr. Soames, and you have got your man. When I add that he possesses a large and very blunt knife, you have an additional aid." Neither Soames nor Watson could understand how the length of the pencil could be deduced from the cuttings, even after Holmes held up "a small chip with the letters NN and a space of clear wood after them." Finally, he explained that the NN represented the end of a word: "You are aware that Johann Faber is the most common maker's name. Is it not clear that there is just as much of the pencil left as usually follows the Johann?"

If using a knife to sharpen a black-lead pencil was difficult, using one to sharpen a colored-lead pencil was even more so. Colored leads contain a good deal of wax and thus cannot be baked into hard ceramic rods, and sharpening them was once a perennial problem. When the Blaisdell Pencil Company developed a pencil that could be sharpened without a knife or device of any kind, the concept proved to be especially useful for colored leads. The lead was wrapped with paper in a spiral fashion, and a controlled amount of the paper could be unwound after just nicking it with a knife or a fingernail. Since the lead itself did not have to be cut, there was also no waste.

Sandpaper or a fine file are excellent for pointing pencils after the wood or paper has been removed, but they are awkward and dirty to handle away from the drafting table or desk. Around 1890 inventors began devising many alternatives to the pocketknife, and those that worked best resembled the small hand-held devices that children still tend to use, but they apparently broke the lead as frequently as they pointed it. In its 1891 catalogue, Dixon offered a small pencil sharpener that resembled a finial for a lampshade but in fact had a conical hole into which the pencil was inserted and twisted against an inwardly projecting blade. The new sharpener had a patented stop to prevent the lead from breaking, and Dixon provided an uncharacteristic apology under the description of the sharpener: "We have spent a great deal of money in the endeavor to get a pencil-sharpening machine that could be sold at a reasonably low price, and at the same time be simple in mechanism, neat, clean and useful. Except for the above little device we know of nothing that we can recommend."

In an 1893 catalogue Johann Faber devoted a whole page to its newly patented Acme pencil sharpener, which consisted of a brass case into which a replaceable steel blade was fitted. It was "not bulky" and could be "carried easily in the waistcoat pocket." Faber claimed that the sharpener was so carefully made and accurately adjusted that "a needle point" could be achieved. The customer was implored to "Try it, please!" As late as 1897 A. W. Faber was offering "lead pointing tools" that consisted of a filelike surface embedded in a piece of wood alongside a "wiping surface" that may have been little more than a piece of felt. The company's pointer for artists' leads appears to have functioned like the Dixon model.

At about the same time that effective pocket pencil sharpeners were being introduced, larger machines designed to be screwed onto a table or desk were being developed. One of the first appeared about 1889. The Gem pencil sharpener, which was made for a Boston firm, consisted of a circular disk of sandpaper against which the pencil was pressed and rotated by a fixture geared to the same crank handle that turned the sandpaper. The Gem even would "point a red or blue pencil perfectly, which all will appreciate who have tried to sharpen these pencils." The Gem was apparently still in use almost a quarter century later, for it was included in a survey of mechanical pencil sharpeners that appeared in *Scientific American* in 1913. But the penknife was far from being made obsolete, for the introduction to the survey noted that "it takes a keen knife and no little skill to sharpen a lead pencil quickly and without waste, and yet nearly everyone likes to have it neatly treated."

The Gem, an early mechanical pencil sharpener employing a rotating disk of sandpaper

Quicker and more efficient alternatives to the "primitive method" of hand whittling were considered necessary, especially for such places as schools, offices, and telephone exchanges, where many pencils had to be kept in "good condition." At a time when there was a growing movement to eliminate all wasted effort in the workplace, it was estimated that a mechanical pencil sharpener could drastically reduce the ten minutes it took a worker to sharpen a pencil the old way: "Borrowing neighbor's knife, two minutes; sharpening pencil, three minutes; washing hands on company's time, five minutes." Among the devices that would make the office of the

early twentieth century more efficient were many kinds of pencil sharpeners that had multiple blades attached to a rotating
wheel designed to whittle whatever was held in its path. Their
advantages were largely offset by how fast the blades became
dull. It was estimated that about one thousand pencils could
be sharpened by one model before the blades had to be replaced at a cost of about sixteen cents.

Other early mechanical pencil sharpeners operated on a
milling principle, whereby a rotating cutter sliced off wood
and lead at an angle to the pencil's length to form a neat conical
point as the pencil was turned against the rotating cutter.
Pressing the pencil with too much force against the cutter
caused the pencil lead to be broken instead of sharpened, however. The single sharpener that seemed to have the greatest
potential was one that had two beveled cutters that not only
rotated but also revolved around the stationary pencil as they
sharpened it. Since the cutters acted on opposite sides of the
pencil point, they did not tend to bend it and therefore did
not tend to break it.

The Automatic Pencil Sharpener Company of Chicago first
made a whittling type of sharpener, which it patented around
1908, but the double-cutter model soon became Apsco's standard product, fourteen models being made in the 1920s.
Apsco advertisements leaned toward overstatement, once
claiming that the pencil was "actually the most important part
of our lives," and used graphic hyperbole when they showed a
mutilated pencil, "sketched from life—sharpened with a knife
—the average job of an average man," beside a picture of
perfection, "the same pencil sharpened by an Apsco Automatic Pencil Sharpener." Another advertisement with the
same pencils asked the question: "Which do you use? Man
made points or 'automatic made'?" When not appealing to
neatness, Apsco discussed efficiency. Above a cartoon of a
man whittling his pencil over a wastebasket was a reminder
that the activity was something the time clock could not check.
At "five minutes to do a five-second job," the knife was too
inefficient to be still used in 1927, according to the pencil
sharpener company.

While Apsco's advertisements may have overstated other
reasons, there were plenty of sound technical reasons to use its
machines. *Consumers' Research Magazine* consistently gave
the Apsco Giant, which later became the Berol Giant Apsco,
the highest ratings among pencil sharpeners, although noting
in 1979 that no instructions were provided beyond "install

with screws provided." Around 1940 the company actually had published an instruction unit for the classroom, and "How to Sharpen a Pencil" contained separate steps for hand and automatic feed models that even kindergartners must have thought beneath them.

Pencil sharpeners may finally have seemed no more to need improvements than instructions, but that is not to say that all pencils could easily be sharpened. Off-center leads were virtually impossible to sharpen, for the lead was constantly bent even by opposing cutters. And colored leads, which often were broken in the wooden shaft, tended to leave loose pieces in the machine to revolve with the cutters and prevent another pencil from being sharpened. Mechanical sharpeners could be opened up to be unclogged, but the electrically driven pencil sharpeners introduced in the 1940s tended not to have their cutters so easily exposed for cleaning.

At least one person was concerned with the extent to which electric pencil sharpeners were being used during the energy crisis of the mid-1970s. José Vila, a New Yorker, saw not only a waste of electricity but also a waste of pencils in sharpeners that "eat up your pencils," and he invented a sharpener consisting of concentric plastic cylinders surrounding a metal shaft tipped with a cutter, all connected through gears and a spring. The assembly was set in motion by the force of inserting the pencil to be sharpened, and as little effort as a single push on the pencil was enough to point it. The device was patented in 1980.

Perhaps one of the largest and most precise pencil sharpeners was developed in the 1950s by engineers at the Eagle Pencil Company. The two-ton machine was capable of sharpening many pencils to cylindrical points whose diameters were identical to within one ten-thousandth of an inch. With this machine comparative tests for smoothness, durability, and opacity of leads could be made with confidence. It was also capable of producing fine needle points for strength tests. Factory-sharpened pencils for consumers are not necessarily pointed with such precision equipment, however. They are often just rolled along a rotating drum covered with sandpaper, hence the characteristic scratches converging on a squat point. In advertising its new machine in 1956, Eagle acknowledged that pencils were also still sharpened with knives and razor blades.

Throughout the twentieth century, manufacturers of drafting pencils have displayed proudly the stubby ends of their

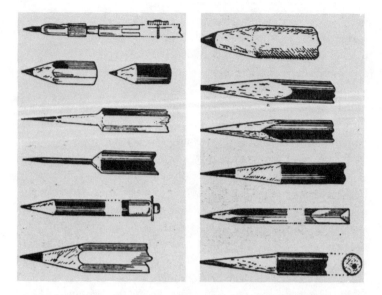

The range of points achievable with the Iduna 2, a universal pencil-sharpening machine

products to emphasize how hard it was to part with them. The American Pencil Company advertised its Venus this way for decades, and Staedtler's ads have featured the last inch and a quarter of its Mars, with the final *s* nicked off and visible on one of the chips beside the pencil, as if a clue for a sleuth. These companies know that even today many engineers, architects, and draftsmen still use a knife to point their pencils down to stubs. But they use the knife only to cut away the wood, being careful not to cut into the lead, which is generally pointed on sandpaper. Cutting into the lead not only dulls the knife but also nicks the lead and weakens it. The suggested procedure varies, but the following from a 1925 textbook is not atypical: "To sharpen the pencil properly, use the knife merely for removing the wood. Do this in such a way as to leave ⅜ in. of lead exposed, and taper the wood back for nearly an inch further. To sharpen the lead use fine sandpaper or emery cloth. Keep the pencil sharp enough to prick the finger."

The long lead and severely cut wood are to minimize any obstruction of the drawing work by the pencil itself. As the text for the U.S. Navy Draftsman 3 course put it: "If you use a pencil with a long, exposed lead, you will be able to see around it to the line as you draw." Some authors have even gone so far as to suggest exactly how the wood should be cut, saying, for example, that "it is better to begin the 'cuts' on the hexagonal edges of the pencil than on the hexagonal faces." Others have even recommended unsymmetrical cutting and pointing procedures. In the 1930s Koh-I-Noor offered a small pocket sharpener called the Tutior Juwel that resembled Johann Faber's but had two blades, one capable of stripping the wood off the lead and the other of pointing it. The stripping blade could be adjusted to give "the long lead of the draftsman or the shorter lead of the sketch artist." The Tutior Juwel could be supplemented with sandpaper to achieve any desired point. Mechanical sharpeners that only removed the wood from drawing pencils were common in drafting offices by the late 1930s, and various hand-operated pointing devices that retained the graphite dust made pencil sharpening neater. In one model, the pencil itself fits into an eccentric hole to serve as a crank handle to turn the point around an abrasive surface inside the body of the device.

The importance of pencil sharpening to engineers and architects was demonstrated in 1920 when a new magazine published by the Architectural Review, Inc., was called *Pencil Points: A Journal for the Drafting Room.* But to some pencil collectors, a sharpened pencil is a ruined pencil, as was certainly the case with the most valuable pencil known to a late-nineteenth-century manufacturer. It was "a cheap-looking affair" owned by a New York lawyer, but the wood came from an ancient tree preserved near the remains of a mastodon in a marl bed in Orange County. The knob on the end of the pencil was made from the mastodon's tooth, and the manufacturer did not expect the pencil ever to be used for writing or drawing.

For every collector who has wanted to preserve an unsharpened pencil, there have been countless inventors who wanted to have their pencils and use them too. In the seventeenth century pieces of plumbago used for writing and drawing were held in a variety of elaborate brass and silver Baroque designs, including one dating from 1636 that pushed the lead out by means of a compressed spring. While this may be said to have been the first propelling pencil, the first mechanical pencils are generally dated from the early nineteenth century. Sampson Mordan was an English engineer who devised locks and pens and, in 1822, patented an "ever-pointed" pencil. In 1833 the American James Bogardus, who was a watchmaker specializing in diesinking and engraving, patented his "forever pointed" pencil. As the lead was worn down in these first

BY HIS MAJESTY'S ROYAL

𝕷𝖊𝖙𝖙𝖊𝖗𝖘 𝕻𝖆𝖙𝖊𝖓𝖙.

S. MORDAN & Co's
PATENT EVER-POINTED
PENCILS,

ARE upon a principle entirely new, and which combines utility with simplicity of construction. The Black Lead is not inclosed in wood, as usual, but in a SMALL Silver Tube, to which there is attached a mechanical contrivance for propelling the Lead as it is worn. The diameter of the Black Lead is so nicely proportioned as NOT TO REQUIRE EVER TO BE CUT OR POINTED, either for fine Writing, Outline, or Shading. The Cases for the Drawing Table or Writing Desk are of Ebony, Ivory, &c. ; and for the Pocket, there are Silver or Gold Sliding Cases, varying in taste and elegance. The Black Lead is of the finest quality, and prepared (by an entirely new chemical process) of five distinct degrees of hardness, and contained in boxes properly lettered for Artists, &c. and at the same time is perfectly suitable for all the purposes of business.

.. .. The Case .. ..

The Patentees desire to remark, that "S. MORDAN & CO.'S PATENT," is stamped on each of their Patent Pencils ; an attention to which, on the part of purchasers, will tend to check any attempt at imposition.

DIRECTIONS FOR USE.

Hold the two milled edges between the finger and thumb of the left hand. Turn the case with the other hand to the right, and the lead will be propelled as it is required for use ; but if, in exhibiting the case, or accidentally, the lead should be propelled too far out, turn the case the reverse way, and press in the point ; which of course in practical use, will seldom or ever be required.

The Black Lead Points are of five distinct sizes, as well as of five degrees of hardness, and contained in boxes marked as follow :—

The	V H	(very hard)	is very small in size..Seldom required
The	H	(hard)	is small............ Hard and black, for fine Drawing
The	M	(medium)	is of a medium size.. For general purposes
The	S	(soft)	is larger............Black for Shading
The	V S	(very soft)	is largest .... Very black, for deep Shading

The Cases are respectively marked with a corresponding Letter.

Attempts have lately been made to impose upon the Public an imitation of their Patent, in which the Lead being attached to the propelling wire, for the useless purpose of drawing back the Lead it is therefore in constant danger of breaking, while in those recommended to the Public by the Patentees, the Lead may be propelled as required for use, without incurring the slightest risk of breaking.—"MORDAN & CO.'s PATENT" is stamped on each case, and no other Patent has been granted for Pencils. They may be had of most of the respectable JEWELLERS, SILVERSMITHS, CUTLERS, AND STATIONERS IN THE UNITED KINGDOM.

MANUFACTORY,
No. 22,
Castle Street, near Finsbury Square,
LONDON.

An 1827 advertisement for one of the first mechanical pencils

mechanical pencils, more could be "propelled" from the tube containing it. The pencils had the obvious advantage of never having to be sharpened, and thus they provided a clean and fixed-length alternative to the ever-shrinking wooden pencil. Many variations on the basic principles of Mordan's and Bogardus's inventions, often cased in silver and gold, were introduced throughout the 1800s, including precursors to many of the basic mechanisms that are still used today. But in spite of the fact that many models were available with companion toothpicks and ear spoons, the mechanical pencil would not seriously threaten the wood-cased kind for almost a century.

The first mechanical pencils were more like novelties and pieces of jewelry than serious writing instruments. They tended not to have the right size, balance, weight, or surface finish to make them suitable for extended periods of writing, and their relatively thick leads did not give anywhere near the fine point possible in a sharpened wooden pencil. Furthermore, the leads tended to give a little both sideways and lengthwise in the barrel, and a play of even a few thousandths of an inch could be distracting to the writer.

The Eversharp pencil was different. It had the length, diameter, and feel of a real pencil, and its "rifled tip" kept the lead from slipping. However, in the beginning Eversharps were made piece by piece, which resulted in a costly and inconsistent product that worked against the pencil's potential selling points. Thus in 1915 the Eversharp Company was looking to acquire some used machinery in order to manufacture the pencils itself. It approached the Wahl Company, a Chicago firm specializing in precision adding machine attachments for typewriters. Wahl's employees were largely trained in watchmaking, and the machinery they used was suited to make Eversharps, but it was not for sale. However, the chief engineer agreed that his company could manufacture the pencil, and Wahl started producing it in 1916. The following year the financially pressed Eversharp Company was bought out by the Wahl Company.

Wahl stressed the mechanical features of the Eversharp in its advertising, which included "a split cross-section, showing how the lead went in, pointing out the rifled tip, describing how the plunger worked, and showing where the rubber was concealed." By the 1917 holiday season, orders could be filled only by working the factory overtime and reorganizing the manufacturing processes to coax two and three times as much output from some machines. Soon 35,000 Eversharps were being made each day, in a continuous process much as Model T Fords were being made. Although this was all happening at a time when materials and labor costs were increasing, the specialized machinery designed by Wahl's engineers enabled the company to keep the price of the pencil constant.

An Eversharp mechanical pencil from the early 1920s

One unexpected complication of growth was the curtailment of Wahl's supply of lead for the Eversharp. Wahl had been buying the lead from a conventional pencil manufacturer, and that company appears to have put a limit on the volume it would supply when the mechanical models began to threaten wood-cased pencil sales. Thus Wahl decided to make its own lead, but according to the company's general manager, writing in 1921: "Lead-making, we learned, is a secret process. Like the manufacture of German dyes before the war, it has always been surrounded by a certain amount of mystery." But Wahl had committed itself to making pencil leads, and so a company chemist "set to work just as if there had never been any lead made in the world before. He did discover that two general processes were used, but he did not learn the details." According to the chemist, Robert Back, writing in 1925 but not revealing any secrets himself, making pencil leads was one of the oldest ceramic industries. Yet "the literature contains little not found in encyclopedias," although of late contemporary trade journals had been giving "narrative descriptions of processes for making the lead sticks."

In the scientific and technological climate of the twentieth century, however, having a general idea of a process is a big advantage in setting up a research and development program to discover details. While precise secrets of other companies might never be found out, Back could find and keep his own secrets. Making leads for mechanical pencils presented special problems, for the diameter of the lead could not be increased to gain strength, as it could for wood pencils. The necessity of making small-diameter leads, which around 1920 meant a diameter of 0.046 inch (or 1.17 millimeters), restricted the use of flake graphite because it lessened the ability of the clay to bind together. Thus the proportion of finer amorphous graphite had to be increased, but it resulted in a lead that wrote less smoothly.

The manufacture of colored leads, which were not hardened by firing in a kiln, presented even more difficult problems. In addition to getting the ingredients right, the leads had to be made to within a 0.001-inch tolerance, so that the advantages of Eversharp's rifled tip would not be offset by a loose lead or one that did not fit. The diamond extruding dies had to be checked regularly for wear that would increase the lead diameter. Another complication with colored leads was associated with the wax with which they were impregnated to increase their writing quality. In order to keep the leads from sticking together, the wax had to be absorbed from the surfaces by tumbling the leads with sawdust. Yet a certain amount of wax coating was necessary on colored leads so they would not absorb moisture and thus soften and swell, as they were apt to do "in a man's pocket during the warmer months of the year."

At first Wahl's lead was "a somewhat inferior product," because the trick of getting "just the right tensile strength" eluded the developers, but eventually the problems yielded to analysis and by 1921 some twelve million Eversharps and sufficient lead to supply them had been distributed. The pencils sold extensively as gifts, and one boy was reported to have received nine Eversharps for his birthday. Finely engraved pencils were selling for as much as sixty-five dollars, even though they had the same essential mechanism as the one-dollar model.

Wahl recognized the importance of taking every opportunity associated with being first with a new product, for it expected the Eversharp to be imitated. Within about five years the company had almost one hundred competitors, many of which copied the Eversharp very closely. Advertisements for the original pencil emphasized less and less the mechanical features, which were no longer unique, and concentrated on "producing in the minds of consumers the idea that *the* mechanical pencil was the Eversharp." With increasing competition at home, the company looked to foreign markets. In considering questions such as "whether Canton, China, and Canton, Ohio, are much alike in their fundamental reactions to sales efforts," Wahl came down on the side of basing its "appeal on beauty and economy and efficiency and other universal selling points." Furthermore, it decided not to translate its pencil's English name.

In the early 1920s domestic advertising for mechanical pencils generally echoed the theme of efficiency contained in sharpener ads and the waste associated with other pencils. Eversharp ads in business magazines reported the results of a survey indicating "a rapidly growing interest in pencil costs," and elaborated: "The average yearly cost of wood and paper pencils was found to be $1.49 per employee—yet only two inches of such pencils are actually used! You would not tolerate such waste in all office supplies." Employers were told that they could "save two-thirds of pencil cost and add to efficiency by furnishing lead for individually owned mechanical pencils," and that the "user of an Eversharp loses no time in sharpening it." The success of the company's lead-making efforts was touted with: "Eversharp leads are smooth, strong, and fit Eversharp like ammunition fits a gun."

Another manufacturer of mechanical pencils claimed that "a seven-inch wood pencil loses six inches of itself in the sharpener." An executive from a wood-cased pencil company reportedly soon visited the mechanical pencil manufacturer and performed a demonstration: "The visitor took a new pencil and a sharpener, made a point, broke it with his thumb, laid it on the desk; and then repeated the process until the pencil was sharpened to nothing. All the points laid in a line on the desk equalled the full seven inches of the original lead in the pencil." Of course, there was much lead cut away to make the little cones lying on the desk, but the mechanical pencil manufacturer agreed to change its advertising, perhaps to the theme that was used frequently by one of Wahl's most visible competitors in the early 1920s.

The advertisements of the Ingersoll Redipoint Company consistently identified its president as "Wm. H. Ingersoll, formerly of Robt. H. Ingersoll & Bro." The ads also featured "the deadly parallel" between a seven-inch wood pencil and a Redipoint, with the former consisting of a wasted three-inch stub, two inches whittled away, creating "muss, lost time, waste," and only two inches of lead to write with, at a cost of five cents. The Redipoint, on the other hand, had none of these disadvantages, and its two-inch leads, "about double the length in the ordinary mechanical pencil," cost only one cent each. The advertising seems to have gotten to the pencil-using public, for some people began to complain that sharpeners were a conspiracy of wood pencil manufacturers.

This same theme was to be the subject of a parody of an antitrust investigation by the Department of Justice and the Federal Trade Commission, in which several years and a million dollars were spent, ending in a massive report uncovering conspiracy between the manufacturers of wood-cased pencils and pencil sharpeners. Instead of a sharpener that left a good cylinder of lead, as in a mechanical pencil, an "iniquitous type" of sharpener was "forced on the public for the express purpose of shortening the life of the lead pencils which the people are forced to buy and thus forcing the purchase of more lead pencils." The "lumber magnates of the Pacific Northwest" contributed to the funds of the trust, according to the parody, thus causing "182.6836 percent" more wood-cased lead pencils to be used than were required by the people of the United States.

Whether seen as humorous or serious, competition drove large wood-cased pencil companies to market their own mechanical pencils and "thin leads" to fit others. The American Lead Pencil Company, for example, tried to sell its Venus Everpointed mechanical pencil by associating its leads with the successful Venus drawing pencil. New companies started up specifically to market inexpensive mechanical pencils. In 1919 Charles Wehn, a pencil salesman, saw a demonstration of an unbreakable, imitation tortoise-shell comb, and he got the idea of making pens and pencils out of the stuff, which he found out was Pyralin, a new material made by Du Pont, and which was much less expensive than the hard rubber or metal that was being used. He designed the pencil "for tomorrow," something lightweight, balanced, colorful, and inexpensive. By 1921 Wehn had opened the Listo Pencil Company, named after the Spanish word for "ready," in Alameda, California, and Listo mechanical pencils became available for as little as fifty cents.

Foreign and domestic competition in supplying leads continued to grow, and in 1923 the Scripto Manufacturing Company in Atlanta began to make a mechanical pencil that would sell for ten cents, in order to have a captive market for the products of its parent company, the only independent pencil-lead maker left in America after World War I. The planning strategy was to make an "excellent" pencil, but one stripped of nonessentials and needless variation. Raw materials alone

would cost almost two cents per pencil, and so if the company
wanted to sell the pencil wholesale for about a nickel, it had
only a little over three cents out of which to get the manufacturing, advertising, accounting, overhead, and profit. Since
making mechanical pencils requires a precision comparable to
watchmaking, and since there were no such precision industries in Atlanta, tool and die makers had to be imported from
the North. When the machinery was built, a decision had to
be made about what kind of labor to employ for jobs like
packaging the leads and pencils. According to Scripto's vice
president, writing in 1928: "We decided that we must employ
negroes, for the wage scales are even lower than for white
female labor. Then someone got the idea: why not use black
female labor?" Of two hundred employees, almost 85 percent
were black women. At first the pencils cost more than twelve
cents to make, but by 1928 the company was making money
on the pencil it sold for five cents. In 1964, when the work
force was still largely black women, Martin Luther King and
other civil rights leaders called for a nationwide boycott of
Scripto products, which ended when a union was recognized.

As mechanical pencils became better and less expensive,
they posed more of a threat to the wood-cased model. While
the larger pencil companies could expand into foreign markets
with their traditional models, smaller companies and distributors did not have that option. The Lo-Well Pencil Company
of New York at one time would offer among other "free gifts"
a Gem pencil sharpener with every order for one or more gross
of pencils. Amidst the increased concern for cost cutting during the Depression, at least one bank was reported to have
issued each new employee a mechanical pencil and one tube of
leads. When that tube was used up the employee was responsible for replacing it.

By World War II, mechanical pencil companies were facing
stiff competition not only at home but also abroad. In the late
1930s, Germany, Japan, and France were among the largest
exporters to Argentina, for example, and models produced in
the United States had to be offered for about seven or eight
cents apiece in order to be competitive with the Japanese. In
1946 Scripto advertised itself as the maker of "the world's
largest selling mechanical pencils," but its least expensive
model cost twenty cents.

Eversharp, remembered as "the first mechanical pencil that
was a writing tool instead of a gadget," had retained its num-
ber-one position in terms of dollar sales, but the expansion of

its manufacturer, the Wahl Company, into fountain pens "never quite clicked." In the meantime, Wahl's share of the profitable but highly competitive refill-lead business had shrunk, and late in 1939 the company got a new board of directors who were determined to meet the challenge of the "smart merchandising" that had enabled Sheaffer and Parker to achieve the largest sales of pen-and-pencil sets. Eversharp was also being challenged by Autopoint, which had captured a large portion of the sales of mechanical pencils for office use. And as if such competitive challenges were not enough, the Federal Trade Commission had only recently taken issue with advertising claims that some Eversharps were "guaranteed not for years, not for life, but forever." The FTC said that rather than a timeless guarantee, the company was really only promising to make any necessary repairs for the flat rate of thirty-five cents as long as it remained in business.

In 1944 *Consumers' Research Bulletin* reported that all mechanical pencils were "of low quality, except those which are sold at inordinately high prices." The "fairly good pencils" that could be bought for about twenty cents before the war were unavailable, and "pre-war quality" could be had for no less than a dollar. The only recommended mechanical pencil was the Autopoint, some models of which could take "fine-line" or "extra-thin" leads with a diameter of 0.036 inch (about 0.9 millimeter), and it was judged "easily the best in its field" of pencils with screw-feed mechanisms, even though the mechanism tended to jam when colored leads were used. An Eversharp repeating pencil, made as early as 1936, was advertised to hold a six-months' supply of lead fed through a clutch mechanism when a button was pressed at the top of the pencil. But this model was judged by the *Consumers' Research Bulletin* to be of intermediate quality because it was available for use only with "square-type" leads, whose 0.046 inch thickness made them "too large to give satisfactory writing performance as compared with a well-sharpened old-fashioned wood pencil." While the Eversharp operated like today's familiar pencils, the larger lead was necessary because fine-line lead was not strong enough to withstand the forces of the clutch mechanism or of writing. Other push-button-type pencils intended for writing as opposed to drafting were not recommended, and neither was the Norma, a "bulky and rather unwieldy" but widely sold novelty pencil that held simultaneously in its thick barrel as many as four different colors of leads.

As competition among mechanical pencils and between

wood and mechanical pencils continued into the 1950s, the Federal Trade Commission was prompted to issue rules to "promote the maintenance of fair competitive conditions," while the companies repeated familiar advertising claims. Autopoint, for example, argued that mechanical pencils were no more expensive than the wood pencils needed to write the same amount, and furthermore "each wood pencil averages thirty sharpenings, each taking at least a minute. That's a half-hour, and on a dollar-an-hour wage rate, adds fifty cents additional cost per pencil, exclusive of sharpener cost." Aside from the fact that wood-cased pencil makers generally claimed only seventeen sharpenings per pencil, in repeating the decades-old argument Autopoint did not mention how much time it took to change the lead or remove a jammed piece from one of its pencils, though in the 1920s it had been a selling point of mechanical pencils that it took twenty seconds to change the lead.

By the 1970s about twenty million pounds of plastics were being used annually by over two hundred firms in the United States to manufacture some two billion writing instruments, and the suppliers of plastics were themselves competing for business, asking such questions as "Can a premium engineering resin make it in a market already satisfactorily served by cellulosics and polystyrene?" Over sixty million mechanical pencils were being sold each year, and the hottest new seller was the pencil using an even finer "fine-line" lead with a diameter of about 0.02 inch (0.5 millimeter), something that had been available for drafting pencils as early as 1961.

Most of the ultra-thin lead pencils were soon of Japanese origin, some with lead as thin as 0.01 inch (0.3 millimeter), but the traditional German pencil manufacturers have also mastered the new technology. Since ceramic leads are not strong enough to be made so thin, the new leads have been made possible only by incorporating plastics in a polymerization process. But while the polymer leads are strong enough to withstand both the forces in the pencil and those of gentle writing, there are limitations and shortcomings that *Consumer Bulletin* identified in 1973 and that users still find objectionable. The very thin leads break easily, wear down quickly, and have to be replaced before they are entirely used up. For such reasons, no fine-line mechanical pencil was then recommended as highly as the Cross, using 0.036-inch (0.9 millimeter) lead that was considered "a reasonable compromise between a satisfactorily fine writing line and resistance to breakage in use."

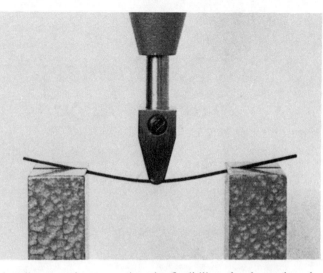

A bending test demonstrating the flexibility of polymer-based leads

By the mid-1980s, Scripto had introduced its Yellow Pencil, a throwaway plastic pencil using 0.5-millimeter lead, and the company expected sales in excess of ten million per year. In all, over one hundred million mechanical and automatic pencils, as those fed through the top have come to be designated, were sold in 1985, and Faber-Castell and others began selling truly automatic pencils—ones that feed ultra-thin lead by the action of writing itself.

18/ The Business of Engineering

*I*t has often been said that an engineer is someone who can do for one dollar what any fool can do for two. Like all witty sayings, this one has an element of truth to it, as long as it is understood metaphorically. And pencil making, our metaphor for engineering, provides an excellent example of what engineering economy means for manufacturing success. As long as a good English or French or German pencil could not be had easily in New England in the first part of the nineteenth century, then any homemade or scratchy pencil was better than none. But the earliest pencil makers also realized that anyone who could engineer a better pencil, whether or not at a dearer price, would have a distinct advantage. Pencil users, especially artists and architects and engineers, who depended upon their pencils for more than scratching notes or marking wood to be cut, would pay a premium price for a quality piece of black lead in a good wooden case. John Thoreau and his son Henry David knew this, and they knew that success would be theirs if they could make a pencil as good as any imported one. Success did come to the Thoreaus, even though their pencils sold for a higher price than any other domestic brand, but the Thoreau pencil business could thrive only until the German pencil industry invaded America around the middle of the nineteenth century. The situation then changed, as Edward Emerson recalled:

A friend who attended, in 1849, a fashionable school in Boston, kept by an English lady, tells me that the drawing-teacher used to direct the pupils to "ask at the art store for a *Thoreau pencil,* for they are the best"; for them they then had to pay a quarter of a dollar apiece. Henry Thoreau said of the best pencil when it was achieved, that it could not compete with the Fabers' because it cost more to make. They received, I am told, six dollars a gross [or about four cents apiece] for good pencils.

In other words, the Thoreaus could not offer a pencil in 1849 for less than one the Germans could make, ship, and sell at a profit. While the Thoreaus were no fools, and while they were for all intents and purposes acting as good engineers in making the best pencil for the price that they could, the Germans were acting as better engineers. By the late 1840s they had certainly mastered the Conté process and were exporting their pencils in vast quantities. And because of the magnitude of their operation and the machinery they employed and because of the resulting economies of scale, Faber was doing for four cents what the Thoreaus might not have been able to do for less than eight. Since the Thoreaus had the basic understanding of how to make an excellent pencil—an ersatz German pencil, which by then had become the world standard—they might have been able to succeed in making it cheaper, or at least cheaper than a foreign pencil, by expanding their family business into a major manufacturing concern and thus enjoying all the economies of scale that the Germans did. However, such an expansion would have required raising capital, investing in new machinery, and increasing output by a great amount.

Henry David Thoreau certainly never seemed predisposed to throw himself into such a major business venture, and his father does not seem to have demonstrated ambitions to be anything but a modest businessman. However, if they had wished, the Thoreaus could probably have raised the capital to expand in the mid-1840s, for they had something few if any others in America possessed at the time—the secret formula for making fine pencil lead out of graphite dust. And, according to Horace Hosmer, the senior Thoreau, at least, had the respect of the community: "I know what the average man of that time was, and I know that he, Thoreau, was cleaner, finer, more intelligent and the light which he hid under a bushel was worth more in gold and silver than all the real and personal property in Concord which was taxable. His great sin was in hiding his knowledge of the foreign process of Pencil making." Hosmer repeated his assertion on the occasion of an interview, as the interviewer noted: "The father Thoreau *very* secretive as to his process."

However, the question of whether to attempt to capitalize on the secret of pencil making may have been rendered moot when technological developments made the business of supplying pure graphite itself so lucrative that the manufacture of pencils was reduced to nothing but a front. According to Edward Emerson, the recently discovered electrotyping process being used around Boston was still a secret in the late 1840s, "and a man engaged in it, knowing the Thoreau lead was the best, ordered it in quantity from Mr. John Thoreau, the latter guarding carefully the secret of his method, and the former concealing the purpose for which he used it." At first Thoreau received ten dollars a pound for ground plumbago; later the price fell to two dollars, but he was selling five hundred pounds a year. When the Thoreaus found out what the plumbago was being used for they "sold it to various firms until after the death of Mr. John Thoreau and his son Henry, when the business was sold by Mrs. Thoreau." Because of the lucrative trade in plumbago and the growing competition from Germany, "after 1852 few pencils were made, and then merely to cover up the more profitable business, for, if the secret were known, it might be destroyed." It was not only the Thoreau pencil business that was affected by foreign competition. Francis Munroe took over his father's prospering pencil concern in 1848, but it was depressed five years later, when "pencil-makers from Germany had established themselves in New York, and with their great skill and an abundance of cheap, imported labor began to push the New England manufacturers pretty hard with their competition."

In this environment, Munroe left pencil making in Massachusetts to enter the lumber business in Vermont. The stories of the Munroes and the Thoreaus and their pencils illustrate in microcosm the often conflicting objectives of real-world engineering and business: making pencils as fine as possible as an end in itself; making pencils better in quality or price than any other available; making pencils to make a living for the pencil makers; making pencils secretly in order to have an advantage over the competition; making pencils overtly to conceal a more profitable business; making pencils for the social and cultural good of artists, engineers, and writers of all kinds. There is no such thing as pure engineering—making a perfect pencil, whether in the artifact or in the abstract—for that would be nothing but irresponsibility or a mere hobby. Engineering, far from being applied science, is scientific business. Edwin Layton has put it succinctly:

> The engineer is both a scientist and a businessman. Engineering is a scientific profession, yet the test of the engineer's work lies not in the laboratory, but in the marketplace. The claims of science and business have pulled the engineer, at times, in opposing directions. Indeed, one observer, Thorstein Veblen, assumed that an irrepressible conflict between science and business would thrust the engineer into the role of social revolutionary.

While Henry David Thoreau seems never to have aspired to being for very long a professional anything, let alone a professional engineer, he does seem to have had a social conscience that asserted individual rights above all. And this trait seems to have dominated his thinking. The stereotypical characterization of the engineer as one who is a "pencil head" or "lead head," with no thoughts but calculations, is at the other end of the spectrum from a Thoreau. Needless to say, the overwhelming majority of real engineers, whether or not professional, fall somewhere in between.

Engineering and business are constantly coming together in mutually beneficial and blissful ways, and the marriage is called industry. The designs engineers put on the backs of envelopes tend to remain there if there is no business interest in financing the realization of those designs. And businesspeople find themselves undersold and their products obsolete if they do not invest in engineering. The small amount of truth contained in the witticism of the engineer and the fool is really much less threatening to industry than the competitive reality: what one engineer has done for one dollar, another engineer will soon do for ninety-nine cents. And it is of boatloads of pennies that fortunes are made.

In the summer of 1921 a young physician, Dr. Armand Hammer, traveled from New York to Moscow to make business arrangements for shipments of medicine and chemicals from his family's firm to the Soviets, who were experiencing grave postwar shortages. He was also planning to do hospital relief work among refugees from drought- and famine-stricken Volga towns. On a trip to the Ural Mountains he not only observed the famine firsthand but also saw Russia's idle industry and economic stagnation. Valuable stockpiles of minerals, gemstones, furs, and the like had accumulated during the European blockade, but there seemed to be no Russian initiative to exchange them for grain or other much-needed commodities. Hammer promptly arranged for American grain shipments to Russia, and after the ships were emptied in Petrograd they were to be reloaded for the return trip with a cargo of Soviet goods. Thus began a long and profitable business.

Such decisive action attracted the attention of Lenin, who summoned Hammer to the Kremlin. According to Hammer's own account:

The two countries, the United States and Russia, as Lenin explained, were complementary. Russia was a backward land with enormous treasures in the form of undeveloped resources. The United States could find here raw materials and a market for machines, and later for manufactured goods. Above all, Russia needed American techniques and methods, American engineers and instructors. Lenin picked up a copy of *Scientific American*.

"Look here," he said, running rapidly through the pages, "this is what your people have done. This is what Progress means; building, inventions, machines, development of mechanical aids to human hands. Russia today is like your country was during the pioneer stage. We need the knowledge and spirit that has made America what she is today. . . ."

"What we really need," his voice rang stronger . . . , "is American capital and technical aid to get our wheels turning once more. Is it not so?"

Armand Hammer certainly agreed, and Lenin went on to explain that Russia hoped to accelerate the process of economic recovery by offering industrial and commercial concessions to foreigners, and he asked Hammer if he was interested. Hammer recalled that a mining engineer on the train had discussed asbestos, and Lenin asked, "Why don't you take an asbestos concession yourself?" When Hammer expressed concern for what he imagined would be long negotiations before closing such a deal, Lenin assured him that the red tape would be cut. It was cut, and in "an incredibly short time" Hammer became the first American concessionaire in Russia. Then suddenly he found himself enjoying not only previously unavailable fine wine but also previously unknown luxurious accommodations in the government guesthouse across the river from the Kremlin. Within about three months of Hammer's setting foot in Russia, Lenin and other Soviet officials had signed the terms of the concession agreement, which granted Hammer buildings, protection of property, freedom to enter and leave Russia at will, expeditious transportation privileges within Russia, access to radio and telegraph stations to transmit telegrams, and the virtual elimination of the threat of any work stoppage.

It was evident to Hammer that Russia needed great numbers of tractors to mechanize its agriculture, and he talked about this with one of his uncles, who had had a Ford agency in southern Russia before the war. While Hammer's uncle did not think that Henry Ford thought much of the Bolsheviks, a meeting in Detroit was arranged between Hammer and Ford. Although Ford appreciated the potential of the Russian market for tractors, he had preferred to wait for a new regime to come to power. However, when he was convinced by Hammer that that would not happen in the foreseeable future, Henry Ford gave young Armand Hammer the agency for all Ford products in Soviet Russia. Hammer took back to Moscow a Ford car and a Fordson tractor and film depicting the Ford works. He also obtained Henry Ford's commitment to invite Russians to learn about automobile and tractor manufacturing in American plants. After about two years of exporting and importing, Hammer was informed by Leonid Krassin, chief of the Foreign Trade Monopoly Department, that the Russians were going to conduct their own foreign trade but other enterprises were still possible. Hammer suggested selling English ships to Russia, but Krassin discouraged that because Russia was reorganizing its shipyards to produce what it needed "more cheaply than the English will sell." What Russia needed, Krassin told Hammer, was industrial production: "Why don't you interest yourselves in industry? There are many articles which we have to import from abroad that ought to be produced here."

Hammer said he would think about it, and he "pondered deeply about Krassin's suggestion." But he had a difficult time deciding exactly what kind of industry to pursue—until the choice was thrust upon him "by an accident." According to Hammer:

> I went into a stationery store to buy a pencil. The salesman showed me an ordinary lead-pencil that would cost

two or three cents in America, and to my astonishment, said the price was fifty kopeks (26 cents).

"Oh! but I want an indelible pencil," I said.

At first he shook his head and then appeared to relent. "As you are a foreigner, I will let you have one, but our stock is so limited that as a rule we sell them only to regular customers who buy paper and copybooks as well."

He went into the stockroom and came back with the simplest type of indelible pencil. The price was a rouble (52 cents).

I made further inquiries and found that there was an immense shortage of pencils in Russia as everything had to be imported from Germany. Before the War there had been a small pencil factory in Moscow run by some Germans, but it had ceased production. Plans were on foot, I found, to remodel and enlarge it as a State pencil factory of the Soviet, but at that time, in the summer of '25, they had not advanced beyond the project stage. Accordingly, I decided that here was my opportunity.

Hammer's suggestion to start a pencil factory was welcomed, but with some skepticism because of the conventional wisdom that pencil making was a "German monopoly." The state factory, which "had produced nothing that could write like the pencils the Russians had been importing from Germany's great Faber plant for generations," was having a difficult time starting up and the supporters of it opposed the opening of a competing factory run by an American. However, Hammer guaranteed, by putting up $50,000 cash, to begin pencil production within twelve months and pledged to produce "a million dollars' worth of pencils in the first operating year." Since the Russian government had set itself the goal of teaching every Soviet citizen to read and write, such an attractive proposal could not be turned down. Approval came within three and a half months, "record time for Russia," and the deal was closed in October 1925. Hammer "did not know the first thing about manufacturing pencils," so he went to Nuremberg to learn.

Just as Hammer seems to have put business before pleasure in Russia, he also put business before engineering—almost to the point of jeopardizing the business itself. Deciding to make pencils, even in the face of the Russian bureaucracy, was easier than actually making pencils. And learning how to make a good pencil was certainly easier said than done, for while pencil making in Nuremberg in the 1920s used the latest of modern machinery, it also kept trade secrets, and discovering them was no easier than it had been a century earlier. After describing physical Nuremberg, the medieval town "where the toys come from," Hammer gives a twentieth-century social and political description from the point of view of someone who is trying to raid its technology:

> Now the little old town is surrounded by up-to-date pencil factories all owned by the Faber family, or its offshoots and connections. Largest of all is the factory of A. W. Faber . . . at a little town named Fürth, a few miles outside Nuremberg.
>
> No prince or feudal baron ever ruled his estates more completely than the firm A. W. Faber rules Fürth. Its word is law and everything—municipality, police, public utilities—is under its control. Years ago the firm decided that a railroad, or even trolley-cars, might bring in undesirables, might make its workers discontented and interfere with the even tenor of their service to the House of Faber, so the railroad passed Fürth by and the stranger within its gates must come by carriage or auto or on foot.
>
> The other pencil factories were strongholds, too, second only to Fürth. Most of the workers had been employed for generations, father to son, a long line of patient craftsmen, each perfect in his job, but knowing little beyond it. In their jealous eagerness to retain the monopoly of pencil-making, the Fabers had been careful never to let any of their subordinates know more than one part of their complicated organization; knowledge of the whole was reserved for members of the family and a few trusted adherents.

After he had spent a week in Nuremberg, Hammer knew no more of the pencil business than when he arrived, and had he been able to cancel his agreement with the Soviets he might have done so. A pencil factory, like any factory, could be looked at as a great stationary engine or a machine whose input is energy and raw materials and whose output is pencils or other products of manufacture. From one point of view the factory may be seen to exist to provide a livelihood for the inhabitants of a town like Fürth, but from another point of view the wages and any satisfaction or sense of accomplishment the workers may enjoy might be seen as a by-product of the business of making pencils. So if a pencil factory was really a large and complex pencil-making machine, and one that, in addition to having satisfied workers, had to operate efficiently enough to make a pencil in Russia cheaper than one could buy a Faber there, then the factory's design and assembly was a matter of engineering for an engineer to carry out. But in and around Nuremberg, Hammer could find no available help.

However, just when it seemed most hopeless, through a local banker he was put in touch with "an engineer who held an important position in one of the principal pencil factories." The engineer, a Faber pencil master named George Baier, had actually once accepted an offer to build a factory in Russia, but the hostilities had interfered with his plans and he was not permitted to leave Russia or return to Germany until after the war. Eventually he did come home to Nuremberg with his Russian wife, but not to a warm reception, and it was years before he could rejoin the local industry. This treatment did not make him loyal to the German pencil barons, however, and Baier, who was making $200 a month from Faber, accepted Hammer's offer of $10,000 a year plus a bonus of a few cents on each gross of pencils manufactured.

Baier told Hammer of other maltreated employees, including a foreman who, after twenty-five years' service in Germany, had accepted a position in a new pencil factory in South America. But apparently the Nuremberg police did not allow him to leave, and he was confined to the city for ten years, unable to work in the industry there. If such a person were finally allowed to leave, he would take few secrets, for his experience would be with ten-year-old machines, processes, and pencils, and hence his value to the South American factory would not be what it might have been. Such human tragedies are not the product of a technology per se but of the management of the technology, and the businessman Armand Hammer and the engineer George Baier knew this. Thus they were able to recruit similarly disillusioned German workers not only by offering higher wages and bonuses but also by promising all the comforts of a familiar home life in Moscow, complete with schools and German beer, though Russian beer was to prove quite palatable. After two months, the staff necessary to start up the factory was finally selected, and their passports were obtained in Berlin, where the pencil industry held considerably diminished influence. According to one account:

The pencil masters and their families were sprung from their cloistered life in Nuremberg and Fürth on the pretense that they were taking vacations in Finland. Hammer had Russian visas waiting for them in Helsinki. The machinery left Germany almost as surreptitiously. At Baier's suggestion and Hammer's insistence, it was sent to Berlin from its point of manufacture with the understanding, by the manufacturers, that a new pencil factory would be built there. Suspicions about its eventual destination were not aroused, even though Hammer requested that all of it be delivered to Berlin disassembled and that the companies concerned with making the machinery assign an expert to the task of reassembling the myriad parts. Once arrived in Berlin, each bit was numbered, then shipped to Moscow.

Since the Russians insisted that his pencil factory also produce steel pens, Hammer next went off to Birmingham, England, which in the mid-nineteenth century had itself been called "toy-shop of the world," to recruit anew. However, he found the same situation as in Nuremberg, "a closed industry, with most of its workers trained from childhood under semi-feudal conditions." This time Hammer immediately advertised in the local newspapers for an engineer, who in turn would make it possible to hire skilled workers.

Upon returning to Moscow, Hammer looked for a suitable location for his factory and its workers' village. He found his site in an abandoned soap works on more than a square mile of land on the outskirts of the city near the Moscow River. Renovations on the existing buildings and the construction of cottages began quickly, and soon the machinery specified and ordered by the engineers was being set up according to their plans. Before beginning production, Hammer had imported pencils at the rate of about two million dollars' worth per year, but in less than six months after he had first visited Nuremberg, his Moscow factory was producing pencils—about six months ahead of schedule. At first American cedar was used, but when Siberian redwood became available Hammer used it instead.

The demand was so great that there was no problem in meeting the goals of the agreement. In the first year the factory produced two and a half times the million dollars' worth promised, and in the second year the price of a pencil was reduced from twenty-five cents to five. Hammer enjoyed a virtual monopoly when pencils could no longer be imported into Russia, and when output increased from fifty-one to seventy-two million pencils in one year, his factory could export about 20 percent of its production to England, Turkey, Persia, and the Far East. Successful production was rewarded in profit, of course, and the Moscow business, whose stationery bore a likeness of the Statue of Liberty, enjoyed first-year earnings of over 100 percent on a capital investment of a million dollars. Although the capital was Hammer's, the profits were split fifty-fifty with the Soviets.

Among the variety of pencils made, the most popular brand was called Diamond and packed in green boxes marked A. HAMMER—AMERICAN INDUSTRIAL CONCESSION, U.S.S.R. The pencils were trademarked with a crossed hammer and anchor, suggesting the crossed hammers that marked some Faber pencils. Hammer was once told by Nikita Khrushchev that he had learned to write using the pencils, as did a succession of Soviet leaders, including Leonid Brezhnev and Konstantin Chernenko.

But neither loyalty nor production and profit came without rewards or incentives to the workers. At first, the Russian workers' "slowness and laxity nearly drove [the] German foremen to distraction," until Hammer instituted a new means of motivating the workers:

> Under the spur of piecework it was a common thing to have men come into their shop a half hour before the whistle blew in the morning in order to turn up their machines so as to start full speed "right off the gun." Now our German foremen were able to report that instead of lagging behind the production rates they had known at home, most of our Russian workers were beating German records. Wages, of course, advanced correspondingly, but so did our profits, and we never had cause to regret the piecework basis.

Job applications soared, and the business prospered to the point that by the end of 1929 the single pencil factory had grown to a group of five factories, manufacturing not only pencils but also related products. According to one contemporary British observer, working in the factories promised many Russians a means for concealing their non-Bolshevik backgrounds. But there was no guarantee of anonymity:

Professors, authors, generals, former captains of industry, ex-government officials, and ladies of noble birth, sit side by side at the cutting machines and lead filling machines, at the trimming and painting plants, and in the packing-rooms, with humble industrial workers. Their only ambition is to sink their individuality, and to destroy all records of their past, in order that they may keep their jobs. Nevertheless, the government agents and spies are constantly tracing their lineage and former careers, and insisting on their being turned into the streets to make room for the genuine proletariat.

As the introduction of capitalism through profit sharing and piecework began to come under fire in the Russian press, there were also other indications that the business climate was changing. The difficulty of obtaining financing in a worsening world economy and the new attitudes of the Soviets toward foreign interests made it a good time to begin negotiating with the Russians for the sale of the factories to the government. Besides, complaints in the press about the large profits of the plants had forced Hammer to drop the prices of his pencils even further. In 1930 a deal was closed, and under Soviet control the Moscow plant came to be called the Sacco and Vanzetti Pencil Factory, after the Italian immigrant workers Nicola Sacco and Bartolomeo Vanzetti, who were executed in 1927 for murders that occurred during a 1920 robbery in a Massachusetts shoe factory, and who had been the cause of worldwide socialist protests.

Although Hammer had stressed maintenance and care of the hard-won pencil-making machinery, when the Russians took it over they seem to have neglected its maintenance, and as a result a fatal accident occurred at the plant some years later. And in 1938 six executives of the factory were brought to trial on charges of falsifying records of output in order to show fulfillment of production plans. However, the charges that millions of "imaginary pencils" were entered into production reports were later dropped when it was discovered that the executives were merely following the orders of superiors, who had determined that each month's production of pencils should include those made in the first working day of the following month.

Whereas in the Sacco and Vanzetti factory imaginary pencils seemed to appear on production reports, sometimes real pencils disappeared into Russians' pockets. According to one story, during treaty negotiations between the United States and the Soviet Union some years ago, pencils stamped "U.S. Government," which had been abundantly distributed around the table at the beginning of each session, were mysteriously nowhere to be seen after the sessions ended. As it turned out, the Soviet negotiators were putting the American pencils in their pockets because such good writing implements could not be obtained so freely in Russia. Thus the pencils were both tangible evidence and forceful symbol of the technological superiority achievable in the West.

Of course, pencils disappear at all meetings, even when there are no ideological overtones, and so this story may be apocryphal or propagandistic. But even if it is, the story still calls attention to a reality. While they may be used and even coveted by negotiators as a treaty itself is being drafted, pencils are nowhere to be seen when it comes time for government leaders in a formal setting to sign the final official version. That is a time for politicians and pens, for who has ever heard of a President handing out souvenir pencils after signing a bill or a treaty? And the same is true of the relationship of engineers to business. While pencils are used to design a factory's machinery and processes, the contracts for their products are invariably signed in ink.

19/ Competition, Depression, and War

With its growing importance, American wood-cased pencil production came to be recognized as an accurate indicator of the country's industrial health, but as European manufacturers began to reenter the world market in the wake of World War I, the prognosis was mixed. Although 90 percent of American distribution was in the hands of the Big Four, as the Eberhard Faber, Dixon, American, and Eagle pencil companies had come to be called, in 1921 they were asking for increased tariffs on imported pencils to counter cheap production costs in Germany and Japan. In addition to the then existing 25 percent ad valorem rate, an additional duty of fifty cents per gross was suggested. A vice president of the New York importing firm of A. W. Faber, testifying against the increase, declared that the Big Four had built up their "control" of the market in spite of a low tariff. A representative of the American firms countered with the charge that A. W. Faber was controlled by German money, which was denied. In fact, the high cost of machinery, which had enabled the pioneering American pencil companies to grow to the size they had, was also what made it hard for new companies to get started and compete.

Competition had become fierce worldwide, and in Argentina, for example, where practically every type of major European and American pencil was sold, the prices of American products had to be kept down to compete with the German ones, and thus yielded less profit to dealers. Furthermore, an overstock had caused prices to drop, with common No. 2

American pencils selling at retail for only sixteen cents per dozen. And the volume of American exports to Argentina was volatile at best. In 1920 exports were at a record high of more than $250,000, twenty-five times the volume before the war, but then dropped to $75,000 in 1921 and to less than half of that the next year.

In England in 1924, pencil production was valued at about nine times what it was in 1907, but taking inflation into account, this represented only a small increase in volume of output. The British began to debate whether they should allow imported pencils to be sold without bearing an indication of their origin. In testimony before the Board of Trade, a Mr. Kirkwood claimed that "the Japanese engineering industry worked its employees sixty hours a week" for low wages. The president of the board countered that he had no official information as to the rates paid Japanese workers but cited the monthly report of the Tokyo Chamber of Commerce to challenge Mr. Kirkwood's assertion. Mr. Kirkwood then offered "proof" that he hoped would cause the government to take action: "I hold in my hand a dozen pencils which were sold to me in London yesterday for a penny." The Board of Trade's "standing committee respecting pencils and pencil strips," which is British for pencil leads, finally would recommend that pencils could not be sold legally unless they were "durably stamped, printed or impressed, in a contrasting colour not less than one inch from either end of each pencil." The last stipulation was to prevent "Japan," say, from being printed at the very end of the pencil, where a dealer could remove it easily by sharpening.

In the increasingly competitive pencil market, manufacturers were turning not only to legislative but also to engineering help. When the Standard Pencil Company set up a new factory near St. Louis, it installed the first electrically heated oven, designed to bake 1,350 pounds of leads at a time. The cost was to be less than six cents per pound at the rate the new factory was first expected to operate. This amounted to a savings of one-third over the cost of operating gas ovens, and as Standard's pencil production would increase to use the electric oven at capacity, the costs were expected to drop further.

In time other pencil companies would reduce their production costs in other ways. The General Pencil Company of Jersey City installed diesel engines to generate all its own electricity at less than half the cost of what it had been paying the local power company. When Eberhard Faber found that it had to shut down its painting operations on days when humidity in Brooklyn rose too high because water was condensing on the freshly lacquered pencils and ruining the finish, a dehumidifier was designed and installed, and not only did pencil production increase but less expensive lacquers could be used in the controlled environment.

While technical advantages could increase production volume and decrease its cost, manufacturers also realized that making a lot of a product efficiently does not necessarily sell any of it. Pencil makers, like others whose success lay in volume, offered special assortments of fast-selling pencils in attractive counter displays, to keep both the manufacturer's name and the product before the customer. An alternative was to create a demand for a particular pencil, so that the customer would specifically ask for it. All the big pencil companies had long advertised their top pencils, but in the late 1920s the Eberhard Faber Company attempted to use advertising in a new way.

One of the problems faced by pencil makers and dealers was handling the enormous variety of pencil styles necessary to match the competition. No dealer could display them all, and no manufacturer could advertise them all. It was Eberhard Faber's plan to standardize the demand for pencils so that

A display of Dixon Ticonderoga pencils pictured in *Back to School,* another Norman Rockwell painting done for the company

dealers could meet 90 percent of it through the pencil assortment in a newly designed counter cabinet. As Eberhard Faber II explained it in 1929, his firm had surveyed a sample of its 25,000 dealers and found that "eight items account for nine-tenths of the volume and many of our dealers are carrying pencils which they will not sell once a year, if at all." The program was "primarily educational," Faber continued, and "every pencil need of the modern office and the individual is covered by the new Eberhard Faber pencil buyers' chart." According to Faber, "fifty percent of the average dealer's business is in five-cent tipped pencils, with colored, ten-cent untipped, ten-cent tipped and copying pencils following in that order."

Faber's idea may have been a good one, but the timing was poor. The Great Depression changed buying habits, which necessarily also changed production and advertising practices. In the pen-and-pencil-manufacturing industry, of which lead pencils made up about a quarter of the value of products, the number of workers dropped by almost 30 percent and the value of goods produced by over a third from 1929 to 1931. In 1932 the sales manager of the Eberhard Faber Pencil Company wrote to his account representative at the J. Walter Thompson advertising agency explaining the "severance of our business relationship." It was not the fault of either the agency or its representative, but rather was because Faber was "not in a position to do a sufficient amount of advertising to warrant your continuing to serve us." In spite of the dark business climate, the company's five-cent tipped yellow pencil floats proudly in a bright cloud on the letterhead, and the sales manager's signature, which slants

The Eberhard Faber Mongol in 1932, as it appeared on the company's letterhead

optimistically upward, is signed decisively in pencil, perhaps a Mongol.

In 1931 Johann Faber, A. W. Faber-Castell, and L. & C. Hardtmuth were operating their European plants at only 60 percent of capacity, so to eliminate competition among themselves and reduce costs, the Big Three, as they were known, combined into a single holding company chartered in Switzerland. The consolidation had been suggested by the success Johann Faber and Hardtmuth had experienced in jointly operating a factory in Romania. The new trust included a Koh-I-Noor subsidiary in Cracow, Poland, Johann Faber plants in Brazil, which before 1930 were Brazil's own largest producers, and Johann Faber's American subsidiary in Wilmington, Delaware. Although this last had not yet begun production, it was the aspect of the merger that most concerned U.S. pencil producers, for once Johann Faber had gotten a foothold in Brazil, the firm had gained practically full control of the pencil market there, in addition to German pencils supplying over 90 percent of Brazil's imports.

At the time, imported pencils amounted to less than 5 percent of American production, and the production of America's Big Four alone exceeded the entire capacity of the European Big Three. But American factories were operating at only two-thirds capacity, and all companies were looking for opportunities to expand their foreign markets while at the same time protecting their domestic ones. American companies were exporting pencils to over sixty countries, with exports greatly exceeding imports. But exports had peaked in the 1920s, and they declined markedly in 1932 when three of the Big Four opened factories in Canada, theretofore the largest importer of U.S. pencils. The sensitivity of the manufacturers at the time can be seen in the fact that the expansions to Canada were prompted by a tariff rate increase from 25 to 35 percent.

High-priced pencils made in Germany and Czechoslovakia continued to dominate imports from those countries, but overall in decreasing numbers as the Depression continued. The shift in consumption from high- to lower-priced products had also caused good five-cent pencil styles, like the Mongol, to lose sales to an "imitation five-cent group" of pencils that often sold at three for ten cents, as well as to pencils selling for half that price. What most alarmed the American pencil producers, however, was that in 1933 Japanese imports in these price categories jumped from only a few thousand to about twenty million, and they had captured 16 percent of the market. Pencil making was not new in Japan, for at least forty producers existed as early as 1913. At first the Japanese used small machines and much hand labor, but after about 1918 they had copied all the modern German machinery. Thus, by the Depression, they had a "well-organized trade and export organization" that enabled them to bring pencils into the United States valued at only twenty-three cents per gross.

The Japanese pencils, which looked like medium-priced, metal-tipped American pencils, were often stamped with a name or trademark that enabled them to enter the United States at lower rates of duty than if they had been stamped to identify their origin. While patent and copyright suits were threatened by domestic manufacturers, whose pencils at the time were selling wholesale at about a dollar or two per gross, what the Americans really wanted was some tariff protection. After all, they were working under the limitations of a presidential reemployment agreement, whereby they had raised the hourly pay rate of their workers. There were complaints that the Japanese pencils were of inferior quality, and there seems to have been some sound basis for concern about how the pencils were represented. In Argentina, for example, an importer of Japanese pencils refused to pay for a large shipment when the pencils were found to contain less than a half inch of lead, with the rest solid wood. The importer claimed that the pencils were not like the sample he had been shown, but in defense the Japanese manufacturer cut some samples open to show that they indeed contained mostly wood. The court ordered the importer to pay.

Acting in accordance with provisions of the National Industrial Recovery Act, the U.S. Tariff Commission recommended in a 1934 report to the President that pencils valued by the importer at less than $1.50 per gross should have duty rates applied on the American selling price of the competitive domestic pencils. No direct action was taken, however, for within months of the report an informal agreement was reached between the State Department and Japanese interests whereby they would limit exports to the United States to eighteen million per year. American manufacturers of other goods threatened by Japanese imports, such as cotton rugs and matches, joined pencil makers in sharply criticizing the agreement, arguing that it was out of all proportion to normal imports, and feared it to be a bad precedent.

The Tariff Commission's convincing if not fully successful case against Japan was greatly facilitated by the considerable information supplied by the Lead Pencil Institute, which was organized in 1929 to collect and disseminate data on production and distribution. In 1933 the institute had ten members, who together manufactured 90 percent of all American pencils. Thus the institute effectively represented the entire industry, which at that time included thirteen firms. They were, roughly in order of size:

Eagle Pencil Company, New York, N.Y.
Eberhard Faber Pencil Company, Brooklyn, N.Y.
American Lead Pencil Company, Hoboken, N.J.
Joseph Dixon Crucible Company, Jersey City, N.J.
Wallace Pencil Company, Brentwood, Mo.
General Pencil Company, Jersey City, N.J.
Musgrave Pencil Company, Shelbyville, Tenn.
Red Cedar Pencil Company, Lewisburg, Tenn.
Mohican Pencil Company, Philadelphia, Pa.
Blaisdell Pencil Company, Philadelphia, Pa.
Richard Best Pencil Company, Irvington, N.J.
Empire Pencil Company, New York, N.Y.
National Pencil Company, Shelbyville, Tenn.

By 1934 the Big Four accounted for no more than 75 percent of domestic pencil production, and the entire industry was feeling the combined effects of foreign competition, increased wage rates and other production costs, and lower demand due not only to the Depression but also to the increased use of recording machinery, mechanical pencils, and fountain pens in business. While the Big Four were not as threatened by Japanese pencils as were the smaller domestic producers, about half of whose business was in pencils retailing for less than five cents each, all were concerned.

One way of alleviating some of the pressures on the industry as a whole was to reduce the staggering number of different styles and finishes in which pencils were being manufactured. Variety added to costs by requiring adjustments to machinery and increased supplies of raw materials and stocks of finished products. The Department of Commerce had formalized a procedure whereby Simplified Practice Recommendations could be developed in conjunction with producers, dealers, and consumers of a certain type of product so that there could be some agreed-upon standards of quality within which fair competition could operate. In 1934 the Bureau of Standards

issued what was essentially a draft of "Simplified Practice Recommendation R151-34 for Wood Cased Lead Pencils." Distinguishing characteristics of pencils with erasers were stated as
follows, in order of increasing quality: (1) natural finish, inserted eraser; (2) nickel tip and white eraser, in loose bulk
packaging; (3) short gilt tip and red eraser; (4) long and decorated gilt tip and red eraser. The last category included the
nationally advertised brands of yellow pencils, and the recommendation was to limit the use of those characteristics, including the yellow color, to high-quality five-cent pencils.
Elaborate and complicated groups, classes, and types of pencils were defined in terms of how they were packed and displayed for sale, whether or not they had erasers, and what
kinds of distinguishing characteristics they had. Standard
lengths and diameters for wood cases, ferrules, erasers, and
leads were specified, as well as what kinds of lead degree,
shape, finish, color, and stamping were allowed in each
category.

The Big Four would naturally be pleased to have their yellow five-cent pencils protected by such an agreement, but
some of the smaller pencil manufacturers might have found
it restrictive. For example, in early discussions of the code,
the elimination of the six-inch-long pencil was proposed,
making the seven-inch standard. However, the six-inch
pencil had provided a use for slats and even finished longer
pencils whose ends were imperfect. By cutting off the bad
inch, the smaller producer could salvage something. It appears
to have been the potentially restrictive nature of the code, in
conjunction with price increases, that set off increased Japanese imports in 1933.

The Simplified Practice Recommendation for pencils seems
never to have gotten more formal than the mimeographed
draft. Although it was indexed in the government's monthly
catalogue in anticipation of its being printed, notices to readers
of volumes of Simplified Practice Recommendations explained
that the only one missing, R151-34, on lead pencils, had "not
reached a point of sufficient stability to warrant issuance in
printed form." The chief of the Division of Simplified Practice
of the National Bureau of Standards had stated in his September 28, 1934, memorandum, transmitting what he had no
doubt hoped would be the final draft of R151-34, that "if you
find that the final changes and corrections . . . affect your acceptance, please advise us." Apparently the division was so
advised, for as of May 15, 1937, the recommendation was still

not available in printed form from the Superintendent of Documents, but "in mimeographed form only, from the Division of Simplified Practice."

Any steps toward standardization were suddenly interrupted in 1938 when the Federal Trade Commission charged the thirteen pencil makers who then accounted for 90 percent of domestic production with price fixing. The commission found that in 1937 the Lead Pencil Institute was reorganized into the Lead Pencil Association to end a two-year price war, and the new association made it possible to maintain uniform prices and standardize terms and conditions of sale, thereby restraining competition. The manufacturers were ordered to cease consulting with each other for the purposes of developing a standardization program whose principal effect would be to limit the variety of products. The successor Pencil Makers Association came to be identified as "a clearing house for trade names [with] no legal authority to prevent the use of names on its register." But while a domestic association could not fix prices or share production figures in ways that violated the antitrust laws, an exemption from those laws existed for an association formed solely for the purposes of export trade. Thus in 1939 the American, Eberhard Faber, and Eagle pencil companies filed papers to join together in the Pencil Industry Export Association.

The Depression had caused other complications for an already troubled pencil industry. Reductions in wages and working hours led to strikes, with the Eagle Pencil Company being struck in 1930, 1934, and 1938. The last strike resulted after reduced demand and increased competition had prompted Eagle to cut working hours and eventually to propose cutting the hourly wage. At the time of the strike the factory was operating only twenty-four hours per week. The strike was an especially visible one, for the Eagle plant occupied the block between Thirteenth and Fourteenth streets and between Avenues C and D, an area of Manhattan "heavily populated with labor adherents." Although the pencil company and a subsidiary, the Niagara Box Factory, together employed only nine hundred workers, the picket lines drew crowds of sympathizers numbering in the thousands. There were clashes with police, nonstrikers were pulled from their machines, and bricks and eggs were thrown. After seven weeks, the strike ended in August 1938, with an agreement whereby the strikers would be reinstated and employees hired during the strike would be dismissed. In a settlement a year earlier the American Lead Pencil Company had become the first in the industry to adopt a closed shop, with an agreement that had included wage increases and guaranteed a five-day, forty-hour week. By 1942 about half the workers in the broader pen-and-pencil-manufacturing industry were unionized.

The American pencil industry was still confronting the economic ups and downs of the 1930s when a new factor came into play. World War II had cut off supplies of the best graphite, that from Madagascar and Ceylon, and inferior kinds from Mexico, Canada, and New York had to be used. The best clay, from Germany and England, had to be replaced with supplies from South America. And wax from Japan had to be replaced with domestic substitutes. But Pearl Harbor created one of the most immediate of the war's effects, for on December 8, 1941, the Mikado pencil was renamed Mirado to disassociate it from its Asian connotations.

In anticipation of shortages, some pencil companies stockpiled materials. The American affiliate of L. & C. Hardtmuth was charged in 1942 with having imported large numbers of pencil leads from Germany and Czechoslovakia before the shipping lanes were closed and then misrepresenting their pencils as wholly American-made. Other companies may have been anticipating guilt by association when they emphasized the importance of their pencils for the war effort. In full-page advertisements for its Winner Techno-tone drawing pencils, A. W. Faber, Inc., presented wartime vignettes. In one, a new battleship was being launched:

Remote from the champagne christening stood a man with a pencil behind his ear.

Suddenly a mighty cheer and the sleek new battle wagon slid down the ways . . . but the man wasn't listening. With trained engineer's eyes he watched every detail of the short journey, making rapid pencil notations and sketches . . .

Back in the drafting room, many men and many pencils elaborated these sketches into drawings and blueprints . . . blueprints for mightier ships, for improving ship ways, blueprints for Victory.

Many ships, planes, tanks, and guns begin with A. W. Faber WINNER Techno-TONE drawing pencils.

Advertising increased during the war also because of record business. Over one and a quarter billion pencils were produced in the United States in 1942. However, critical materials shortages soon led to a ban on the use of rubber or any kind of metal for pencils. Manufacturers had generally anticipated this restriction, and the American Lead Pencil Company had begun experimenting with plastic ferrules in 1940. By 1944 the company was displaying them on its Venus and Velvet pencils, and was planning to recommend them to postwar customers. Plastic ferrules were on the market as early as 1942, and erasers made out of rubber substitutes would also be held in paper or cardboard ferrules. Dixon's wartime use of yellow-banded green plastic ferrules on its Ticonderogas introduced the now-familiar color scheme for the pencils.

Dixon Ticonderogas from the early 1940s, equipped with typewriter erasers and a point protector, and marked by two yellow bands around a gilt ferrule

In early 1943 the War Production Board limited output of wood-cased pencils to 88 percent of the 1941 level, and it estimated that only two-thirds of the 1939 consumption of wood-cased pencils would be required to meet essential civilian needs in the event of total war. Since neither the domestic wood nor the inferior graphite being used to make pencils was considered scarce at the time, the restrictions were believed to have as their object the conservation of transportation and manpower in raw materials production. The mechanized factories used only about three thousand unskilled workers, three-fourths of whom were women, and so pencil making itself could not have had a significant impact on the wartime labor force.

In wartime England, the manufacture, supply, and price of pencils were tightly controlled. The Board of Trade ordered that, as of June 1, 1942, pencils were to be made in a limited number of degrees and were not to have polished finishes. This was not an altogether bad development, according to *The Economist,* which evidently felt that pencils had gotten out of hand anyway:

> Few will regret the passing of the fancy pencil of "novelty" shape and colour; the advertisers' gift pencil of low quality lead and high quality paint and polish will be a small loss even to a drab war time world. . . . The more recent development of the pencil cannot be regarded as progress in either a utilitarian or an aesthetic sense. The pencil with the rubber top which always fell off and was lost, the inefficient, propelling pencil for which refills were often not obtainable and the empty cases of which clutter up all our desks, the bridge pencil, the point of which seldom survived the rubber, will all pass unwept. If the raw materials of pencils are to be limited, the needs of the accredited pencil users must be safeguarded. The draughtsman must have his hard pencil to draw the lines of infinitesimal marking. The staff officer must have the coloured pencils to mark his map in such a way that he knows at a glance the course of the battle. The censorial and editorial blue pencil must not be crowded off the market by the free distribution pencil advertising Buggin's Beer. If the interest of legitimate pencil users is fully safeguarded the control of pencils is both timely and welcome.

When the war ended and pencils became unregulated, the frivolities that *The Economist* lamented returned to the free marketplace. Pencil making, like everything else after World War II, was to use more plastic, and science, technology, and engineering were to play increasingly important roles.

20/ Acknowledging Technology

The shortages of pencils created by World War II were not expected to be alleviated immediately after hostilities ended. In order to supply the government, the armed forces, and the war industries with their increased demands in the wake of Pearl Harbor, American and British manufacturers had to cut back on deliveries for civilian uses and had to curtail production of more frivolous styles of pencils. Existing stocks had to be allocated and distributed in ways that appeared equitable, so that manufacturers would not lose the loyalty of the dealers and distributors who would have to be relied upon again once peace returned.

Since it was expected to be years before Germany and Japan could resume their prewar levels of pencil production, countries that had depended on imported pencils had to look to other sources. The Netherlands, for example, which used about thirty-five million pencils per year in the 1930s, had never had a domestic pencil industry and had relied upon Axis countries for over 70 percent of its pencil imports. After the war a pencil industry was organized in Holland not only to supply domestic needs but also to make pencils for export. Raw materials were to come largely from within Holland, and Ceylon graphite was to be obtained from Britain, which at the time held large stocks. The latest woodworking machinery was to be bought from the United States.

Less legitimate ways of dealing with pencil supplies were employed elsewhere. As late as 1949 an army corporal from Thief River Falls, Minnesota, was sentenced to six months at hard labor for attempting to smuggle more than thirty thousand dollars' worth of German pencils into France using an army truck. In 1951 the Federal Trade Commission ordered Atomic Products of New York to stop selling mechanical pencils without disclosing that they were made in Japan. It was years after the war before the British could again purchase all the pencils they had once enjoyed, no matter where they were manufactured. While in 1942 *The Economist* had been happy with the disappearance of novelty pencils, *The Times* was equally happy with the return of variety in late September 1949: "This week pencils of all kinds, colours, shapes, and sizes will be back in the shops, and one of the minor delights of life will be restored to us again."

The return of pencils to the free marketplace was not taken for granted by the manufacturers, at least in America. Although the war years had brought more orders than could be filled, pencil companies "with an eye to the competitive future" continued to place advertisements in magazines, sometimes even stepping up and intensifying their promotional campaigns. In 1945 the Eagle Pencil Company introduced Ernest Eagle, a cartoon spokesman, whose function was to tie together the company name and the products it manufactured. Marketing pencils was traditionally believed to be especially difficult because "pencils have always been sort of unromantic, inanimate objects about which it was difficult to get anyone very much excited."

Such a view was confirmed as early as 1927, after Eagle had studied selling patterns in some large cities. According to the findings, most people would ask for a hard, soft, or medium pencil in a stationery store and be satisfied with any of a number of different brands offered at a fair price. Eagle wanted the customer to ask for an Eagle pencil, and set out to develop a campaign to effect that. The German pencil maker's tradition of "featuring its factory, years of experience, or the general background of egotism" was rejected. Eagle felt that the public was interested in itself, not in the patriarch of a fourth-generation pencil empire, and the subject of the new advertising campaign was to be the pencil user and the pencil user's unconscious mind.

The widespread habit of scribbling and doodling was focused on as something both inherently interesting and associated with pencils. Handwriting analysis was experiencing a vogue at the time, and Eagle engaged a graphologist to analyze pencil scribblings. Advertising copy was designed to call attention to patterns in scribbling and to link the habit with Eagle's Mikado pencils. For ten cents plus the head of the Mikado that appeared on each box of one dozen pencils, anyone could send in a sample of scribbling and receive in return the graphologist's personalized analysis. The reply would also contain enclosures advertising Eagle's entire line of stationery products.

During the Depression, it had taken more than frivolities to sell pencils. Eagle's new advertising manager, Abraham Berwald, had admitted on taking the job that he knew the pencil mainly as a tool used for jotting down an "O.K." on things that crossed his desk. All pencils and pencil advertising looked pretty much the same to him, and he was not at all sure that there was anything about a pencil that he could advertise successfully. Eagle's president, Edwin Berolzheimer, hired Berwald anyway, inviting him to take his time coming up with a new idea, "mosey around the factory, pry into this and that and see what you can find—keep thinking about what a pencil needs to justify its being advertised nationally."

In his moseying, Berwald became familiar with what was essentially the engineering department of the Eagle Pencil Company. Its chief was Isador Chesler, who had worked in Thomas Edison's laboratory, and whose job it was "to experiment with materials and methods, to develop new processes, to do things better." In other words, Chesler was working as an engineer. Since Berwald had found that pencil advertising was qualitative, reflecting the industry practice of testing by methods that relied on rules of thumb and experience, he asked Chesler if it would be possible to devise some quantitative testing procedures whereby claims of quality could be backed up by numbers.

While Chesler and other pencil engineers might not have needed anything beyond their quick scribblings on the back of an envelope to tell whether a batch of pencil leads was up to par, the challenge of quantifying exactly how good they were would have been a welcome one. A machine reminiscent of Edison's first phonograph was made to carry a large drum of paper upon which a Mikado pencil lead was pressed. As the drum revolved, a line was traced whose length could easily be determined by multiplying the drum's circumference by the number of revolutions it took to wear out the pencil lead. From then on consumers could be told not just that an Eagle pencil lasted a long time, but that when they bought a Mikado they were getting "thirty-five miles for a nickel."

Having quantified the amount of writing a pencil could do, Chesler was next asked, "How can we prove by actual test that Mikado pencils have stronger points than other pencils—that they will stand more writing pressure?" The engineer developed a device similar to a weighing scale, on which the pencil was pressed at a uniform writing angle until the point broke, the strength being read off the scale's dial. While the Mikado did well in the first quantitative tests against the competition, its superiority was not enough to advertise convincingly. This disappointing result naturally raised the question of whether Mikado leads could be made stronger. Since it was established that pencil points were weakened by poor adhesion of the lead to the wood and by the poor support offered by splitting wood, Chesler set out to improve binding and reduce splitting. This led to the development of chemical processes whereby the waxed lead is coated so that glue adheres better to it, and the wood is impregnated so that its fibers better resist being split. The resulting "Chemi-sealed" Mikado pencils were compared with various other five-cent pencils by an independent testing laboratory and found to be significantly stronger.

In the mid-1930s Eagle ran full-page magazine ads with the headline "Buy Pencils on Facts." Twenty years later Abraham Berwald was still with Eagle, signing his letters in pencil. One such letter came to the attention of a writer for *The New Yorker,* who wanted to find out "whether it is the policy of the outfit to transact all its written affairs in this informal, if in the circumstances quite suitable, fashion." Berwald said that he and a few other old-timers had stuck faithfully to pencils, but "the pups among the company's staff don't go in much for this display of loyalty." He told the interviewer that when Eagle first marketed indelible pencils in 1877, they were used routinely, until displaced by the typewriter and fountain pen, to write business letters and checks. Such checks were still good, Berwald insisted, as long as there was no specific prohibition against using pencil. He also pointed out that Eagle's vice president, Henry Berol, a descendant of the founder, but one who used the Americanized version of the family name, was the only person in the firm allowed to use magenta leads. Any memo or annotation in magenta was then unmistakably Berol's. Some pencil companies advertised the advantages of assigning a different color of lead pencil to each executive for the same reason.

Before the writer could leave the office, Berwald demonstrated that he had not lost his interest in or his flair for testing

Eagle products. He first described how old-fashioned colored leads were so brittle that they could hardly be sharpened without breaking. He then held up a handful of uncased carmine leads and announced that in the old days they would have "smashed into six or seven pieces" each if dropped on the floor. Berwald then threw one of the new leads on the floor, to show that it would not break. When Berwald finally began to throw leads all about his office, exclaiming that "we have made colored leads flexible," the visitor moved to the door.

But Berwald wanted to demonstrate one last thing. Since Eagle had been advertising that its Turquoise drawing pencils could be sharpened to a needle point, the old skeptic wanted to convey exactly what that meant. He called in a young associate, who proceeded to wind up a phonograph and insert a finely sharpened Turquoise pencil point into the playing arm. Soon a "scratchy but stirring" rendition of "The Star-Spangled Banner" brought Berwald to his feet, and after a respectful silence when the music ended, he announced that the visitor was the first witness to the demonstration of what it meant to have a needle point.

By 1953 American consumption had reached nearly 1.3 billion pencils annually. There were twenty-three firms, but the Big Four still dominated and were the only companies in the industry making everything that goes into a pencil—the leads, slats, ferrules, and erasers. But that was apparently not all that the giant pencil companies had in common. In early 1954 the government filed suit charging the Big Four with violating the Sherman Antitrust Law. The suit charged that at least as early as 1949 they had been conspiring to fix prices, rig bids, and allocate sales of pencils to local government agencies and large industrial users. At the time of the suit the Big Four had annual sales in excess of $15 million and accounted for 50 percent of domestic and about 75 percent of export sales of pencils. The companies entered pleas of no defense and were fined $5,000 each. A consent decree, whereby they agreed to abstain from the illegal practices, was also accepted by the court.

Among the results of price fixing appears to have been the increase of the price of the "five-cent pencil" to six cents, its price in 1953. Rising costs could no longer be absorbed without cutting quality, according to Eberhard Faber, which soon raised to seven cents the price of its Mongol, which was being advertised as "the first well-known brand of yellow pencil." In the mid-1950s the Mongol was promoted as "America's standard of quality—and today's Mongol is the smoothest-writing, blackest-writing, longest-wearing pencil you can buy." Since quality had traditionally been the theme of Eberhard Faber's advertising, the company stated that it did not want to resort to a hard-sell campaign to maintain its share of the market as costs continued to rise.

Although Faber was not selling directly to consumers, having made a decision in 1932 to sell exclusively through distributors, it did not feel that it could afford not to communicate directly with the individual pencil buyer. In 1956 Faber launched a major advertising campaign, in which the price of Mongols was listed as two for fifteen cents, an intermediate step to the ten cents each that was considered inevitable. In two-page, four-color advertisements, which were unprecedented for the industry, the Mongol was declared to be the consumer's best buy at "2,162 words for one cent." A footnote explained the actual writing tests conducted by a testing laboratory, did the cost calculation, and concluded that "even *more* savings" would result from large lots. The twin-packs carried "Advertised in Life" stickers, and "people were buying two pencils instead of one because they were packaged that way." The ad writers seem to have learned the value of quantification, which Berwald had used so successfully for Eagle, but they used it only with regard to price. After comparisons of the old numberless kind—"writes blacker with less bearing down," "needs less sharpening"—they were most likely alluding to Eagle's claims with their double-entendre punch line: "The pencil with the strongest points in its favor is Mongol!"

Eberhard Faber was enjoying 15 to 20 percent of the total American pencil business, and it was in the process of moving from Brooklyn to Wilkes-Barre, Pennsylvania, where it was expected to have "the world's most modern pencil plant," putting out 750,000 pencils per day. If the sales force could sell all Faber could produce, gross revenues would reach $7 million in 1957. The eastern sales manager was optimistic, planning to tell his customers how a pencil was made, and the company's first non-Faber president, Louis M. Brown, saw the rising national economy as good news for the pencil business. He said, perhaps having the experience of Faber's most recent success in mind, that "tomorrow doesn't just happen, it is planned today. And the planning is first done with pencils."

Amid such growing competition, other pencil companies took less drastic steps, and put their efforts into redesigning their packaging, as Dixon did "to distinguish its brand and drive home its quality story." By 1957 the Big Four had been joined by a fifth large pencil producer, Empire, and once again imported pencils were threatening the American industry. In Washington the Lead Pencil Manufacturers Association was testifying before the House Ways and Means Committee against a proposed bill to extend the President's authority to cut tariffs by as much as 50 percent. Japan was cited as the principal threat, having recovered its prewar industrial strength, aided in part by the 50 percent tariff reduction that had been in effect since 1945. From a cottage industry at the beginning of the twentieth century, the Japanese pencil industry had grown into a world competitor.

Early in the century there were also small pencil factories in India and there was some hope by the government that the industry could give rise to a "stream of prosperity." However, according to one contemporary native observer, it was no simple matter to expand a fledgling cottage industry into a major industry to make pencils in large quantities:

> Wood, graphite and clay are its main raw materials and to a superficial observer, it looks as if we had plenty of these in India. In a way, we have, in a way, we have not. Wood of an inferior quality we have in abundance; for a superior quality, we must take resort to plantations or import. As to graphite, we have any number of mines in Madras, Travancore and Ceylon; but the refining has got to be done in foreign countries, because the industrial uses of graphite are unknown in India. Clay, we have in any amount, but it requires some expert knowledge to find out the right sort. I know of at least two factories that began work without making any distinction between the right sort of clay and the wrong one and had to spend thousands of rupees in late experimenting. A thorough preliminary inquiry has got to be made regarding wood, graphite and clay, if the pencil industry is to be successful in India.

Such a program of research was to begin in the 1940s but was interrupted by the war, and thus India's dependence on imported pencils continued. In 1946 the United States was exporting $4 million worth of black-lead pencils, and 1947 promised to see increases of 50 percent or more. At the time, the Philippines, Hong Kong, and India were the principal outlets for the American products, with India by far the most important market for all kinds of writing instruments. The Controller of Printing and Stationery asked the National Physical Laboratory in New Delhi to draw up purchase specifications, since none were readily available, even from the countries of foreign manufacturers. India was well aware of its large consumption of foreign goods, and from 1946 to 1948 alone the country imported enough pencils to last for forty-five years as based upon earlier rates of use.

While the Indian investigators recognized the wide variation in raw materials, special techniques, and secret processes employed in the manufacture of pencils, they argued that the purchaser was really interested only in how well the finished pencil worked. Among the characteristics that were considered important were quality of writing, reliability of grading, wear of the lead, resistance of the mark to chemical reactions over time, and the straightness and whittling quality of the wood.

A search for indigenous woods for pencil making had been going on in India well before 1920; by 1945 eighty different woods had been identified as promising, and they were tried by various pencil factories. First-quality Indian pencils, like their English and German counterparts, had to rely on foreign woods such as American and East African cedar. As of the mid-1940s, the Forest Research Institute at Dehra Dun had declared truly suitable only one Indian wood, a juniper found in Baluchistan, but the remoteness of the source and the high occurrence of knots, dry rot, and slanted grain in the timber did not make it an economical prospect. Furthermore, the trees grew very slowly and very crooked.

Of eighteen other woods identified by 1945 as suitable for second-grade pencils, deodar was considered "moderately suitable, but costly." However, by the early 1950s, when India was consuming almost three-quarters of a billion pencils annually, deodar was reconsidered and came to be the preferred alternative to foreign woods. Further research on the seasoning of deodar had proven that its light yellow color, "which the public is not accustomed to in pencils," could be inexpensively changed to "a pleasant violet colour" by immersing the pencil slats in dilute nitric acid, a treatment that incidentally also improved the cutting quality of the wood. Thus the Forest Research Institute finally declared in 1953 that "not only is deodar satisfactory for making first class pencils, but it is also superior to the East African cedar, a timber on which the Indian pencil industry is largely dependent." The latter wood had also shown a marked tendency to warp in the humid Indian climate, and the two halves of wooden pencil cases tended to separate because of the incompatible warping of the mating slats.

While the wood problem was being so thoroughly considered by the Forest Research Institute, the investigators at the Physical Laboratory concentrated on what the wood enclosed. Although wartime conditions made it difficult to obtain all grades of pencils from all the different manufacturers, some testing of leads commenced in the mid-1940s. Among the first things to be measured was the electrical resistance of pencil leads. Such measurements could be made on untipped pencils without doing any damage to the pencil itself. Since graphite is a good conductor, the resistance of a pencil lead could be measured by inserting it in an electrical circuit containing an ohmmeter. Leads separated within the wooden case could be identified by a broken circuit, and even partially broken leads could be detected by excessive resistance. Quality control and consistency of grading could be correlated with the resistance of unbroken leads.

The strength of an uncased pencil lead, or slip, was tested in an apparatus looking much like the scales found in a doctor's office. The slip was placed in a frame that supported it at two points about a half inch apart, and the lever arm bore down on the slip's midpoint. This configuration is known technically as three-point bending, and as weight was added further out on the lever arm, the pencil slip was bent in much the same way it would be during writing. As expected, the strength of pencil leads increased as the hardness increased, a necessary quality, since harder leads tend to be pressed more severely when writing or drawing.

Other qualities considered important to the Indians were the blackness, wear, and writing friction of pencil slips. To quantify blackness, the mechanism of a traveling microscope was modified to enable closely spaced parallel lines to be drawn under constant pressure. The paper, blackened with a fixed number of lines, was then inserted in a box equipped to measure the amount of light reflected by the pencil marks into a photocell. A galvanometer reading could be correlated to the grade of pencil, and uneven blackness in a series of lines from the same pencil could indicate a poor mixing of graphite and clay.

The wearing quality of a slip was measured by drawing lines in the same microscope-like apparatus, with the drawing paper replaced with sandpaper. This accelerated the rate at which the slip was shortened, and the number of millimeters of lead required to draw a certain distance on the sandpaper could be measured. Such a wear measurement would be an indication of how long a pencil would last. The expected result, that softer pencils wear down faster, could be quantified. The friction between a pencil slip and a piece of paper was measured on an apparatus consisting of a paper-covered trolley pulled at a steady rate under a weighted pencil slip, with the amount of force needed to pull the trolley being measured. The Indians argued that the greater the friction, the greater "the strain experienced when a number of pages have to be written in pencil at a stretch."

After approval by its Engineering Division Council, the Indian Standards Institution in 1959 issued its "Specification for Black Lead Pencils," which was mainly to aid the development of a relatively young and indigenous pencil industry. The committee responsible for the standard included engineers and scientists from the National Physical Laboratory and the Forest Research Institute, as well as pencil manufacturers and users. The document attempted to reduce the number of pencil grades by eliminating the 5B, 3B, B, H, 3H, and 5H grades of drawing pencils, arguing that pencils graded close together, like 5H and 6H, often overlap in hardness anyway. The grades of general writing pencils were recommended to be "Hard, Soft, and Middling." Tests for uniformity, strength, wear, friction, and blackness were incorporated into the standard as recommendations to manufacturers, whom the Standards Institution hoped "would equip themselves with the necessary testing facilities before too long," warning that otherwise the recommendations might become government requirements. While woods for pencils are not specified in the standard, four are listed as comparable to American cedar—deodar, cypress, juniper, and a Nepalese alder. Four additional woods comparable to African cedar are also listed.

The committee that drafted the Indian standard acknowledged consulting the pencil standards of Russia, Japan, and the United States, perhaps with a view to the potential for exporting pencils to those countries. While the Indian standard may have derived some guidance from the foreign standards, it is much more explicit and detailed than most, especially with regard to quantitative testing methods. It benefits little from the American standard, for example, which is largely a guide to writing government purchasing specifications based on pencil style, size, and grade.

It should come as no surprise that a country newer to pencil making like India should have more technically explicit standards than the older pencil-producing countries like England, Germany, and the United States. In these latter, the major companies of the competitive industry developed their own scientific and engineering techniques for testing and controlling their products because such practices were good, if not necessary, for business. What the Indian government laboratories began to do in the 1940s was what the largest American pencil company was already doing.

We know that Eagle developed tests to quantify the wearability and strength of its pencils because it was an advertising decision to go public with those tests. But the rules of thumb and experience that Abraham Berwald found not sufficiently quantified for his promotional purposes were not necessarily less useful means of quality control for the Eagle factories. It was only when Berwald wanted to compare a Mikado with a Mongol, for example, that he needed standardized quantitative measurements rather than the individually self-consistent but collectively incomparable rules of thumb of separate manufacturers. One reporter's account in 1949 of activities at the Eagle Pencil Company showed the prominent role of testing:

Looking in on Eagle's twenty-man research laboratory, one sees a device not unlike that which is used for drilling oil wells. Nearby stand a pressure scale, a mileage meter, a reflectometer and a bowing machine. Some of the "old hands" would have no truck with such "gimmicks," but the competitive advantages that accrue from the use of these machines have repaid the management's costly (but undisclosed) investment in them. The "well drilling" device turned out to be a fourteen-foot-high structure that housed a giant pendulum weighted with a 540-pound bob. Watching its 49,920 oscillations from a single impulse, a technician explained that when the point of a pencil is pressed against a sheet of paper on a platen attachment to the pendulum shaft, the friction of the lead slows and finally stops the motion. Purpose: measuring the smoothness of the lead; the smoother the lead, the longer the swing of the pendulum. It eliminates all guesswork in determining the relative writing smoothness of different lead formulae. On the pressure scale, with the pressure sometimes running as high as five pounds, a pencil is pressed to test the breaking point. The bowing machine bends lead to the breaking point. Now the once-brittle compound has been made so flexible that it will bounce on the floor without shattering and take a point in any sharpener without breaking.

With "thirty miles for a nickel" [sic] as its slogan, the company proved twenty years ago that the Mirado pencil would draw a line thirty-five miles long. Although the slogan is unchanged, laboratory tests show that the pencil will draw a line seventy miles long.

While this rare glimpse behind the scenes in a modern pencil factory was described as a look at "research activities," it is in fact a report on the testing procedures used to back up this one manufacturer's claims. Smaller pencil companies that survive making cheap pencils and that have no aspirations for their products to compete with a quality pencil may not care very much about the wear or friction qualities of their leads, but it is unlikely that there is anything in the foreign standards or the research that led up to them that the larger firms in the older pencil-producing countries did not know long ago. The example of India's crash research program in pencil standardization is the technological equivalent of the biological dictum that ontogeny recapitulates phylogeny.

In keeping with their looking at an engineered artifact in an engineering-scientific age, the Indians published many engineering research papers on pencils independent of the standard itself. In 1958, for example, a paper evaluating clays noted that some poorer pencils scratch the paper because abnormally large clay particles are at the tip of the lead, and thus the clay must be finely divided before being used. The paper reports on the chemical analysis of indigenous and imported clays and several of their physical characteristics, especially those relating to particle size. While the Thoreaus and the Joseph Dixon Crucible Company, for example, may not have quantified their conclusions the way the Indians did a century later, they started from the same premise and had the same objective. Another paper, published in the early 1960s by the Indians, considered the abrasion characteristics of clays, because certain clays caused rapid wear in the dies through which the slips were extruded. Other papers addressed the limitations of the strength test in the standard, because in "mending" a pencil with a knife the lead is struck rather than pressed gradually. This observation led to the development of an impact test and a piece of apparatus that is not unlike the huge pendulum that Eagle had used to test its own leads for smoothness, which is nothing but the absence of friction.

One of the great benefits of the scientific-engineering method is that it does provide a rational means of approaching problems, and solving them, in reasonably short times. What the Indians were able to accomplish collectively and openly in less than a decade had taken Western pencil manufacturers in an earlier and more closed age much longer to realize and develop. And while even as late as the 1920s Armand Hammer had encountered difficulties in trying to import pencil-making technology from Germany, in contemporary America it had been possible to master lead making from a few hints. In the wake of World War II the secrets of pencil making could no more be exclusively held by anyone than could those of the atomic bomb. Indeed, it may well have been the Manhattan Project itself that has provided the paradigm for research and development to all countries with the talent and determination to master a technology, whether it be bomb or pencil making. But Conté's crash wartime program a century and a half earlier to develop a new pencil lead might also serve as a model.

In an advanced technological climate, the decision to start up a pencil factory does not require family or trade secrets, for these or substitutes for them can be bought or inferred by analysis. For those who do not have the resources or technology to compete from scratch in the pencil-making industry, there can still be opportunities. In the late 1960s, when unemployment on the reservation of the Blackfeet Indians was ranging from 40 to 70 percent, Chief Earl Old Person and the heads of other Montana tribes approached the Small Business Administration for help in setting up their own companies. The Blackfeet Indian Writing Company was formed in 1971 to put together wood-cased pencils largely by hand. By 1976 the company was making a profit, and by 1980 it was employing one hundred Blackfeet Indians assembling pencils and pens. By the mid-1980s the company had annual sales in excess of $5 million, and its smooth-writing and handsome natural-wood pencils had won many loyal users. But heavy competition, including that from Japanese and German imports, provided a constant reminder to the Blackfeet that an open technology creates challenges as well as opportunities.

21/ The Quest for Perfection

In his pencil company's 1892 catalogue, Eberhard Faber II stood behind his products with the following statement:

All goods coming from my factories I warrant to be of the very best material, of uniform quality, most carefully finished, and always full count. It is my aim to manufacture perfect goods only.

While it may have been Faber's aim to make perfect pencils, there is no doubt that he failed to achieve that goal. This is not to say that he was insincere in his warranty, for he very well may have believed that his company's best pencil was indeed made of the best graphite, clay, and wood available—without paying an unreasonable price. He may have believed that each and every best pencil was of the same quality—as far as inspectors could tell. He may have believed that the finish given every best pencil was as fine as a paint job could be—within reason. Finally he may indeed have believed that a package of a dozen pencils always contained twelve—and it very well may have. But the goal of "always full count" was really the only one that might have been humanly achievable without qualification.

Productivity and counting go hand in hand, and how pencils are counted depends upon who is counting. Salom Rizk, as a thirteen-year-old Syrian orphan, counted a stubby fraction of a pencil as a cherished possession. More fortunate pencil users tend to count them by full units and maybe even dozens.

Salesmen count pencils maybe by dozens, but hope to count by grosses, the preferred units of manufacturers. Pencil collectors count by the hundreds and the thousands and the tens of thousands. After the war, Camp Fire Girls counted them by the hundreds of thousands when they shipped pencils to the children of war-devastated Europe. An older Rizk, when he was living in America, asked for unused pencils by the millions to send to poor schoolchildren around the world, in a drive he called "Pencils for Democracy." And some countries now measure their outputs in billions, potentially providing several pencils for every man, woman, and child in the world to use for writing and figuring. The simple physical artifact multiplies the power of the individual.

When Ralph Waldo Emerson wished to describe the body as opposed to the mind of Thoreau, the essayist marveled over the "wonderful fitness" of the physical and mental abilities of the surveyor and pencil maker, who could "pace sixteen rods more accurately than another man could measure them with rod and chain." Emerson went on to give further examples of Thoreau's "most adapted and serviceable body": "He could estimate the measure of a tree very well by his eye; he could estimate the weight of a calf or a pig, like a dealer. From a box containing a bushel or more of loose pencils, he could take up with his hand fast enough just a dozen pencils at every grasp."

While this is no doubt hyperbole approaching hagiography, it certainly is also an implicit acknowledgment that Thoreau could measure, weigh, and count so quickly because he had a wealth of experience with the measure of trees, the weight of animals, and the count of pencils. Dexterity in counting pencils was an especially useful talent for a pencil maker, of course, and one of Horace Hosmer's memories attests to the fact that Thoreau's talent was not unique: "Emerson spoke of Thoreau's ability to pick up just 12 pencils at a time as something unusual. Tieing the pencils in doz bunches was commonly done by girls and women, and a fair days work was 1200 bunches. I should say that 1000 of these bunches would be picked up without counting and of the right number."

Elbert Hubbard, who also found this talent remarkable enough to devote to its description almost a full page out of the couple of dozen comprising his 1912 preachment on Joseph Dixon, remembers the accuracy achieved in a pencil factory to be even greater: "One of the needs for which a machine has never been invented is the picking up of twelve lead-

A bundle of one dozen Thoreau pencils in their original wrapper

pencils out of a mass at one motion. In the Dixon works the visitors are surprised and pleased to see scores of bright, healthy, active girls, who reach a hand into a box without looking, and pick out twelve pencils with one grab, ninety-nine times out of a hundred." According to Hubbard, "Joseph Dixon himself used to boast that he could do this."

The process of counting and packing pencils has fascinated many an observer, including the one who visited the Dixon factory in the late 1870s and saw the "counting board":

> This is merely a board, on which are fastened two strips of wood, about four inches apart, having in each strip one hundred and forty-four grooves. Catching up a handful of pencils, the workman rubs them along this board once and back, thus filling all the grooves,—the pencils lying in them as a pen lies in its rack on the ink-stand,—and he has counted a gross of twelve dozen without possibility of mistake, and in five seconds' time.

Another observer, this one at the Cumberland Pencil Company in Keswick in the late 1930s, described a means of counting that did not rely on any mechanical aid. After they were inspected, the pencils were counted in groups of three dozen:

> Taking a bundle of pencils in both hands the operative grasps a number of them in her left hand and behold— she holds exactly three dozen. She does it in a twinkling. I tried, but found it very difficult. It was done by holding the pencils in a hexagonal group, having sides, 5, 3, 5, 3, 5 and 3 pencils in length respectively. . . . Three sides of a group of this nature fit naturally into a medium-sized hand. An advantage of this quick method of counting is that hexagon pencils can be similarly grouped and counted.

This curious fascination with a hand becoming a head is yet another, albeit symbolic, manifestation of the importance of nonverbal thinking in dealing with the world of artifacts. Like a quick, rough calculation, it seems to rely on instinct honed by experience. But the ability to estimate quantities and sizes without explicitly or consciously counting or measuring seems to be something that can develop naturally in those who do not resist it. Thoreau was among other things a born quantifier and measurer, as his accounts in *Walden* demonstrate and to which his graduated walking stick attests, and probably neither he nor the other pencil bunchers made any conscious effort to fit their one hand to the correct number of pencils that would be wrapped in the label being readied in the other hand. Indeed, the very ability to count with their fingers would no doubt have freed their minds of the pencils and pencil wrappers to think of things less tedious than the necessary task at hand.

While Eberhard Faber's full count was thus an achievable promise, the most important aspect of his warranty was what was unsaid. Yet it was exactly what is implicit that can really sell a product. And what really sold pencils like Faber's was the belief that they were among the best that could be had for the price at the time. Those who wanted the top of the line paid top money for their pencils. And those who wanted a less expensive pencil could select an inferior model—which Faber could still claim to be the very best for that price. The claim of absolute perfection for all of his company's goods, the inferior as well as the superior pencils, was simply relative.

Today's Mongol, Velvet, Mirado, or other quality writing pencil is really beautifully made. It contains a strong piece of lead that takes a fine point and is smooth-writing. The wood is straight-grained and sharpens easily. The pencil is nicely finished with a bright paint job and crisp lettering. The ferrule is neatly decorated and holds the clean eraser straight and firmly. In short, the pencil appears perfect, and it is an object we can admire in the way we might admire a new automobile or a new bridge. But if the pencil does appear perfect, the way this year's new car does and the latest great bridge does, then why do these objects ever change? Why should there ever be new models or new designs?

While the exhaustion of the supply of some raw materials and the discovery of new supplies of others can affect not only their availability and cost but also their quality and efficacy, the ongoing and seemingly innate human endeavors of innovation and engineering make perfection a relative term and its goal a moving target. What real inventors and engineers really see in "perfect" artifacts is imperfection. Take any one of the "best" No. 2 pencils available today, for example. While it seems to have the kind of perfection Eberhard Faber II warranted a century ago, on closer inspection and reflection it also leaves some things to be desired.

The writing pencil I have in my hand is a top-of-the-line model from a major American pencil manufacturer. Some no doubt consider it the epitome of pencils, for it can be sharpened to a fine point that makes a uniformly dark line, which the pink eraser can remove without a trace. This pencil's smooth yellow finish and softly rounded hexagonal shape give it a classic look and a luxurious feel. The manufacturer is clearly serious about making this appear to be as fine a writing pencil as money can buy, for on the hexagonal face opposite the one in which the brand name is stamped in gold there is something blind stamped, or impressed, in fine sans-serif letters that can only be read if the light hits the pencil just right: "U.S.A. 'CHEMI-SEALED' QUALITY CONTROL NO 0407."

If I hold this pencil close up and twirl it in my fingers, however, I begin to see that even the most conscientious attempts to control quality must allow for a degree of variability. Just as highway lanes must be wider than our cars to give us some latitude to swerve a bit now and then and to allow us some margin for error when we are driving at sixty-five miles per hour, so quality control in a manufactured product must necessarily admit a range of acceptability and allow for some slight misalignments, albeit only in thousandths of an inch, in the high-speed machines that produce pencils. Thus quality control no more means that every pencil will be exactly the same than that every called strike in baseball will be perfectly centered over home plate. The concept of a strike zone means that a pitch need not be perfect to be a strike, and the reality is actually that not every pitch would be called the same way by different umpires or even by the same umpire in different innings.

The pencil I am holding reveals its blemishes when I look hard enough for them. The wood on one side of the point is rougher than on the other side, and as I twirl the pencil I can begin to make out the ever so fine line between the two halves of the wood case; there are also variations in color, texture, and grain. On this pencil a bit more wood and yellow paint have been removed from one side than the other, suggesting that possibly the lead is ever so slightly off center or that the pencil was not held quite straight or not rotated uniformly during its factory sharpening. At the other end of the pencil, the painted band around the ferrule is actually a bit sloppy and scratched and the eraser is ever so slightly cocked to one side. On the face of the pencil bearing the brand name, the grade designation 2½ is a bit large for the width of the flat surface and so curves around to the adjacent face. But this is admittedly nit-picking and there is certainly nothing about this particular pencil that should have caused a quality controller to reject it. If anyone were to argue that, then our perfectly common pencils would be scarce indeed or cost us dollars apiece.

It is not the quest for perfection in mere appearance and alignments alone that drives inventors and engineers. In making a new pencil or a new machine to make old pencils faster, trade-offs between appearance and economy must be made. But it is often the quest for perfection in performance that is pursued before anything else. If the lead in a pencil is so far off center that it is likely to break when bent by the centering action of a sharpener, then it is clearly inadmissible in a quality pencil. But if the lead is so little off center that it can only be noticed by the super-critical engineer with a magnifying glass, or the pernickety writer looking for Zen in a pencil, then it is no great imperfection. But if the pencil lead, whether centered or not, tears the paper or does not draw a uniformly black line, then that is something else.

Pencils, like automobiles and bridges, are meant to be more than just objects to be admired when they are new. The pencil is designed to be destroyed, its wood to be cut away, its lead to be used up, however slowly. And it is in the process of using rather than just looking at the pencil that its real faults and imperfections become clear. While the pencil has the advantage over the quill pen that there is no need to carry an ink bottle into which the point constantly has to be dipped, the wood-cased pencil does need to be sharpened occasionally. Furthermore, as the pencil is used up by being sharpened, the hand has constantly to readjust to the diminishing heft. While it would be nice if the pencil kept its appearance as it grew shorter (and to many a user the pencil still does seem to be an attractive object at almost any reasonable length), it is the more functional shortcomings that prompt a great deal of innovation and engineering.

Innovation springs from perceived failure. If the wood-cased pencil did not keep its point and if repointing the pencil altered its feel, then what better replacement for it than an "ever-pointed" pencil in a case that did not change its size as the lead was used? An 1827 advertisement explains why the "ever-pointed" pencil was an improvement over the old kind:

> The Black Lead is not inclosed in wood, as usual, but in a SMALL Silver Tube, to which there is attached a mechanical contrivance for propelling the Lead as it is worn.

The diameter of the Black Lead is so nicely proportioned as NOT TO REQUIRE EVER TO BE CUT OR POINTED, either for fine Writing, Outline, or Shading. The Cases for the Drawing Table or Writing Desk are of Ebony, Ivory, &c.; and for the Pocket, there are Silver or Gold Sliding Cases, varying in taste and elegance. The Black Lead is of the finest quality.

The large print in this copy emphasizes the drawbacks of the wood-cased pencil, and this is characteristic in describing innovation. While every early-nineteenth-century pencil user was aware of the constant need to cut and point the wood-cased pencil, that was hardly a reason to eschew its use as long as there was no better alternative. But when inventors wanted to explain why their invention was patentable, the most sensible thing to do was to point out, as in the advertisement, what imperfection in existing devices was eliminated in the new device.

This "new, improved" syndrome, which exists often implicitly in the development and often explicitly in the marketing of everything from breakfast cereals to suspension bridges, is at the heart of all innovation and engineering design. Whether the new pencil is cheaper (not expensive) or has smoother-writing (not scratchy), stronger lead (not easily broken) that is ever-pointed (no sharpening), its advantages are premised upon negating the old's disadvantages. But everyone may have gotten accustomed to the old product, hardly noticing any inconvenience in use or imperfection in manufacture, so its failures have to be emphasized.

This phenomenon occurs repeatedly in such common everyday products as toothpaste and soap, where a "new, improved" version of a familiar brand often appears on the shelf without much ado. Since a manufacturer seldom wants to negate its own superseded product, the new, improved version will not negate the old version so explicitly. Rather, the new, improved version will claim that it "makes teeth whiter" or "cleans better," but of course this means that the old toothpaste did not make teeth as white as the new and that the old soap did not clean as well as the new. It is only when a product is trying to displace a competitor that its shortcomings are broadcast explicitly. A curious dilemma arose recently when a breakfast cereal originally named Just Right, presumably because it had the perfect mix of ingredients, came out with a new recipe. Since it would have been a clear contradiction to have a "New,

Improved Just Right," the advertising campaign had to take a humorous approach and hope no consumers took its message too seriously or thought about it too carefully.

While there would seem to be benefits for all—manufacturer, seller, and buyer—in the development of "new, improved" products, there can be periods of difficult transition. When a new model of the Eversharp was introduced in the mid-1920s, dealers still had over a million of the old models on their shelves. Only a cleverly conceived promotional campaign aimed at the dealers convinced them to stock the new model without asking the manufacturer to allow them to return the old for credit. At the back of its 1940 catalogue, Dixon recorded its policy on innovation by announcing that the company "reserves the right to make improvements in its products without incurring obligations on goods sold previously."

Another common characteristic of innovation in products and designs, or at least in the marketing and promotion of them, is the need to educate the potential customer in how to use the often more complicated replacement. Who today can imagine having or needing instructions on how to sharpen or use a wood-cased pencil, for we seem to learn these things as children learn to speak. But was this so when cabinetmakers first enclosed plumbago in cedar cases? Was their purpose or use obvious? Manufacturers of the first mechanical pencils certainly seemed to take little for granted and published "directions for use" right in their advertisements: "Hold the two milled edges between the finger and thumb of the left hand. Turn the case with the other hand to the right, and the lead will be propelled as it is required for use; but if, in exhibiting the case, or accidentally, the lead should be propelled too far out, turn the case the reverse way, and press in the point; which of course in practical use, will seldom or ever be required."

While such instructions may be clear to a twentieth-century veteran mechanical pencil user, they may have been about as clear in the early nineteenth century as most computer user manuals are today. When read carefully, with close attention to detail and to grammar, the directions reveal their own imperfections. What finger and thumb? Turn the case how? And so forth. Nevertheless, the novelty of the mechanical pencil is made clear in these directions by the anticipation that its proud owner will be "exhibiting the case." But whatever its owner's intentions, it is much more likely that the use of the mechanical pencil would have been learned by a hands-on fiddling than by a close reading of instructions, much the way we learn to use personal computers more by trial and error than by reading manuals.

The operation of mechanical pencils was really much less complicated than the directions made it out to be, of course, and the advantages of such pencils, whether used by the exhibitionist or the sullen writer, were real. The almost precious and yet easily broken lead that soiled clothes and hands could now be retracted into a case when not in use or when carried in the pocket. Thus great numbers of mechanical pencils, sometimes termed automatic, propelling, and repeating pencils, appeared in the nineteenth century, perhaps reaching a peak, or nadir, in the late Victorian period, when pencils disguised as gold charms were being sold. The ever-pointed pencil seemed to be enjoying its heyday at a time when another and then more recent innovation, the bicycle, was being advertised for its own advantages as "an ever saddled horse which eats nothing."

While a pencil may not appear to consume much of anything but a writer's time, for the manufacturer one of its essential components—the wood case—consumed forests of red cedar. Thus the metal-cased pencil could also have advantages for the pencil maker, especially during times when supplies of wood for pencil cases were diminishing in quantity and rising in price. A related example may be the introduction of New Coke, perhaps motivated more by the supply-and-demand dynamics of the sugar used in the old Coca-Cola than by any lack of perfection in it. The phenomenon of New (Improved?) Coke being so poorly received by drinkers of old Coke even led to the unusual technological development of a presumably superseded product being brought back, while perhaps not exactly as it was, at least in a purportedly classic form.

The mechanical pencil, along with its relative, the metal pencil case into which a replaceable wood-cased pencil stub with a threaded-brass end could be inserted and slid out for use or secured for carrying in pocket or purse, fell into a period of less imaginative design in the twentieth century. Improved techniques of bonding wood to improved pencil leads that took and held a sharper point perhaps helped displace the metal pencil cases and mechanical pencils, whose rather thick leads were not suitable for fine work. The widespread use of reasonably priced and good wood-cased pencils in schools and offices

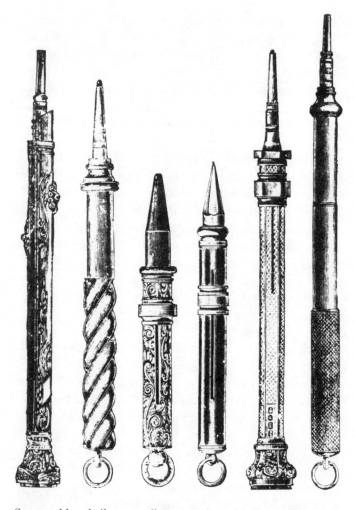

Some gold and silver pencil cases and mechanical pencils from around the turn of the century

equipped with efficient sharpeners no doubt also kept the mechanical pencil from taking over. Besides thick leads, older mechanical pencils also had other disadvantages: they required a relatively large capital investment and their mechanisms were prone to break, thus needing maintenance generally beyond the ability of the pencil user.

New fine-lead mechanical pencils, many of which come in inexpensive models that can be reloaded by inserting an entire cartridge full of new leads, have found a tremendous following among all kinds of writers, and their novelty is emphasized by the rather elaborate instructions for use and refilling that are contained on some of their packages. These pencils may be preferable for extended writing because they really never do need sharpening, because their lead seems to write much more smoothly (and quietly) than most wood-cased pencils, and because they have a constant weight and length. But the new automatic pencils are also rather delicate, and a heavy writer can wear out several of them in a year. Thus the choice of pencil is still not totally clear-cut, and for writers and editors who can have the luxury of freshly sharpened pencils of a preferred length standing at the ready on their desks each morning, almost the way rifles are stacked at the ready for soldiers, the old wood-cased model may indeed still be the weapon of choice over the plastic automatic.

The mechanical pencil has not been the only challenge to the "perfected" wood-cased writing instrument. The development of the fountain pen advanced throughout the nineteenth century, and Alonzo Cross, who founded his company in 1846, developed a "stylographic pen" in the late 1860s. This innovation with an ink-depositing needle point was widely copied and advertised by the end of the century. A ball-point pen was patented as early as 1888, but it did not become a practicable invention until the 1930s. The first ball-point pens sold in America cost $12.50 each in 1945, but they were prone to skip, leak, and blot, and so it was not until after a new ink was developed in 1950 that ball-point pens came into widespread use. But even when their price dropped the way that of pocket calculators has in more recent memory, the pencil was not displaced. According to one admirer, writing when the ball-point pen was still news:

> The wood pencil seems to be a tool that can't be put out of business. It has never been seriously threatened by the invention of fountain pens, mechanical pencils, ball-point pens or typewriters. Today's business executive . . . still has a row, stack or glassful of pencils within easy reach. One company president has three or four dozen freshly sharpened pencils placed in a jar on his desk each morning. When he uses one he pushes it aside—he may write a long memo or simply jot down a man's name, but he won't touch that pencil again until it reappears the next morning freshly sharpened.
>
> Another big wheel has an even dozen freshly sharpened pencils placed on his desk each morning. They are all new. He refuses to use a resharpened pencil, must have the full seven-and-a-half-inch length of a new one. His

daily discards go to vice-presidents and other lesser lights around the office. In contrast, there's the operating head of one distinguished old firm who insists on short pencils. He likes them about five inches long and his secretary must scurry around swapping new pencils for ones that have lost a couple of inches in trimming. Thomas Edison demanded them even shorter than that: three and a half inches long, so they would fit, lying down, in his right-hand vest pocket. He persuaded a pencil factory to turn out short pencils just for him.

The Eagle Pencil Company did indeed make a special pencil for Edison, but it was four and a half inches long, "vest-pocket high," according to another source. Yet even in spite of such inconsistencies in the lore, the anecdotal staying power of the pencil is supported by more objective evidence. While pencil manufacturers had some concerns about the ball-point pen, introducing competing "liquid graphite" products in the mid-1950s, wood-cased pencils were enjoying all-time record sales in the 1960s, and American output alone approached two billion pencils a year. Even today, pencil manufacturers see no end to the demand for the classic pencil. If no one else, then temperamental executives and writers like Ernest Hemingway, who reportedly got himself in the mood for writing by sharpening dozens of pencils and then, like Virginia Woolf and Lewis Carroll, sometimes wrote standing up, will never give up their quirks or their quotas. According to Hemingway, "wearing down seven number two pencils is a good day's work," and John Steinbeck declared that an electric sharpener was necessary lest he waste time pointing all the pencils he needed for his day's work.

But if writers write with pencils they seldom write about them as other than tools. There is a slight poem by Carl Sandburg entitled "Pencils," in which the poet assures us that "eager pencils" will stop opening and ending stories only when the stars come to a stop. While such a sentiment is no doubt reassuring to pencil manufacturers, as a poem on this subject Sandburg's is in a class by itself in the English language. Steinbeck and Hemingway seem to have paid more attention to the pencil than most writers. In *A Moveable Feast* Hemingway described writing in his notebook with a pencil in a Paris café and being distracted by a girl who came in and sat nearby: "I watched the girl whenever I looked up, or when I sharpened the pencil with a pencil sharpener with the shavings curling into the saucer under my drink." While the girl was apparently waiting for someone else, Hemingway could think that she belonged to him and his story, as "all Paris belongs to me and I belong to this notebook and this pencil." But Hemingway's is not a romance with the pencil. For all the author's fetishism we know nothing about the pencil's brand or color or size or grade or quality.

Perhaps no writer has ever admitted to thinking more about his writing instruments than Steinbeck, whose journal entries attest to his seeming obsession with pencils—their points, their shapes, and their sizes. In letters to his friend and editor, Pascal Covici, which Steinbeck composed while writing *East of Eden,* he said of his characters: "They can't move until I pick up a pencil." But which pencil he picked up depended on his mood and on the weather, for he recognized that a damp day affected the leads. At one time he confessed to blunting sixty pencils a day, and he regularly asked his editor to send more. Those the Nobel laureate seemed to like best were the Eberhard Faber Blackwing and Mongol "480 #2⅜ round," but neither was always right for his writing mood, and on one occasion he confessed:

> For years I have looked for the perfect pencil. I have found very good ones but never the perfect one. And all the time it was not the pencils but me. A pencil that is all right some days is no good another day. For example, yesterday, I used a [Blackwing] soft and fine and it floated over the paper just wonderfully. So this morning I try the same kind. And they crack on me. Points break and all hell is let loose. This is the day when I am stabbing the paper. So today I need a harder pencil at least for a while. I am using some [Mongols] that are numbered 2⅜. I have my plastic tray you know and in it three kinds of pencils for hard writing days and soft writing days. Only sometimes it changes in the middle of the day, but at least I am equipped for it. I have also some super soft pencils which I do not use very often because I must feel as delicate as a rose petal to use them.

The pen, on the other hand, is the more familiar symbol of the writer, and the writer is the pen personified. Even a bel-letrist like John Middleton Murry, who published a book of essays entitled *Pencillings,* did not write about the pencil. In a note to his book Murry even disowns its title, revealing that the one he had chosen was too long for the editor to fit in the space allotted to the newspaper column in which his "little essays" originally appeared. Murry further misleads the pencilophile with the title of the book's first essay, "Chiaroscuro." It is no paean to the pencil, but is in fact Murry's lamentation about the incomprehensible aspects of contemporary literature. As if it were not bad enough that there is no pencil in *Pencillings,* Murry reminds us that the pen is celebrated more mightily than the pencil, adding insult to injury by writing an essay extolling the pen. In "The Golden Pen," he waxes poetic about the writer's foil:

> The pen of my dream is a golden pen; it glides over a great sheet of white paper like crisp parchment; it is dipped into a crystal well of ink blacker than a raven's breast; and the lines it traces are as fine as those which Indian artists draw with an elephant's hair. And it seems to me that if all these things were mine, the thoughts of my brain would be as clean and fine and definite as they. An idea would rise before my mind like a bubble. I should only have to trace the outline. The bubble would break, the dust of its rainbow colouring would float down, settle on my ink before it was dry, and be imprisoned in it for ever.

There has been little like Murry's hyperbole published about the pencil. The gold pen, the crisp parchment, the crystal well, the black ink—the nouns enhanced with approval and pleasure—extend the sensations of using something of beauty and value. But for all its polished prose, which Murry may have achieved only after much rubbing with an eraser on a draft done in pencil, his celebration of the pen gives no hint of an appreciation of the technological achievement embodied in an artifact the way a more overt cultural achievement is embodied in the literary artifacts that were usually the objects of Murry's study.

The millennia of technological innovations that led through craft and engineering to the artifact that is Murry's dream pen are no less a part of our cultural heritage than the literary innovations that led through the world's literature to his prose poem. Yet the conjunction of cultures is rare. Even Thoreau, who with his very own hands made some of the finest pencils of his time, did not sing their praises. This, however, may have been a self-imposed silence, for he was not so reluctant to brag about his handiwork on the common cabin near Walden Pond.

Paeans to the pencil have been largely anonymous, and often corny. But one writer, at least, in personifying the object, had prose as purple and aspirations as grand as Murry's:

I am the pencil, the first chronicler of new-born thought. I come from the sleeping graphite beds, and the balsamic frills of kingly cedars. In my heart, I carry the black carbon of Pluto's world—half-brother to the diamond.

I memorandum the business of continents and strike the trial balance in the traffic of nations. I am the hub in the wheel of theory—the keystone in the structure of fact.

I note the doings of the world in the dizzy hours of the morn while presses wait like couchant beasts to fling my efforts to sleeping millions. I am man's best friend. I am his only confidant. . . .

I am the cosmopolitan, known in every . . . town and hamlet where the brain of man connives. I am the pencil and my mission is service.

The mission of engineering is also service to mankind, but perhaps one reason that engineering and technology are not conspicuously thought of as such active parts of our inherited culture is the very nature of innovation. While pens and pencils can evolve along parallel paths, and while wood-cased pencils and mechanical pencils can coexist, because writers as well as their instruments come in all shapes and sizes, only the most improved and perfected versions of artifacts seem likely to survive a given age. We no longer seem to want to use an early-nineteenth-century wood-cased pencil, even one containing the finest English graphite, assuming we were able to find such a thing. Our late-twentieth-century models, with their uniformly graded lead that does not break off in the pencil sharpener, are willy-nilly preferred. And our latest mechanical pencils are certainly better than the old heavy models with thick, brittle leads. No, we generally do not want to use old instruments when there are new, improved versions to serve us better. If we collect old pencils or pens it is often for sentimental reasons or for their appearance or curiosity or intrinsic value, made all the more so by their very scarcity.

It is different with literature and art, especially, for the new does not need to displace the old. James Joyce's *Ulysses,* although structured after Homer's *Odyssey,* does not supersede it. We still use Homer, in the sense that we read and enjoy him as many, many generations have. And in our greatest art museums centuries of paintings hang under the same roof—not as material collections of historical curiosities but as art collections of cultural achievement that can be no less moving to the twentieth-century viewer than they were to the painter's contemporaries.

What seems to distinguish the appreciation of art and literature from that of invention and engineering is the question of function. Whereas an artwork can succeed by conveying a sense of unity and by evoking an emotional or aesthetic response, an artifact of technology tends to be judged first by its functional success. A pencil must write, and a beautiful pencil that does not write loses something of its beauty as a pencil. Not only must an artifact have the unity appropriate for its physical functions, which usually makes it meaningful in the context of its own technological tradition, but a successful artifact will function in some way better than its predecessors. And in the world of artifacts, price is a curious component of function. So Thoreau's expensive pencils were successful in America, and Hammer's inexpensive ones in Russia.

A pencil that has a flaw, such as the early American ones that were filled with scratchy lead or the early Soviet ones that were imported and expensive, will be easily displaced by one with a smoother lead or one more easily bought. And the old, inferior pencils soon tend to disappear entirely, for they are not of much use or value, and they are certainly not considered objets d'art. Their flaws, whether realized or appreciated before, have been made all too obvious and inexcusable by the qualities of the new, improved pencils. In literature, however, even when the users of a work—its readers and critics—discover egregious flaws in its fabric, the published work is not expected to be revised or improved. No author would think of explicitly rewriting the novel of another, leaving alone the elements that succeeded and improving all the parts that the critics found to fail.

Even when a simple factual error is made in a work of art, the original is not necessarily revised to correct it. Thus John Keats's sonnet "On First Looking into Chapman's Homer," a poem in which the discovery of the Pacific Ocean is wrongly attributed to Hernando Cortez, remains in anthologies today as written and first published, even though the famous lines are incontrovertibly wrong historically:

> *Then felt I like some watcher of the skies*
> *When a new planet swims into his ken;*
> *Or like stout Cortez when with eagle eyes*
> *He star'd at the Pacific—and all his men*
> *Look'd at each other with a wild surmise—*
> *Silent, upon a peak in Darien.*

Simply changing "Cortez" to "Balboa" does not necessarily improve the poem the way replacing scratchy with smooth lead does a pencil. Since "Cortez" and "Balboa" have a different number of syllables, the scansion of the line, and hence the poem itself, would be altered. Furthermore, "stout Cortez" has a quality of sound in its place that fits the poem there, and the sound unity of "stout Cortez" ties it with "eagle eyes" at the end of that line to reinforce its subtle internal rhyme. In short, the line as Keats wrote it has so many interrelationships among its words that changing one could start a chain reaction that might destroy the whole poem. The poem as it stands has a metrical and evocative unity that makes it a classic that will not be displaced even if a modern-day Keats succeeds in crafting a superb sonnet that corrects the factual shortcoming.

Poetic license has a long tradition of well-deserved respect, and no engineer would call for equal rights for artifacts. If a pencil or a bridge is seriously flawed and the engineer discovers that, then it will be fixed or rejected. If the flaw is minor, it may remain in the original artifact, but when a new lot of pencils is manufactured or a similar bridge is designed for another location, the flaw should be corrected. If it is not, then a competing pencil company will make a better pencil or another bridge designer will design a better bridge, and eventually the new, improved artifact will displace the old. This is not to say that a product of engineering is any less of a whole than a poem, for changing one detail in an engineering design can also threaten the integrity of a whole machine or structure. Indeed, as television and the newspapers have recently reminded us, seemingly trivial changes supposedly made for the better and seemingly trivial details not given due attention can cause the collapse of an elevated walkway in a crowded hotel lobby or the explosion of a space shuttle in cold weather.

To recognize that truly improved artifacts displace the old

is also to recognize that the unity of an artifact of engineering is judged by its performance not only aesthetically and intellectually but also functionally and economically. None of this is to say that John Keats was any less of a perfectionist than Eberhard Faber. Both could say with integrity, "It is my aim to manufacture perfect goods only."

22/ Retrospect and Prospect

*I*n 1938, on the occasion of the addition of Konrad Gesner's book to the exhibit on the history of the recorded word, *The New York Times* editorialized about the evolution of the pencil since Gesner's first reference to one. The typewriter, it was feared, was driving out "writing with one's own hand," in both pen and pencil, which the editor clearly preferred, and he concluded with the concern that "libraries of a century or two hence may be searching for the last reference to pencils." Almost half a century later it was the computer that was going to be the end of the pencil, but that too is not likely to come to be. Pencils are being manufactured worldwide at the rate of about fourteen billion per year, and reports of the pencil's impending passing have been so greatly exaggerated that its staying power has come to be the subject of amusement.

To open a seminar on the importance of back-of-the-envelope calculations even in the age of the computer, a visiting professor of engineering showed a cartoon from an Australian magazine. In the background, a bunch of young students are sitting glumly before computer screens in a classroom, and the only student who is not so occupied is obviously enjoying himself by drawing with abandon at a desk without a screen. In the foreground, a very sad student is looking up at his teacher, who is saying, "I'm afraid you'll just have to wait until it's your turn to play with the pencil. . . ."

The pencil has been described as a generic word processor by Philip Schrodt, who calls the pencil point the character insertion subunit and the eraser the character deletion subunit. His clever parody first appeared in 1982 in *Byte* magazine, which purported to reproduce the "documentation" of the new product that incorporated "the essentials of a word processor in a sublimely simple form." After first congratulating the purchaser, the imagined manufacturer sings the praises of his product, while implicitly pointing out the shortcomings of competing products: "We are sure that you will find this word processor to be one of the most flexible and convenient on the market, as it combines high unit reliability with low operating costs and ease of maintenance." The documentation, in the manner of its genre, is not "sublimely simple," of course, and this is how a brand-new pencil is sharpened:

> To initialize a word-processing unit, carefully place the character insertion subunit into the left side of the initialization unit and rotate the word-processing unit approximately 2000 degrees clockwise while exerting moderate pressure on the word-processing unit in the direction of the initializer. Check for successful initialization by attempting a character insertion. If the insertion fails, repeat the initialization procedure. The word-processing unit will have to be reinitialized periodically; do this whenever necessary. (*Warning:* do not attempt to initialize the word-processing unit past its character deletion subunit. Doing so may damage both the word processor and the initializer.)

This parody has been followed by others, including an elaborate book-length pastiche, *The McWilliams II Word Processor Instruction Manual,* by the prolific author of serious personal-computer and word-processing manuals. A shorter, but no less clever, parody by Terry Porter, has reversed the order of things and described "The Pencil Revolution":

> Have you considered that the most important development of the age has been the introduction of the "personal" or "home" pencil, which has prompted a series of changes we should call the "pencil revolution"? . . .
> Once pencils became prevalent in schools, many conscientious parents concluded that home pencils should be purchased to ensure that their children would remain up to date. There was alienation and fierce controversy when it was learned that the pencil purchased for educational purposes ended up being used for games. Eventually,

interactive pencil games were invented, including Tic-Tac-Toe and Hangman.

While parodies and cartoons can be great fun, they often mask not a little truth. Modern engineering, epitomized by the personal computer, has brought us wonders beyond the wildest dreams of our parents. But to those who are not engineers, engineering appears also to be full of jargon and empty of fun. Furthermore, many of the latest products of engineering seem excessively complicated, difficult to get to work the first time, full of dangers to themselves and to ourselves that we have to be forewarned against, and accompanied by a great deal of tedium. And if these are the images of the personal computer, then what can the layperson think of less personal technologies like the generation of electricity, the processing of waste water, or the manufacture of steel? "Ho-hum"?

Engineering can be its own worst publicist, because the more it succeeds in making a product or a service that is reliable and efficient, the more engineering itself becomes virtually invisible and seemingly humdrum. Were physicians so successful in keeping people well and lawyers so successful in keeping them agreeable, those professions themselves might be less successful in maintaining their own status. It is our need of a doctor when we are ill and of a lawyer when we are wronged that gives them power over us. If the doctor cures us, the doctor is as a god. If the doctor fails us, God has taken us. And if the lawyer wins our case, the lawyer is a hero. If the lawyer fails us, we just did not have a good enough case. But the engineer is always dealing with us through engineered artifacts and systems. If they work they are taken for granted, almost as if given by nature; if they fail, not nature, but the engineer has failed us.

Engineering, like all professions, must necessarily contain esoteric knowledge. It is, in part, what defines a profession. But the aims, ideals, and, to a certain extent, the essential features of what makes a profession need not be inaccessible to the uninitiated. The Hippocratic oath is not a doctor's shibboleth, and the courtroom drama is not a secret ceremony. Every citizen knows these public characteristics of those professionals as they relate to people. But the engineer deals with things and with processing things. The artifact is almost always an intermediary between the professional engineer and the layperson. When an engineer does deal directly with people, it is often more as a businessperson. So if people are to understand the engineer as an element of society and as one bound to its culture, they must understand what it is that the engineer does, and in some general sense how it is done, even if it is generally done privately with a pencil at a drawing board.

Something so seemingly simple as the pencil itself, along with a consideration of its manufacture and use, provides one vehicle for conveying this understanding. The world of pencil making is a microcosm. Just as the pencil can be equated in parody with the latest high technology, so can the pencil and its history seriously instruct us by analogy and suggestion in the ways of engineers and engineering. The very commonness of the pencil, the characteristic of it that renders it all but invisible and seemingly valueless, is really the first feature of successful engineering. Good engineering blends into the environment, becomes a part of society and culture so naturally that a special effort is required to notice it. By looking closely at the origins and development of something so ubiquitous as the pencil, we are better able to appreciate the achievement of a great bridge or an efficient automobile. And we can do so without having or needing the detailed esoteric knowledge of the structural or automotive engineer. We can know that the bridge or the automobile was conceived first by a human mind and given its first embodiment as a concept in a human mind or in a sketch done by a human hand and not as a bunch of numbers given by equations in a computer. We can know that a natural gas supply system or a can of soda delivers energy or refreshment on demand without exploding in our faces because some engineers worried about how their designs might go wrong. But we can also know that these things are not perfect, because no artifact is perfect.

Understanding the development of the pencil, though it has spanned centuries, certainly helps us to understand also the development of even so sophisticated a product of modern high technology as the electronic computer. Understanding the obstacles that must be overcome in identifying, obtaining, and processing the right materials for pencil leads, which sometimes have the right properties for writing and sometimes do not, helps us to appreciate the triumph of the silicon chip. Realizing how complex the development of so simple and common an object as the pencil can be cannot help but cause us to marvel before the computers on our desks, but at the same time we can appreciate that the synthesizing of it all is implicit in the point of a pencil.

Not all pencil makers have been professional engineers, of course, but all the problems they solved in advancing the art of pencil making have been problems of engineering. When craftsmen made pencils the way they had been taught, they were acting as craftsmen, but when they departed from tradition, as the young William Munroe did, and evolved new and improved artifacts, they effectively were acting as engineers. Modern engineers, with the benefit of mathematical and scientific tools in place of joiners' and cabinetmakers' tools, have been able to adapt pencil making to changing conditions of materials and supplies, economics, and politics in a much more rapid fashion than was once dreamed. Indeed, the single most significant advance in pencil making in the last four centuries was the development by Conté of graphite-clay composition lead, and this was really a natural extension of his work with clay and graphite in an environment that nurtured the research and development on a crucible for use in molding cannonballs. Like Conté, all engineers are potentially revolutionaries, but revolutionaries with a technological tradition.

Pencil making began as a cottage industry and not as a fully developed, deliberate extension of the cabinetmaker's art. Similarly, some of the most creative developments in computer hardware have also begun in garages—the cottages of the automobile age. And software development still depends largely on computer hackers, some of them living and working literally, if we are to believe the stories, in cabins and shacks not unlike Thoreau's at Walden Pond. And, again if we are to believe the stories, the work habits of some hackers are not unlike Thoreau's on-again, off-again involvement in—one might almost say his forays into—the pencil business. But Thoreau's lack of a commitment to the conventions of the workplace did not at all imply a lack of commitment to the product of his workplace. For all their unconventional, or even too conventional dress, engineers must not be judged by their clothes or personalities any more than the pencil should be judged by the color of its wood case. What matters ultimately is the artifact and how it functions in society and in the marketplace, what engineers do not with abandon but with responsibility. If the pencil does not write, it will not sell; if it writes better than any other, not only will it sell but it can also command a higher price.

Thoreau understood pencils and the marketplace, and in the late 1840s he understood that the marketplace was becoming crowded not only with American but also with foreign manufactures. His father and he knew, too, that the secret of pencil leads that might be inferred, if not exactly extracted, from an encyclopedia could not be kept secret for very long, and indeed it was to be broadcast to the millions of people attending the Great Exhibition in 1851 and to anyone willing to read books about the Crystal Palace and its contents. Although Thoreau could no doubt have made further improvements in his own pencils in order to keep ahead of the competition, his temperament was not that of a career pencil maker and so he and his family got out of the pencil business in favor of selling pure graphite to the emerging business of electrotyping, a business that had a new secret.

Engineering does not deal just in secrets, of course, but those are among the realities of private industry. Corporations necessarily must have some advantage if they are to recover the investment that they make in retaining consulting engineers and in maintaining an engineering staff to carry out the research and development required to manufacture a new product or even to continue to make an old product when the traditional supplies of materials become unavailable. Once a newly developed product is released to the marketplace, it also becomes available to the competition.

A cottage industry like the one in which John Thoreau and his son engineered a new pencil for America did not maintain a formal laboratory or possess the formal knowledge of chemistry that might have enabled them to take a few French pencils apart and analyze them for their ingredients and clues to their manufacture. Rather, the scholarly Thoreau had to stick his nose in books and sniff out whatever trails might be there. Other, less literate contemporaries of his, also having no laboratories or chemistry or chemists in their employ, might have depended upon an oral tradition to learn that the best pencil leads were made of graphite baked with clay. Somebody already working with those materials, such as a Joseph Dixon, a maker of crucibles and other graphite products, might be expected eventually to elaborate on a hint and come up with his own secrets.

The removal of the pencil industry from New England cottages to New York factories meant the beginning of a new era. As physical plants grew, a relatively small investment in a research and development department was not a luxury but a necessity to protect the much larger investment. A research department would naturally be staffed with engineers and scientists whose job it would be to understand the pencil and its industry as the microcosm that it is and to understand the ways in which pencils are manufactured and can be manufactured better. Part of the job of such a group of people is to be prepared to answer any questions and solve any problems that might arise on the floor of the factory or out in the marketplace. Why does this wood splinter during cutting? Why do the points of these pencils break so easily when they are being sharpened? Why do these pencils not write as smoothly as the competitor's?

Another job of the research and development department is to initiate or "perfect" new products in order to capture a greater share of the market with a "new, improved" pencil. Ideas for such new pencils may come from the sketchbooks or notebooks of engineers, out of the dreams of the company president, or out of the suggestion box in the lunchroom. But wherever they originate, they will only be made possible by a careful selection of materials and the deliberate development of a process suitable for production in volume. If the company is not willing or able to invest in new machinery, this may seriously restrict how well or even just how the new pencil might be made. And making a pencil that is not the very best that one might be able to make at the time might be to make a failure. There are clearly lessons for today in the story of pencils.

Some manufacturing industries seem to have become invaded in recent years by a process known as "reverse engineering," whereby "research and development" consists of taking apart the new product of a political or mercantile competitor and determining how to make reproductions of it. But the idea is at least as old as the pencil. While the practice may provide short-term advantages, a company's exclusive dependence upon it could lead to long-term decline and to the eventual extinction of the company. While reverse engineering can also contribute to incremental innovation, its too-narrow application can be self-limiting. A research and development staff that did nothing but take apart pencils would come to know only about pencils as they were being made. When a competitor who was developing entirely new manufacturing processes, anticipating the exhaustion of certain supplies of raw materials, for example, came out with a revolutionary new pencil, perhaps incorporating new materials, its reverse engineering would be an engima and not just a matter of disassembly and analysis. Puzzling over the new pencil could cause the copycat company to fall further and further behind its competitors, who might already have on the drawing board the president's latest dream, whose "forward engineering" might include deliberate booby traps for the reverse engineers.

Most of our children use pencils and personal computers with equal facility nowadays, and we all should feel similarly comfortable with any new artifact, no matter how it was engineered. For if it is not an elaboration upon something with which we are already familiar, there should be a familiar older artifact through which we can approach the new with confidence. Thus we should be able to achieve an understanding, if not of an innovation's inner secrets, then certainly of its achievements, promises, and possibilities, and those of its industry. Even though the pace of the personal-computer industry has been frantic, to say the least, its dynamics are little different in principle from those of something seemingly so simple as the pencil industry. That is what makes parodies and satires ring so true. And while the apparent simplicity and commonness of an artifact can mask its achievements and complexity, the story of its origins and history reveals them. The story of a single object told in depth can reveal more about the whole of technology and its practitioners than a sweeping survey of all the triumphant works of civil, mechanical, electrical, and every other kind of engineering. The grand tour of engineering leaves little time for sketching many details.

While pencil making is a near-perfect metaphor for engineering as well as providing in its own right an excellent case study of technological development, there is no end of common objects whose close scrutiny rewards us with an understanding of the rest of the world as well. The biologist Thomas Huxley used a lump of carpenter's chalk to relate to the workingman's world of water wells and lecture-room slates the scholar's worlds of microorganisms and geology and their implications for Darwinism. The chemist Michael Faraday, in a course of lectures delivered before a juvenile audience at the Royal Institution, used a common candle to illuminate the world of chemistry, telling his young listeners that "the child who masters these lectures knows more of fire than Aristotle did." And the aviator and author Anne Morrow Lindbergh, "searching for a new pattern of living," took freshly sharpened pencils down to the beach and, after a while, felt her mind come to life and found treasures of thought in shells tossed up as gifts from the sea. Others, being perhaps of a less literary bent, have been no less moved or inspired by something of their special choosing. As a grain of sand contains a world within it, so objects as apparently plain and simple as straight pins can hold a multitude of lessons and meanings to prick the mind into activity.

Adam Smith opened his *Inquiry Into the Nature and Causes of the Wealth of Nations* with a classic description of straight-pin making in order to explain the effects of the division of labor:

> One man draws the wire, another straights it, a third cuts it, a fourth points it, a fifth grinds it at the top for receiving the head; to make the head requires two or three distinct operations; to put it on is a peculiar business, to whiten the pins is another; it is even a trade by itself to put them into the paper. . . . Each person . . . might be considered as making four thousand eight hundred pins in a day. But if they had all wrought separately and independently, and without any of them being educated to this particular business, they certainly could not each of them have made twenty, perhaps not one pin in a day.

Charles Babbage, known as the father of the modern computer because of his pioneering work on calculating engines, also used pin manufacturing to explain the benefits of the division of labor, "perhaps the most important principle on which the economy of manufacture depends." For "pin" we can read "pencil," of course, and the Nobel economist Milton Friedman has commented that in the 1980s the pencil provides a neat lesson in free-market economics: the "magic of the price system" gets thousands of people to cooperate so that we may buy a pencil "for a trifling sum." Perhaps we tend to overlook such an achievement because it is so close at hand, as Henry David Thoreau demonstrated by forgetting to mention the very thing he made and was using to draw up a list of supplies for making an excursion into the wilderness more civilized. But keeping in mind the complexity of the simple pencil can help us to simplify the more complex aspects of technology and society.

The story of the pencil and the industry that produces this small and inexpensive yet powerful and indispensable object is truly the story of a microcosm. It is the story of the familiar thing that we can hold in our hand and admire, press against paper and test its mark, twirl about in our fingers and see its seams and its blemishes, break apart if we wish and see at the same time its simplicity and its complexity. The pencil in our hand can be the automobile in our garage, the television in our home, the clothes on our back. When we know the story of the pencil in the world, we know the mythical proportions of technology: the discovery of plumbago and its exploitation led to the transmutation of a chemical relative of coal into black gold. When we know the story of the black lead mines of Cumberland, we know the finiteness of resources: the unchallengeable supremacy of the English lead pencil is a thing of the past. When we know the story of the French pencil, we know the value of research and development: Conté's efforts of two centuries ago continue to make the modern pencil possible. When we know the story of the pencil in the nineteenth century, we know the inescapable business of technology: the German pencil might have dominated the world market at mid-century, but the ultimate triumph of the American pencil could serve as inspiration for the future.

Appendix A
Appendix B
Notes
Bibliography
Illustrations
Acknowledgments
Index

Appendix A

from "How the Pencil Is Made," by the Koh-I-Noor Pencil Company

The graphite, however pure, is apt to have foreign matter mixed with it, so the first thing to be done is to carefully clean it. The gravitation process is one commonly used. In this, the graphite is mixed with hot water until in a fluid state when it is fed into the first of a number of tubs, usually six, set on steps, the first the highest, the next one step lower, and so on down the line. The fluid is kept in motion by an agitator and pours from the top through a fine mesh sieve to the next tub and so to the last. The sieves grow finer of mesh until that through which the fluid is fed to the last is about 200 meshes to the inch. In each tub, the impurities, which are heavier than the graphite, sink to the bottom so that the material finding its way into the last tub is pure and entirely free from all foreign matter.

The clay, the finest of which for lead pencil purposes is found in Czechoslovakia, is cleaned in the same manner as the graphite. As grit, however minute in size, is fatal to a good lead, and as this thorough washing process insures freedom from it, this extreme care is justified.

The fluid graphite is pumped into a filter press which squeezes the water out, leaving the graphite in large square cakes. The clay is dried in the same manner.

After further drying, the graphite with the addition of clay, which is proportioned according to the degree of lead to be made, is mixed with water and thoroughly milled. The quality of lead to be made determines the time of the grinding. The longer the grinding, the better the lead. It is interesting to note that the average time required to grind Koh-I-Noor lead is approximately two weeks.

The mixture is now ready for forming into leads. It is put into heavy iron cylinders in the bottom of which is a die, usually of sapphire, of the diameter of the lead to be formed. Under great hydraulic pressure, this mixture is forced through the die, coming out like an endless, round shoestring. This is taken, broken off in correct lengths, and laid out on flat iron plates where it is straightened, dried and cut to pencil lengths. These leads are packed in crucibles and sealed up in a furnace where, in a temperature of more than 2000°F., they are tempered. After a gradual cooling, the leads are "prepared" by immersing in a bath of hot oils and wax. This process has an important effect on the smoothness and general marking ability of the lead. Now with a final drying and cleaning in sawdust, the lead is ready to be put in the wood.

The thoroughly seasoned cedar wood is cut into bolts from which are sawn the "slats" which are worked into pencils. These slats are slightly longer than the length of a pencil, the thickness of a half pencil, and usually the width of six pencils.

The slats are planed and grooved for the leads in one operation. The grooves are deep enough to cover one half the diameter of the lead.

After brushing the grooved slats with glue, the leads are laid in and a similar grooved slat is fitted over the first. A quantity of these are placed in a frame, pressed carefully under hydraulic pressure, locked in the frames, and set away to dry.

The pencils are cut apart in shaping machines by cutters revolving at high speed. Passing through once, half of the pencils' circumference is formed. By turning the block and repeating the operation, the pencils are fully formed. In this state they are known as plain cedar pencils and, while qualified to do the work of finished pencils, there are many operations necessary before they are ready to be offered for sale to the users.

From the shaping machines, the pencils are taken to be sandpapered. A fine paper reduces every slight unevenness of surface, leaving them with a velvet-like feel.

Now the color is applied, usually by varnishing machines. The pencils are fed automatically from a hopper, through a bath of color, then through a disc of felt which smooths the color on and removes the surplus. An endless pin-belt receives the pencils, carries them over a heated compartment, emptying them, dried, into a basket at the other end of the machine. This operation is repeated over and over again until there is a good covering of color. Then the finish coat is applied and the pencils are taken to the sizing machine.

Varnish has accumulated on the ends of the pencils. This is quickly removed in the next operation by two large, rapidly revolving sandpaper-covered drums set opposite each other with a space between, through which moves an endless belt carrying the pencils. The drums turn toward each other, and the sandpaper surface, coming in contact with the ends of the pencil, removes all varnish and at the same time reduces the pencils to exactly the same length.

In America, most people want an eraser on their pencil. As the eraser is usually held by a metal ferrule, there are several operations necessary to put this on. First, a shoulder is cut or pressed on one end of the pencil, usually by an automatic machine, the ferrule is fitted on, prick-punched, securing it to the wood, and then the rubber plug is inserted, sometimes by an automatic machine and often partly by hand.

Stamping is the next process. This is done by different methods, according to the quality of the pencils. Where pure gold leaf is used, as on the finest Koh-I-Noor goods, the stamping is often done with hand presses. The gold leaf is cut into narrow strips which are applied to the pencil.

The stamping material may also be in the shape of a large roll, automatically fed over the pencil. Sometimes on cheaper goods, bronze powder is used. In any case, the stamping die, which is of steel and heated, generally by electricity, is brought into contact with the pencil under some pressure. The stamping material lies between the pencil and the die and is thus pressed into the painted surface. Any surplus stamping material is then wiped from the pencil, leaving the lettering sharp and clear.

The branded pencils, after a careful inspection and cleaning, are now ready to be boxed.

Appendix B

A Collection of Pencils

Sometime after I began to think about the pencil as a symbol of the engineer in society and about pencil making as a paradigm for the engineering process generally, I began to collect pencils of all sorts. These were to be my totems and my tools for writing about engineering, and their origins and history were to be those of engineering itself. But I found that getting to the roots of something so common as the pencil was no easy task. Very early in my efforts to learn about the first pencils and pencil making from the usual sources of scholarly books and articles, not to mention the oral tradition of pencil makers, I realized that reliable information about pencils was no easier to find today than it had been in centuries past. Thus, in lieu of information about them, I accumulated the artifacts themselves and hoped to have so great a variety and such great numbers of physical pencils about me that I could even sacrifice some now and then to break and take apart whenever I had a question or a hypothesis about what made them tick.

Soon I realized, however, that really old pencils are as hard to come by as information about old pencils. It was bad enough to learn that when antique dealers bought old toolboxes, they kept the tools but threw out the carpenter's pencils with the wood shavings and sawdust. But I also soon learned that when they bought old trunks owned by engineers who had followed construction jobs around the world, they kept the drafting instruments but threw out the pencils and erasers. When they bought the contents of artists' and architects' studios, they kept the drawings but discarded the drawing pencils. When they bought anyone's estate, they kept everything but the pencils, or so it seemed.

When I realized how difficult it was to find even an early-twentieth-century wooden pencil in local antique shops, I placed a classified ad in *Antique Week* under the category "pens-pencils wanted." The ad ran in several consecutive issues of the newspaper, which has a circulation of 65,000, and I received a total of five replies. One offered an old pencil display case from a stationery store, one a child's pencil box made in Japan, one a salesman's demonstration kit of how a pencil was made, one some canceled stock certificates from pencil companies, and, finally, a retired schoolteacher and self-styled "collector" offered some honest-to-goodness wooden pencils. She meticulously listed what she had to offer, ranging from the rare (an eighty-year-old box of drafting pencils in mint condition) to the common (several short, thin advertising pencils that magazine publishers once stuffed in envelopes and mailed out by the millions). It is a good thing I did not rely on finding pencils through classified ads. My family was an enormous help, and my brother especially has been able to find for me a number of very interesting old pencils at antique shows and flea markets in the New York metropolitan area, where pencils were once made by the billions.

I have learned that I am not the only person interested in these things. There is an American Pencil Collectors Society, with over three hundred active members and a monthly newsletter, and a British Writing Equipment Society, with over three hundred members worldwide. One of the principal activities of the American collectors is exchanging pencils imprinted with their unique membership numbers, which in late 1988 reached as high as No. 1333, and trading mostly for advertising pencils, which are never sharpened or used. One pencil collector in North Dakota has amassed over twenty-five thousand examples in seventy-five years. Yet his collection is far from the largest, and the hobby knows no numerical or political limits. There is a collector in Russia with many pencils from the Armand Hammer factory that flourished in Moscow in the late 1920s. A Texas man collected only used pencils, and especially prized those sent by famous people in response to requests. Although he thought counting was bad luck and in the late 1930s was probably far from his goal of a million stubs, he did note that of the thirty-one pencils governors had sent him, twenty-nine were yellow and the erasers were generally in good condition.

Some people have collected not pencils but theories about other people's pencils, believing that they are expressions of the user's personality. Pencils also follow fads. When King Tut's tomb was uncovered, for example, the Ramses pencil was introduced. This was shaped like an obelisk and was colored red, green, and blue on polished black. The pencil was imprinted with an image of the Sphinx and was topped by a pyramid-shaped eraser. Such novelties were once as ubiquitous as they now are rare.

Finely printed and illustrated old trade catalogues, as rich in their contents as Cumberland wadd holes and saved in ways that their wares seem not to have been, document the variety of pencils that have been manufactured over the years. Booklets published in the

1890s by the Joseph Dixon Crucible Company, for example, indicate that that firm alone was then making "over seven hundred styles" totaling over 30 million pencils a year. Fifty times that many were made in all of America in the 1950s, but in only half as many styles.

A great variety of pencils are manufactured to this day, and many can be found as souvenirs in museums and tourist attractions, especially in Britain, where the pencil still seems to be regarded as an object of some value and is often of some deliberate and restrained design. Some of the most attractive pencils can be bought in South Kensington. They are of a slender, rounded-triangular shape, with a fine enamel finish and a gold-painted eraserless tip, and they are imprinted "Victoria and Albert Museum" and "Science Museum," in typefaces appropriate to each. These pencils feel comfortable in the hand, and the lead takes a good, sharp point that writes smoothly.

In the United States, finely crafted souvenir pencils do not seem so common as novelty pencils, but novelties are certainly not limited to any one country and museums do not have uniformly good taste. My son once brought back from a school field trip to a museum of lapidary art a pencil attached to a plastic tube filled with little polished stones. My daughter has given me a two-foot-long, one-inch-thick pencil that actually writes, and my wife has brought me a pencil with a dentist's mirror where we expect an eraser. In Amsterdam I found a short pencil painted convincingly to look like a filter-tip cigarette, and everywhere I have found common pencils painted to look like anything but common pencils. Many of these latter seem not to be designed to be used, however, for their wood casings often seem to be made of very poorly matched halves and their leads are not always centered, suggesting inferior quality beneath the gloss.

Like all tourist attractions, the John Hancock Observatory has a gift shop, and, as in virtually all gift shops, one can buy a souvenir pencil. In fact, a variety of pencils were available on the day that I visited. There was one of the new automatic kinds with the very thin lead, perhaps made in Japan but suitably stamped to identify it as a souvenir of the Boston tower. There was also for sale a handsome pencil with a natural wood finish imprinted in tiny capital letters with only the single word "TAIWAN." While this pencil had little to distinguish it as a souvenir of Boston, it was certainly something to amuse children, for as they twirl the pencil while writing or drawing, its four-colored lead, clearly visible through the eraserless end of the pencil as a neat pinwheel of purple, green, red, and yellow, lays down a line or a word in a spectrum of colors. But instead of being near the blunt end, where virtually all of today's Western pencils are imprinted, "TAIWAN" is located very close to the sharpened tip. And since visiting Boston, I have bought other wood-cased pencils made in Taiwan and Japan; they too are imprinted, and very faintly, away from the eraser end. Thus, after a first sharpening, only "WAN" or "PAN" remains visible to give a hint of the pencil's provenance, and after another sharpening there is no evidence of where these very attractive pencils come from. One pencil that I bought recently has a striking ferruleless black eraser and is also imprinted "LEADWORKS" in large letters in the conventional location. When I first saw this pencil, I assumed it was made by some young American craftsperson in a Vermont cottage heated by a wood stove—until I saw the faint image of the word "Japan" near the end to be sharpened.

In spite of possibly misleading us about their place of manufacture, these pencils are functionally superb. The multicolored Taiwanese pencil is fun to draw and write with, its round shape makes it easy to twirl with such control so as to generate more shades of color than I can count, and the lead in the Leadworks is as smooth as any, its black mark easily erased by the curious black eraser. Thus, like an imported automobile, the pencil's performance makes it competitive in our market. Unfortunately, while my first impressions were of the finish and feel, I have found the wood in both of these pencils unnaturally white and odorless, thus flawing otherwise very attractive and effective pencils and reminding me of their origins in lands where no suitable pencil cedar grows.

The other pencil that I bought in Boston was one of a more common, domestic variety. While the multicolored pencil was sharpened and ready to be used by bored children or curious adults, the souvenir pencil was unsharpened. Indeed, its manufacturer may not have expected it to be sharpened or used, because the lead is visibly off center in this pencil, and trying to sharpen it would likely result in many broken points and lots of frustration. The pencil is round, so that a logo can be printed continuously around the surface, and while no country is claimed as the origin of this pencil, indications are that it is of the inferior kind that are sold to jobbers for imprinting and souvenir sales. For all its properly colored wood, albeit badly mismatched, it is a pencil whose performance would not allow it to compete very well in the world market.

Things are certainly not always what they appear to be, and some pencils are not even designed to be used for writing or drawing. There is an infuriating puzzle made out of a pencil with a loop of string emerging out of the end where an eraser normally would be. The loop on the trick pencil is long, but it is not long enough to pass the pencil itself through. An experienced joker who knows the trick can attach the pencil to a buttonhole on an unsuspecting person's coat, and the fun is supposed to come as the uninitiated tries to get the pencil off.

According to Jerry Slocum, who collects and classifies puzzles old and new, the trick pencil was invented about one hundred years ago by the great nineteenth-century American puzzle maker and chess-problem composer Sam Loyd for the president of a New York life insurance company who wanted a gimmick to help his agents sell policies. When Loyd first showed the puzzle to the executive, he apparently was not impressed. However, Loyd demonstrated the trick's usefulness when he attached it to the president's buttonhole and bet him a dollar that he could not remove the pencil within half an hour—without cutting the string. When he finally admitted defeat, Loyd jokingly agreed to reveal the secret if the president would buy a life insurance policy, and he thus was convinced that his agents could use the trick. While it has been claimed that it is because of Loyd's pencil trick that "to buttonhole" became an expression meaning to grab someone's attention, the *Oxford English Dictionary* documents that usage in England in the 1860s, when Loyd was only in his twenties. But conflicting claims as to when and how words and phrases come into use on different sides of an ocean are as difficult to pin down as when and how pencils first made the transatlantic crossing.

Since the buttonhole trick would not work if the pencil was shortened by sharpening, often a leadless piece of wood was employed. Another trick pencil was also leadless, but for another reason. A late-nineteenth-century writer remembered the "deceptive pencil" this way: "When a child at school, we inserted two needles in the stump of a pencil from which the lead had been removed, and attached a string through a hole in the middle of the pencil. An innocent comrade was asked to hold the pencil between his thumb and forefinger, and to press very hard. Then, on our pulling the strings, the needles were made to protrude and the victim of the joke was forced to let go of the affair very promptly."

While they may amuse some children, leadless or inferior pencils are a risky way to advertise a product. The merchant distributing free pencils with the name and logo of his business or its merchandise might hope that they would not be saved unused like mementos of museums or tourist attractions, but kept in daily use and view. Traditionally, pencils intended to carry advertising have had to be good pencils, lest their breaking lead and crazed finish reflect badly upon their sponsor. Indeed, many old advertising pencils were very well made and made to carry a more permanent message than just something that would disappear into the pencil sharpener. Hence short wooden pencils have been disguised in simulated bullet shells and other cases clever and durable enough for their owners to keep the imprinted token in use for a long time.

Pencil cases made from real cartridge shells were sold in London at the end of the nineteenth century. The shells were inscribed "Remember Gordon," and part of the sales proceeds was to go to the Gordon Memorial College at Khartoum. The case of the Khartoum pencil was "guaranteed to have been actually used by the British Troops at 'the Battle of Omdurman' " and came in models with screw and ratchet actions. Other pencils that saw a different kind of wartime service include one on display in the Pencil Museum in Keswick: it contains a hidden map of Germany and a compass.

The pencil, whether wood-cased or not, is not merely an object of frivolous novelty, serious advertising, emotional fund raising, or survival behind enemy lines. Design magazines seem regularly to list the common pencil as a "paragon of pure design" or a "monument to longevity in design," and report on variations on it. One recent feature on new and notable items of industrial design pictured what the editors described as "at long last, an improvement on . . . the lead pencil: . . . a ribbed wooden body to offer a better grip and a more interesting appearance." However, old-time butchers, whose fingers tended to be wet and greasy, could take such pencils from their aprons when they wanted to tally the bill on the brown paper bag in which the meat would be carried home, and kosher butchers could use a pencil guaranteed free of any pork-derived products. But it is still the most common of all pencils, the yellow, hexagonal No. 2, that is considered the design classic.

If we can't always remember or find old pencils today, whether yellow and hexagonal or any other color and shape, it's not because we tried to lose them when we or they were young, or threw them away with abandon the way Johnny Carson and David Letterman have been wont to do on their television shows. Not very long ago in America a pencil box or a pencil case was as important to schoolchildren as what it contained, and many children imitated the man who seemed always to keep a pencil behind his ear (a practice Egyptians started with their reed pens) or the woman who seemed always to have one stuck in her hair. Among the countless catalogues that came in the mail for the 1988 holiday season, some of the most ostentatious offered imprinted wood pencils and pencil boxes, including a ceramic one from Neiman Marcus containing a dozen wooden pencils imprinted with the purveyor's name and "the ultimate pencil box," handmade and standing on turned pencil legs, from the San Francisco Museum of Modern Art.

Pencils can be as important as toys, and they often have been used as toys. We once wrinkled our faces to hold mustache pencils between the upper lip and the nose, and we scribbled pencil mustaches on posters in the days of erasable graffiti. Young boys made pencil tusks hang from their nostrils, and older girls put pencils under their breasts to test if they needed a bra. We twirled, chewed, tapped, doodled, and sometimes even took notes with pencils during classes, as we would later during meetings.

We use pencils to stir paint, prop up windows, open stubborn plastic bags, dial telephones, and punch holes in aluminum beer cans whose ring openers have come off in our fingers. As calculator buttons grew smaller we used, in an ironic twist, the eraser end of a pencil to tap out our sums. Still later, as parents, we showed our small children how to fit the pencil eraser into the holes left by the broken buttons on a Speak & Spell, and now our children show us how to use a pencil to remove tapes from a videocassette recorder whose eject button has fallen inside. It works because a pencil lead conducts electricity.

Some of us, before our arthritis got too bad, tried to experience the sensation known to the medical profession as Aristotle's anomaly: "When the first and second fingers are crossed and a small object such as a pencil is placed between them the false impression is gained that there are two objects." Apparently, for some people at least, when the pencil touches two parts of the skin that are not ordinarily touched simultaneously by a single object, the one pencil is perceived as two. As our arthritis got worse, our doctors prescribed medicine in containers designed to be opened with a pencil acting as a lever.

The pencil is always an extension of the fingers. With a pencil we can count beyond our ten digits, usually striking out every four marks with a fifth—four vertical fingers made into a hand by a diagonal thumb. We can turn the pages of slick magazines and catalogues more quickly with the dry eraser than the licked finger. We can dial or press telephones that our nails are too long or our fingers too fat to work. We can hold more places in books by sticking pencils where our fingers were. We can point to details that our fingers would obscure. We can exaggerate our gestures. We can make visible what our fingers can only trace in air. We can vote not by raising our hands but by marking our secret ballots.

Indeed, the secret ballot was such an important institution that before the days of computer cards (themselves sometimes to be filled in with No. 2 pencils), special pencils for the voting booth were made with cords and screw eyes attached so that the pencils might be tied down to keep them from being carried away or lost. Pencils with fancy cords were made for dance programs, and flat and ultra-thin pencils with tassels attached have been made for use as bookmarks and book-leaf cutters. Such pencils had no erasers, but in the 1928 American presidential campaign, pencils with oversized erasers in the shapes of the heads of Herbert Hoover and Al Smith competed for votes.

As late as the 1940s all pencils were objects of value, even if they cost as little as a penny. In its 1940 catalogue, the Eagle Pencil Company offered its top-of-the-line Mikado (five cents in its basic style) specially equipped—perhaps for writers like Nabokov, who remarked that his pencils outlasted their erasers—with an "oversize eraser, big enough to outlast the pencil." Other styles of the Mikado, "the biggest selling quality pencil of its type because it is super bonded," came with oversize erasers attached not to the pencil proper but to a detachable protector, which could fit over the sharpened lead when not in use. These are clear indications of how the pencil was regarded not as something to be neglected or thrown away but as something to be used down to a stub, which, unlike nineteenth-century pencils, could still contain lead. While John Steinbeck could not use pencils once he felt their ferrules touch his hand, he did not discard the shortened ones but gave them to his children.

Even today, if a cache of pencils is found in an old desk, there are likely to be two-inch stubs among the unsharpened souvenirs, perhaps, in John Updike's words, "so old their erasers don't erase and even the graphite has gone waxy and refuses to write." But what had accumulated over fifty years is likely to be thrown away in a second by the desk's new owner. In the days of more conservative consumers, pencil stubs were seldom too small to hold, and for the writer who did not feel comfortable using a stub, pencil extenders were sold. These devices function something like a *porte-crayon* and are much like old penholders, in that the pencil stub can be inserted into the end of a shaft, not unlike the way the first lead pencil was in a wooden tube. Extenders were especially common among engineers and draftsmen, whose favorite pencils were priced dearly. The use of an extender also has the advantage that the pencil does not appreciably change its heft as it wears down.

For all the varieties of wood-cased pencils that have been manufactured and sold, they are mostly forgotten today. A recent reprinting of the 1902 edition of the Sears, Roebuck catalogue, from which the publisher "omitted those pages which were mostly repetitious," left out the wood pencils but included a selection of mechanical pencils and novelties. Other contemporary catalogues show many interesting and intrinsically valuable gold and silver pencil cases, and examples of these can be found at many an antique show or flea market. One type was designed to hold a short wood-cased pencil, whose point could be protected when not in use by retracting the pencil into the case. Captain Charles Ryder, the narrator of *Brideshead Revisited,* recalled his father using one of these while reading after dinner in an upright armchair: "Now and then he took a gold pencil case from his watch chain and made an entry in the margin." The metal pencil cases had rings at what would be the eraser end, by which they could be attached to watch chains and the like. Small silver and gold snap rings were sold so that the pencil could easily be removed from the chain, as Captain Ryder's father's seems to have been. Some cases were of a sheath design so that the pencil could be "withdrawn from the sheath for use, without the trouble of detaching it from the chain." Flat pencil cases of a sheath design were also made, to be carried in a waistcoat pocket without creating an excessive bulge.

Such specialized cases required special pencil refills, of course, and "flat cedar pencils" were sold in boxes of six and one dozen. The round cases required small wood-cased refill pencils with threaded brass ends that screwed into the metal case. These pencils were so short and sometimes so slender that they would have been uncomfortable to use without a case, but they were an economical use of black lead and wood. I have a sterling silver case made by S. Mordan & Company that is engraved with the name and address of its original London owner and that has a still-serviceable Koh-I-Noor pencil stub (grade HB) only a fraction of an inch long. Who today but a frugal draftsman would use a pencil down to such a stub? But our pencils, unlike many of those of the Victorians, have lead from end to end, and some engineers and draftsmen, when a good pencil's stub is too small to hold even in a pencil extender, have been known to cut away the last of the wood case and use the lead in their compasses.

While the common seven-inch-long yellow writing pencil may account for the vast majority of all pencils made today, there is no single yellow pencil that is everyone's favorite, and the beauty of a pencil will no doubt always be in the eye and the hand of the beholder. The authors of *Quintessence,* a book whose subtitle is "The Quality of Having *It,*" are firm in their defense of the Mongol No. 2 as "the very best pencil there is," but they were forced to make a choice, for they could not very well show a generic yellow pencil as "the best." While they claim that Mongols have the "perfect mix of clay and graphite" and are "topped with the best erasers," others choose the Mongol for different reasons: "It's covered with wood and full of ideas. Everytime I pick one up something else comes out. And the type on the side is good, and the eraser is pink." Such sentiments could describe other brands of pencil also. The Think Big! store in New York, which specializes in selling large versions of some of the most common items found around the home and office, chooses to take the Dixon Ticonderoga No. 2 ("soft") as its model for a six-foot pencil. Ask others, and they will claim the Faber-Castell Velvet (descendant of the American Venus-Velvet) is the pencil of choice, or the Berol Mirado (once the Eagle Mikado) is the number one No. 2. Still others will shun classic yellow for the steel-black hexagonal Faber Blackwing, a dignified-looking fifty-cent pencil with a distinctive flat ferrule. (The Blackwing's extra soft lead makes it so smooth and easy to write with that the pencil has been imprinted with the slogan "half the pressure, twice the speed.") While all these and still other different kinds of pencils share the quality of being a pencil, at the same time all of them and none of them can claim to be *the* pencil.

Notes

Full references are given in the Bibliography.

CHAPTER 1 *What We Forget*

p. 3 list of essential supplies: Thoreau, *Maine Woods,* pp. 839–40.

4 "in his pocket": Ralph Waldo Emerson, "Thoreau," p. 244.

5 scribers to mark: Bealer, pp. 103–4.

"one dozen Middleton's": Oliver Hubbard, p. 153.

English pencils: Oliver Hubbard, p. 156.

6 fencer's foil: "Andrew Wyeth: The Helga Pictures," National Gallery of Art brochure for the 1987 show.

"I am a pencil": see *New York Review of Books,* January 16, 1986, p. 26.

Emmanuel Poiré: Caran d'Ache, pp. 1–2.

7 "everything begins": Remington, p. 24.

engineers did not feel: Ullman et al., p. 70.

sketched his own hand: see, e.g., Hart, Plate 40. Cf. Carpener, p. 814.

basically right-handed: Carpener, p. 814.

8 "In order that": quoted in Turner and Goulden, p. 170.

9 "If historians": Lynn White, *Medieval Technology,* p. v.

skill with the pencil: Vitruvius, Book I, Ch. i, para. 3.

10 "He writes in atrocious Latin": Drachmann, p. 12.

"He has all": Morris Hicky Morgan, quoted in Vitruvius, p. iv.

p. 11   "the social antithesis": Zilsel, p. 550.

     "probably often illiterate": Zilsel, p. 551.

     "Praised be God": quoted in *Encyclopaedia Edinensis,* "pencil" entry.

12   "the writer has had": quoted in Norton, p. 13.

     "do-it-yourself addict": Remington, p. 26.

     "all the principles": David Shayt, Smithsonian Institution, private communication, September 26, 1988.

13   "an introduction": in Friedel, p. 3.

     "nail picked up": Harding, *Catalog,* p. 10.

CHAPTER 2   *Of Names, Materials, and Things*

15   pencil was named: *Oxford English Dictionary,* "pencil" entry.

     tails of animals: Dickinson, "Besoms," p. 100.

16   precursors of the broom: Dickinson, "Besoms," pp. 99–100.

18   "Mr. Ross ate pencils": Pinck, p. 6.

22   "Made hemp fibres": quoted in Friedel and Israel, p. 134.

23   "They twist and stick": quoted in C. Lester Walker, p. 91.

CHAPTER 3   *Before the Pencil*

24   "what has been": Ecclesiastes 1:9, quoted from the *Oxford Annotated Bible.*

25   letter to his friend: Cicero, *Letters,* in Winstedt's translation, Vol. I, p. 115.

     "If you find": Cicero, quoted in Viollet-le-Duc, *Discourses,* Vol. I, p. 152.

27   metal style and wax tablet: see, e.g., Voice, p. 131.

28   "*His comrade carried a staff*": Chaucer, in Wright's translation, p. 263, with thanks to James Wimsatt.

     early epistles: Astle, p. 201.

     "diplomatic science": Astle, p. ii.

     "in Plautus": Astle, p. 201.

     "iron styles": Astle, p. 207.

     pugillares: Astle, p. 200.

29   "Cassianus was put": Astle, p. 207.

     sharpening slate pencils: *New York Times,* October 23, 1945, p. 16.

31   plummet: see, e.g., Voice, p. 131.

erased with bread crumbs: Meder, p. 58.

Theophilus wrote of an alloy: Dickinson, "Brief History," p. 74; Lynn White, *Medieval Religion,* p. 322.

"the first man": C. S. Smith, p. 106.

32 "crooked and oblique": Beckmann, 3rd ed., Vol. IV, pp. 348–49.

Edward Cocker: *Chambers's Encyclopaedia,* new revised ed., 1987.

"Having a Book": Cocker, reproduced in Whalley, *English Handwriting,* illustration 82b.

"parallel lines marked": Palatino, p. 16.

"goose-quills were used": Oliver Hubbard, pp. 157–58.

33 *productal . . . paragraphos:* Alibert, p. 4.

"the sheet with the black lines": Palatino, p. 17.

34 "Another more important": Pliny, Book XXXIII, xix.

Paper-cased metallic: Mitchell, "Black-Lead Pencils," p. 383T.

five small pieces of graphite: Mitchell, "Graphites," p. 380.

Less ancient reports: *Compton's Encyclopedia,* 1986 ed., "pencil" entry.

CHAPTER 4 *Noting a New Technology*

36 Konrad Gesner: Bay, pp. 53–86. Cf. biographical entry in *Encyclopaedia Britannica,* 15th ed. Some sources spell Gesner's first name with a *C*.

De Rerum: with thanks to Debora Shuger for help with the translation.

37 "father of bibliography": Bay, p. 53.

"German Pliny": Ley, p. 125.

"father of zoology": Ley, p. 130.

"born with a pen": Bay, p. 64.

38 "no more than an inky pencil": Fairbank, p. 85.

"The stylus shown": Gesner, quoted from the translation in Meder, p. 121, note 2.

41 "I remember": Mathesius, quoted in Meder, p. 114.

reprinted, enlarged: Beckmann, 3rd ed., Vol. IV, pp. 352–53.

"more complete": *Chambers's Encyclopaedia,* new revised ed., 1987.

p. 41   *"lapis plumbarius":* Francis White, p. 466.

"all the tools": Palatino, p. 2.

as early as about 1500: Cumberland Pencil Company.

as late as 1565: Fleming and Guptill, p. 5.

43   "unacquainted with the time": Beckmann, 3rd ed., Vol. IV, pp. 353–54.

scientifically accurate: Acheson, "Graphite," pp. 475–76.

A. G. Werner . . . K. W. Scheele: Berol Ltd., "The Pencil," p. 3.

traditional local name: Lefebure, p. 74.

"for its use in scoring": Plot, p. 183.

"The Mineral Substance": Plot, p. 183.

44   "misconception of graphite": Staedtler Mars GmbH, *History,* p. 2.

"To ascertain how old": Beckmann, 3rd ed., Vol. IV, pp. 345–46.

C. T. Schönemann: cf. *Literary Digest,* June 5, 1920, p. 98.

45   "The recorded history": Lefebure, pp. 18–19.

"Reading up wadd": Lefebure, p. 75.

46   "The uprooting": Fleming and Guptill, p. 5.

"Here also is found": Camden, quoted in Voice, p. 133.

Germans were involved: Jenkins, pp. 225–26.

Flemish traders: Cumberland Pencil Company.

47   "it is much more": Imperanti, quoted in Beckmann, 3rd ed., Vol. IV, pp. 350–51.

the term "vine": Voice, p. 133.

Borrowdale lead was widely exported: Voice, p. 133; Beckmann, 3rd ed., Vol. IV, p. 350; Mitchell, "Black-Lead Pencils," p. 384T.

"pointed pencils": Beckmann, 3rd ed., Vol. IV, p. 354.

"I think also": Cesalpino, quoted in Meder, p. 114.

48   streets of London: Voice, p. 133.

"which you would": quoted in Voice, p. 133.

readily stolen and disposed-of: Lefebure, pp. 85, 87.

49   "bomb shell": quoted in Fleming and Guptill, p. 7.

"felony to break into": quoted in Fleming and Guptill, p. 7; cf. *Journals of the House of Lords,* Vol. XXVII, p. 645.

"Le Roy le veult": *Journals of the House of Lords,* Vol. XXVII, pp. 703–4.

CHAPTER 5 *Of Traditions and Transitions*

50 "I believe more": Truman Capote, quoted in Winokur, p. 87.

"I have written": Vladimir Nabokov, quoted in Charlton, p. 43.

"Few articles": Alibert, p. 3.

53 "his square": Ben Jonson, quoted in Voice, p. 133.

"black-lead pen": John Evelyn, quoted in Voice, p. 133.

"black-lead pencils": quoted in Voice, p. 133.

"Bleistift, Blay-Erst": Meder, p. 114.

"in catalogues": Meder, p. 114.

54 microscopic and chemical investigations: Mitchell, "Pencil Markings," p. 517.

55 "PENCIL, an instrument": *Encyclopaedia Britannica,* 1st ed.

56 Bavarian town of Nuremberg: Meder, p. 115. Cf. Staedtler Mars GmbH, *History,* p. 2.

English town of Keswick: Lefebure, p. 79.

"Buy marking stones": quoted in Voice, Fig. 2.

"There is also": Pettus, quoted in Voice, p. 134. Beckmann (3rd ed., Vol. IV, p. 354) identifies fir as the specific kind of deal used for pencils.

58 "Its natural Uses": Robinson, pp. 75–76.

59 "inclined to think": Beckmann, 3rd ed., Vol. IV, p. 355.

"pens of Spanish lead": Meder, p. 114.

60 a Keswick joiner: Voice, p. 135.

"technique of glueing": Staedtler Mars GmbH, *History,* p. 2.

Friedrich Staedtler: Staedtler Mars GmbH, *History,* p. 2.

"art of working in wood": Martin, p. 122.

61 The original process: Voice, p. 135. Cf. Sutton, p. 711.

62 "a mouthful": Lefebure, p. 86.

63 "The lead cutter pounds": quoted in Fleming and Guptill, pp. 8–9.

"the end of an old pencil": Austen, p. 306.

p. 64　"I have not a word": Austen, p. 307.

　　　European graphite: Voice, p. 136.

　　　metallic lead alloyed: Voice, p. 137.

　　　"PENCIL, is also": *Encyclopaedia Britannica*, 2nd ed.

　66　"carry the week's production": Fleming and Guptill, p. 9.

CHAPTER 6　*Does One Find or Make*
a Better Pencil?

　68　"wherewith all the world": *Encyclopaedia Perthensis*, 2nd ed.

　　　"LEAD, BLACK, OF PLUMBAGO": *Encyclopaedia Perthensis*, 2nd ed., "Lead (III)" entry.

　　　"I have seen": *Encyclopaedia Perthensis*, 2nd ed., "LEAD (III)" entry.

　69　"At present": Beckmann, 4th ed., Vol. II, p. 395.

　70　Nicolas-Jacques Conté: McCloy, pp. 78–80.

　　　"every science in his head": Gaspard Monge, quoted in *Encyclopaedia Britannica*, 11th ed., "N.-J. Conté" entry.

　　　experiments with hydrogen: *Historische Bürowelt*, p. 12.

　　　Conté's innovative process: Voice, p. 137; McCloy, p. 79.

　71　early German pencil makers: Staedtler Mars GmbH, *History*, p. 4.

　　　a groove about twice: see Voice, p. 136.

　　　as early as 1790: see, e.g., Fleming and Guptill, p. 8, which dates Conté's discovery from 1790.

　　　Hardtmuth himself claimed: J. W. Hinchley, in discussion appended to Mitchell, "Black-Lead Pencils," p. 389T.

　　　Conté's son-in-law: Boyer, p. 149; *Historische Bürowelt*, p. 12; *Dictionnaire de Biographie Française*.

　　　freeing Continental pencil makers: Staedtler Mars GmbH, *History*, pp. 2–3.

　　　state-owned pencil factory: Staedtler Mars GmbH, *History*, pp. 3–4.

　72　used to some extent in England: Mitchell, "Black-Lead Pencils," p. 384T.

　　　Borrowdale graphite finally did run out: Mitchell, "Black-Lead Pencils," p. 384T.

　76　"CRAFT": Diderot, *Encyclopedia*, Gendzier's translation, p. 85.

77 "wakened hands": Lawrence, p. 448, "Things Men Have Made."

"artists who are at the same time": d'Alembert, in Diderot, *Encyclopedia,* Gendzier's translation, p. 39.

"sordid toil": Agricola, p. 1.

CHAPTER 7 *Of Old Ways and Trade Secrets*

80 German miners: Collingwood, p. 1.

Staedtler history: Staedtler Mars GmbH, *History*, pp. 1–5.

82 families of Staedtler, Jenig, and Jäger: *Die Leistung*, p. 8.

83 "bunglers": *Die Leistung,* p. 10.

"the trade was": Johann Faber, p. 3.

84 Faber history: See A. W. Faber, "A. W. Faber" and *Manufactories.* Cf. Alibert.

85 "Considering": the Hoovers, in Agricola, pp. iv–v.

"In almost all instances": Martin, p. iii.

86 "yet their confidential": Andrew, p. 2.

"a subscription for 1000": Andrew, pp. 3–4.

carbon paper: Andrew, p. 7.

87 "ten parts by weight": *Scientific American Supplement,* April 2, 1898, p. 18558.

88 "A countess": Frary, p. 8.

89 "The industry is restricted": *The Engineer,* May 11, 1917, p. 415.

CHAPTER 8 *In America*

91 "It was an inferior article": Sackett, p. 16.

92 "the goat of Acton": Hendrick, p. xxii.

Leffel's Illustrated News: Hendrick, p. 25, note 1.

"In the beginning": Hosmer, in Hendrick, p. 23.

93 "The first pencil factory": Nichols, p. 956.

"She obtained": Voice, p. 139.

94 David Hubbard: Hosmer, in Hendrick, p. 23.

95 "Boys then had to submit": William Munroe, Jr., p. 147.

96 "Before finishing": William Munroe, Jr., pp. 147–48.

Joseph Gillott: *Scientific American,* September 20, 1873, pp. 177–78.

the innovative Munroe: William Munroe, Jr., pp. 148–49.

p. 97 "And seeing": William Munroe, Jr., p. 150.

Munroe's first experiments: "The Lead pencil," anonymous typescript from the files of Robert Gooch, p. 5.

"But his mind": William Munroe, Jr., p. 150.

98 "In 1812 William Munroe": Hosmer, in Hendrick, pp. 23–24.

100 "He had great difficulty": William Munroe, Jr., p. 151.

101 "This continued until 1819": William Munroe, Jr., p. 152.

Ebenezer Wood and James Adams: Hosmer, in Hendrick, p. 25.

"hand and brain": Hosmer, in Hendrick, p. 24.

102 first machines: Hosmer, in Hendrick, p. 24. Cf. William Munroe, Jr., p. 152; Nichols, p. 956.

hexagonal and octagonal: Nichols, p. 956; Hosmer, in Hendrick, pp. 24–25.

"a gentleman in looks": Hosmer, in Hendrick, p. 25.

"had always encouraged": Allen, p. 8.

CHAPTER 9 *An American Pencil-Making Family*

104 "I dont know": Thoreau, *Correspondence,* p. 186.

106 "the first American school": see, e.g., Schodek, p. 13.

John Smeaton: Turner and Goulden, pp. 276–78.

John Rennie: Smiles, *Selections,* p. 194.

107 Joseph Dixon: *Dictionary of American Biography,* Vol. III.

"he was told": Meltzer and Harding, p. 136.

Francis Peabody: Erskine, p. 187.

108 Dunbar & Stow: Harding, *Days,* p. 16; William Munroe, Jr., pp. 297–98.

"the Lead Pencils": quoted in Meltzer and Harding, p. 138.

"I do a thorough job": private communication, Anne McGrath, curator of the Thoreau Lyceum, August 20, 1987.

"Thoreaux": Thoreau, *Correspondence,* p. 570.

Ebenezer Wood's mill: Harding, *Days,* p. 17.

109 Munroe business faltered: Harding, *Days,* pp. 17, 32, 45.

"greasy, gritty": Meltzer and Harding, p. 136. Cf. Edward Emerson, pp. 32–33.

The warm mixture: Edward Emerson, p. 135.

accepted an offer to teach: Harding, *Days,* pp. 52–54.

110 "the first Polygrade": Johann Faber, p. 3.

111 "Lothar Faber": Johann Faber, p. 3.

Harvard's library: Meltzer and Harding, p. 136.

112 letter to his brother: Thoreau, *Correspondence,* p. 23.

113 "A coarser kind": *Encyclopaedia Perthensis,* 2nd ed., "Lead, black, of plumbago," under the main entry "Lead."

114 obtained some clay: Edward Emerson, pp. 32–33.

his father's suggestion: Edward Emerson, p. 135; Harding, *Days,* p. 56.

mechanical drawings: Moss, p. 9.

"narrow churn-like chamber": Edward Emerson, p. 33.

"The machine spun around": Harding, *Days,* p. 56.

115 began his *Journal:* Thoreau, *Journal,* Vol. I, p. 592.

exchange journal passages: Thoreau, *Journal,* Vol. I, p. 594.

116 carried his diary and pencil: cf. Ralph Waldo Emerson, "Thoreau," p. 244.

"He would make our pencils": Edward Emerson, p. 3.

a memorial tribute: Meltzer and Harding, p. 49.

"improvements in the pencil line": Thoreau, *Correspondence,* p. 114.

machine to drill holes: Harding, *Days,* p. 157.

117 "I observed here pencils": Thoreau, *Journal,* Vol. II, p. 289.

Conté . . . produced round leads: McCloy, p. 79. Cf. Mitchell, "Black-Lead Pencils," p. 384T.

holes in rubies: Beckmann, 4th ed., Vol. II, pp. 395–96. Cf. *Cassell's,* Vol. IV, p. 23.

varying the amount of clay: Harding, *Days,* p. 158.

"IMPROVED DRAWING PENCILS": quoted from the illustration in Meltzer and Harding, p. 137.

118 exchange of letters: *Thoreau's Pencils.*

Thoreau pencils did cost: Meltzer and Harding, p. 139.

a Boston bookstore: Stern, p. 17.

119 labels and advertisements: see, e.g., Meltzer and Harding, pp. 137–38.

p. 119   University of Florida library: with thanks to Maggie Blades. Ives, p. 10, item HDT6. Cf. Ives, p. 11, item HDT9. Marble, p. 37, notes having a "gift-pencil bearing the stamp —'J. Thoreau & Son, Concord, Mass.' "

"J. Thoreau & Son": see, e.g., Meltzer and Harding, p. 137.

120   "JOHN THOREAU & CO.": quoted from Meltzer and Harding, p. 138.

121   invention of raisin bread: Harding, *Days,* p. 183.

Charitable Mechanic Association: *Thoreau Society Bulletin,* Winter 1961, pp. 7–8.

sound thinking about business: see, e.g., Thoreau, *Walden,* pp. 361, 366.

"The farmer is endeavoring": Thoreau, *Walden,* p. 349.

A Week on the Concord: Marble, pp. 157–58.

122   "nearly nine hundred volumes": Thoreau, as quoted in Marble, p. 158.

ground plumbago: Harding, *Days,* pp. 262–63.

"Why should I?": Thoreau, quoted in Harding, *Days,* p. 262.

"Plumbago, Prepared Expressly": see Meltzer and Harding, p. 137.

"My pen is a lever": Thoreau, *Journal,* Vol. I, p. 315 (August 4, 1841).

123   "As I was desirous": Thoreau, *Walden,* p. 549.

considered a joke: Stowell, p. 9.

"What I have observed": Thoreau, *Walden,* p. 554.

124   "H. D. Thoreau, Civil Engineer": Meltzer and Harding, p. 172.

"LAND SURVEYING": quoted from the reproduction in Harding, *Days,* facing p. 461.

"his habit of ascertaining": see *New York Review of Books,* January 15, 1987, p. 48

"he could pace": Edward Emerson, p. 242.

"I so much regret": Ralph Waldo Emerson, "Thoreau," p. 248.

CHAPTER 10   *When the Best Is Not Good Enough*

126   "a copious table": Whittock, title page.

127   "plumbago, is a dark": Whittock, p. 375.

128 One hundred thousand exhibits: see, e.g., Beaver, p. 9.

semicircular electric clock: Hunt, p. 1. Cf. *Illustrated London News,* February 8, 1851, p. 104.

129 "for subjects so important": Hunt, p. 546.

130 "the upper surface": Hunt, p. 547.

131 England, France, Germany, and Austria: *Official Catalogue,* pp. 19, 199, 236, 274.

"It is not generally known": *Tallis's History,* Vol. II, pp. 154–55.

132 "Messrs. Reeves and Sons": *Tallis's History,* Vol. II, pp. 152–53.

133 "The diamond": Hunt, p. 29.

in the north of Scotland: Hunt, p. 39.

134 "artificial plumbago": *Dictionary of National Biography,* Vol. II, p. 1279. Cf. Voice, p. 138.

135 "the somewhat recent discovery": *Tallis's History,* Vol. II, p. 153.

"close examination": Hunt, p. 40.

"best black-lead": *Illustrated London News,* September 2, 1854, p. 206.

"M. Conté, in 1795": Hunt, p. 39.

"for drawing, engineering, &c.": *Official Catalogue,* p. 896.

136 "pure Cumberland black-lead": *Scientific American,* July 27, 1850, p. 356.

"Great Britain and the Islands": *Illustrated London News,* August 5, 1854, pp. 118–19.

a natural host: *Transactions of the Newcomen Society,* 18 (1937–38): 245.

"Against the magnificent background": Marshall and Davies-Shiel, p. 13.

137 even though machinery: Cumberland Pencil Company, p. [2].

"Situated in a slightly": *Illustrated Magazine of Art,* p. 252.

"The men": *Illustrated Magazine of Art,* p. 254.

138 "The fashion of varnishing": *Illustrated Magazine of Art,* p. 254.

140 "And we might conclude": *Illustrated Magazine of Art,* p. 254.

141 "When its commercial value": *Illustrated Magazine of Art,* p. 252.

p. 141 "cut up, pounded down": *Illustrated Magazine of Art,* p. 253.

called a plummet: *Cassell's,* Vol. IV, p. 23.

143 a recent visitor to China: Getchell.

"It was essential": Clark, Vol. II, p. 806.

CHAPTER 11 *From Cottage Industry to* Bleistiftindustrie

145 Staedtler family business: J. S. Staedtler, p. [7].

146 Johann Froescheis: Lyra, "Early Days," p. 1.

"possible to make pencils": *Die Leistung,* p. 10.

German industry: E. L. Faber, p. 7; Staedtler Mars GmbH, *History,* p. 3; *Die Leistung,* p. 10.

"in all towns and cities": *Die Leistung,* p. 10.

147 sixty-three different types: Staedtler Mars GmbH, *History,* p. 4.

"the great advantage": *Die Leistung,* p. 10.

Kreutzer family: *Die Leistung,* pp. 10–11; Staedtler Mars GmbH, *History,* p. 4.

Kaspar Faber: E. L. Faber, p. 8; A. W. Faber, *Manufactories,* p. 9.

148 Georg Andreas: Lyra, "Early Days," p. 1.

Anton Wilhelm Faber: A. W. Faber, *Manufactories,* p. 9.

Lothar Faber: A. W. Faber, *Manufactories,* pp. 10–12.

149 "Since the Exhibition year": *The Builder,* July 27, 1861, p. 517.

150 Jean Pierre Alibert: Alibert, p. 21.

"to be of excellent quality": Alibert, pp. 22–23.

151 "in no way inferior": A. W. Faber, *Manufactories,* pp. 15–16.

earned him further honors: Alibert, pp. 23–26.

"and that it deposited": A. W. Faber, *Manufactories,* p. 16.

"five years more": Alibert, p. 27.

long-term and loyal workers: A. W. Faber, *Manufactories,* pp. 25–26.

152 "unwilling, or cannot afford": Alibert, p. 16.

"He himself dwells": Alibert, pp. 16–17.

153 King Max: Alibert, p. 33.

"who ever after will keep": Alibert, p. 20.

"The first car": Alibert, pp. 36–37.

154 "in spite of the periods": Alibert, p. 38.

155 "I dedicate this album": Alibert, pp. 38–39.

the weak joint: *Scribner's,* p. 807.

placed on the market: Alibert, p. 34.

157 extend the range: Alibert, pp. 27–28.

"purified black lead pencils": *Knight's Cyclopaedia,* "plumbago" entry.

letter designations originated: see Watrous, p. 163, note 17. Cf. Langlois-Longueville, p. 286.

Conté used the numbers: Larousse, *"crayon"* entry.

Both German and French: *Meyers Enzyklopädisches Lexicon* (Mannheim, 1978), *"Bleistift"* entry; Boyer, p. 150.

158 "the appellation 'Siberian Graphite' ": A. W. Faber, *Price-List,* p. 12.

"I call special attention": A. W. Faber, *Price-List,* p. 10.

159 an established name: A. W. Faber, "History," pp. 52–55; A. W. Faber-Castell, "Origin and History," p. 1.

160 "At that time": Johann Faber, p. 5.

Among the problems: Johann Faber, pp. 3–4.

161 Siberian graphite from elsewhere: Johann Faber, p. 14.

Franz von Hardtmuth: Carlo Gherra, in *The Pencil Collector,* 28, No. 8 (September 1984): 1.

162 "the great diamond": Fleming and Guptill, p. 24.

"In goods": Fleming and Guptill, p. 3.

"the original yellow pencil": Fleming and Guptill (New York ed.), p. 20.

163 "natural polished": A. W. Faber, *Price-List,* e.g., p. 13.

"It may make a pencil look well": Smithwick, p. 351.

CHAPTER 12 *Mechanization in America*

164 "Strange to say": Day, pp. 10–11.

165 "Established 1827": Day, p. 11.

Joseph Dixon was born in 1799: Erskine, pp. 186–87. Cf. *Dictionary of American Biography,* Vol. III.

pencils dating from about 1830: *Scribner's,* p. 810.

166 crucibles: See, e.g., Cleveland.

p. 167 $5,000 loss on pencils: Erskine, p. 190. Cf. Nichols, p. 956; *Encyclopaedia Americana,* "lead-pencils" entry.

"solid black lead": Cleveland, p. 35.

"By a private door": Jersey City *Evening Journal,* December 20, 1872, quoted in Cleveland, pp. 33–34.

168 "Certain Germans": Cleveland, tipped-in folder for black-lead pencils.

169 imitation Dixon Stove Polish: Cleveland, tipped-in folder for stove polish.

Among the first machines: see Erskine, p. 190; *Scribner's,* p. 809.

"birthplace of the world's": *New York Times,* January 27, 1974, p. 74.

"only three quarters": *Manufacturer and Builder,* p. 81.

170 "the Dixon Company are": Elbert Hubbard, pp. 18–19, 22.

Dixon's son-in-law: E. L. Faber, p. 12.

171 "became absorbed": *New York Times,* March 4, 1879, Eberhard Faber obituary.

securing cedar tracts: E. L. Faber, p. 9.

172 "the ingenuity of the Americans": Johann Faber, p. 4.

It was the best of times: see *New York Times,* March 4, 1879, Eberhard Faber obituary.

"except during periods": Wharton, p. 159.

173 "the oldest pencil factory": see Eberhard Faber Company, *Story,* pp. 4, 6.

174 "the right of title": E. L. Faber, p. 10.

When Lothar Faber died: Eberhard Faber Company, "Since 1849."

Berolzheimer: E. L. Faber, pp. 11–12.

moved to California: telephone conversation with Charles Berolzheimer, October 25, 1988.

acquired in 1988: Nashville *Banner,* September 17, 1988.

175 Edward Weissenborn: Venus, "100 Years," p. 2.

American Lead Pencil Company: E. L. Faber, p. 11. Cf. Faber-Castell Corporation, "Date Log," p. 1.

"There was a great demand": Hosmer, in Hendrick, p. 59.

"A pencil maker offered": Hosmer, in Hendrick, p. 27.

176 "In 1864 I took": Hosmer, in Hendrick, p. 28.

Hosmer also finished: Hosmer, in Hendrick, pp. 28, 86.

177 "the very convenient method": Beckmann, 3rd ed., Vol. IV, p. 356. Cf. Beckmann, 4th ed., Vol. II, p. 393, where "gum elastic" is replaced with "Indian rubber."

"a substance excellently adapted": Joseph Priestley, quoted in Speter, p. 2271.

tree resin: Dick Walker, p. 83.

"indiarubber": Riddle, p. 538.

metal point protectors: see *Dictionary of American Biography,* Vol. VI, "John Eberhard Faber" entry.

"a lead pencil": *Scientific American,* July 4, 1863, p. 11.

"secured a piece of prepared rubber": Kane, p. 454. For variant spellings of Hyman Lipman's name, see Bump, p. cxcii, and *Encyclopaedia Britannica,* 15th ed.

Joseph Reckendorfer: *Encyclopaedia Americana,* International ed., 1986, "pencil" entry.

"no joint function": de Camp, *Heroic Age,* p. 101.

178 Eagle Pencil Company: See McClurg, p. 123, item no. 140, which shows the patent date to be May 21, 1872.

less than a penny each: Sears, Roebuck *Catalogue,* Fall and Winter 1941–42, p. 590.

90 percent of American pencils: see *Scientific American,* August 22, 1903, p. 137.

"fix pencil marks": *Scientific American,* February 19, 1881, p. 121.

over 700 pencil styles: Dixon Crucible Company, *Pencillings,* pp. [10–11].

"Soon after the appearance": Dixon Crucible Company, *School Pencils,* p. 24.

179 "mouthpiece, adapted": McClurg, p. 125.

"The truth is": Dixon Crucible Company, *School Pencils,* p. 25.

180 Italian soldiers: Wharton, p. 158.

it is the ferrule: Cf. Ecenbarger, p. 17.

CHAPTER 13 *World Pencil War*

182 One observer: *Scientific American,* May 26, 1894, p. 332.

183 duty on imported pencils: *Philadelphia International Exhibition, 1876, Official Catalogue of the British Section,* Part I, p. 243.

if one pencil was missing: *Scribner's,* p. 809.

p. 183 with leads cut from graphite: *Literary Digest,* June 5, 1920, p. 98.

184 simple machine: Smithwick, pp. 350–51.

Germany had introduced: see, e.g., A. W. Faber, "A. W. Faber," p. 4; *Dictionary of National Biography,* Vol. II, p. 1279; Remington, p. 26; Rocheleau, p. 130.

185 grooving two slats: Remington, p. 25.

"ten hands": *Scientific American,* April 12, 1884, p. 226.

186 "The machine separates": *Scribner's,* pp. 807–8.

187 "Johann Faber": Doyle, p. 700.

Bavaria's . . . pencil factories: Stephan, pp. 191–92; *Scientific American,* December 21, 1895, p. 387.

188 "suffering severely": *Scientific American,* December 8, 1900, p. 359.

another consul: *Scientific American,* June 8, 1901, p. 358.

"1000 Workmen": General Imperial Commissioner, p. 437.

American copying pencils: *New York Times,* February 29, 1914, p. 13.

"The Japanese": London *Times,* January 6, 1916, p. 4.

England was the biggest importer: *New York Times,* April 20, 1919, p. 7.

raw materials: *New York Times,* November 28, 1920, p. 19.

189 prices for the American pencil wood: *Literary Digest,* January 17, 1920, pp. 98–99.

Japan: *New York Times,* April 16, 1922, Sect. VI, p.6.

Johann Froescheis: *New York Times,* November 2, 1915, p. 16.

A. W. Faber: U.S. Court of Customs Appeals Reports, Vol. XVI (1929), pp. 467–71.

London County Council: London *Times,* March 1, 1921, p. 17.

"displacement": *Scientific American Supplement*, Nov. 22, 1919, p. 303.

Alien Property Board: A. W. Faber, "A. W. Faber," pp. 11, 13. Cf. Faber-Castell GmbH, *Bleistiftschloss,* p. 98.

191 "excellent and well-known": London *Times,* January 25, 1906, p. 15.

"in the Koh-I-Noor": *New York Times,* November 9, 1906, p. 5.

"yellow pencil": Vivian, p. 6931.

Koh-I-Noor Pencil Company: Vivian, pp. 6925, 6929, 6931.

Manx cat: Vivian, p. 6928.

192 "fine goods": see, e.g., Dixon Crucible Company, *Standard Graphite,* p. 62.

"finer and softer": *Scribner's,* p. 805.

"The coarsest and heaviest": *Scribner's,* p. 805.

193 "For the cheapest pencil": *Scribner's,* p. 806.

Venus de Milo: Venus, "100 Years," pp. 3–4.

CHAPTER 14 *The Importance of Infrastructure*

196 "Cars must come before roads": Henry Ford, quoted in Hammer, *Quest,* p. 106.

197 one of the largest cantilever: Fraser, pp. 133–34.

"Toredos": Fraser, pp. 134–35.

"pulled out a pocketknife": Fraser, p. 135.

198 Saul Steinberg: with thanks to Charles Blitzer, who first showed me a Steinberg pencil.

"the study of architecture": Saul Steinberg, in Rosenberg, p. 235.

English riddle: Taylor, p. 40, no. 101.

199 "Lead pencils are designed": Nichols, p. 956.

201 red cedar was imported: Fleming and Guptill, p. 10.

202 "one seventh of all": Melvil Dewey, quoted in Tichi, p. 67.

"Historical exhibits": Adams, p. 341.

Only one-fifth: Decker, p. 108.

future supply of red cedar: Sackett, p. 46.

203 fallen trees: *Scientific American,* September 13, 1890, p. 160.

rotting trees: *Scientific American,* April 27, 1912, p. 386.

"the supply": *New York Times,* July 8, 1911, p. 3.

"the average pencil": *Scientific American,* September 13, 1890, p. 160.

"in the ordinary": *New York Times,* July 8, 1911, p. 3.

as much as 40 percent: Godbole, p. 21.

editorial: *New York Times,* July 10, 1911, p. 6.

204 "A good pencil wood": Sackett, p. 46.

graded in this way: Russo and Dobuler, p. 15.

205 still buying up old fence posts: Cf. *Deschutes,* p. 2.

p. 206   wood was dyed: Faber-Castell, "Story of the Lead Pencil," p. 4.

impregnated with wax: California Cedar, p. [6].

basswood and alder: Helphenstine, p. 654.

mutarawka: *New York Times,* April 13, 1924, Sect. III, p. 13.

Little St. Simons Island: telephone conversation with the island's resident naturalist, James Bitler, October 25, 1988. With thanks to Rebecca Vargha.

Siberian redwood: *New York Times,* June 9, 1928, p. 21.

"The treatment gives": Sackett, p. 46.

207   "The pencil-using public": Sackett, p. 46.

Cutting triangular pencils: Voice, p. 141.

woodworking machinery: G. S. MacDowell to Smithsonian Institution, letter dated March 1, 1976. Cf. *World Book Enyclopedia,* 1988 ed., "pencil" entry.

208   prefer hexagons: Decker, p. 108.

209   Marc Isambard Brunel: Beamish, especially Ch. VIII; Turner and Goulden, pp. 361–62.

earned a royalty: Beamish, pp. 96–97.

"The tolerances": Nichols, p. 957.

210   "Wait till you hear": C. Lester Walker, p. 91.

average . . . tree yields: *Compton's Encyclopedia,* 1986 ed., "pencil" entry.

210   "many, many thousands": Empire Pencil Company, promotional pencil card, 1974.

211   Pencil Street: Metz, p. 1.

red cedar was still abundant: undated clipping from Robert Gooch.

"triple coextrusion": *Modern Plastics,* April 1976, p. 53.

CHAPTER 15   *Beyond Perspective*

213   "The stylus": Gesner, quoted in Meder, p. 121, note 2.

215   pencil definition: *Webster's New Collegiate Dictionary,* 1961 ed.

216   "Many objects": Ferguson, "Mind's Eye," p. 827.

"If there had been no": Pye, p. 72.

218   "Pencils must be round": Steinbeck, p. 47.

"big flat lead": Fleming and Guptill (New York ed.), pp. 22–23.

Thomas Wolfe: Wharton, p. 156.

"This shape prevents": Israel, p. 352.

220 "In fact, all kinds": Vitruvius, Book VI, ch. viii, para. 10.

221 Perspective drawings appeared: cf. Ferguson, "Mind's Eye," p. 831.

223 colorful cover: *Engineering News-Record,* May 21, 1981.

"*ENR* got almost": *Engineering News-Record,* May 13, 1982, p. 9. See *Engineering News-Record,* June 18, 1981, p. 9, for letters.

224 Although orthographic projection: cf. Ferguson, "Mind's Eye," p. 831.

theoretical foundations: Baynes and Pugh, p. 32; see also Booker.

architectural drawing: Booker, p. 135.

"the usual mode": Binns, pp. vii–viii.

225 "To find the end elevation": Binns, p. 9.

"The object of a section": Binns, p. 14.

227 engineering drawing instruments: V. & E., pp. 1 ff.

Pens made from the quills: V. & E., pp. 3–4.

228 a trade throughout Europe: Dickinson, "Brief History," p. 75.

"the graphite stick": Gautier, quoted in Meder, p. 121, note 15.

George Washington's set: V. & E., pp. 6–7.

red-morocco pocket case: Oliver Hubbard, p. 153.

229 mechanical drafting pencil: A. W. Faber, "A. W. Faber," p. 12.

"A satisfactory drawing pencil": Svensen and Street, p. 24.

graphitic carbon content: Mitchell, "Black-Lead Pencils," p. 388T.

230 twenty-one pencil grades: *Encyclopaedia Britannica,* 15th ed., "writing" entry.

No matter what the designation: Halse, p. 85.

"How the paper": Steinbeck, p. 12.

234 Color continued to be used: Baynes and Pugh, p. 175.

Blueprints were available: Andrew, p. 14.

"three large freight-car loads": V. & E., pp. 12–13.

p. 235 "smooth the path": *Journal of Engineering Graphics,* 24 (February 1960): advertisement bound between pp. 18 and 19.

"any fool can tighten": quoted in Rolt, p. 153.

CHAPTER 16 *The Point of It All*

238 "sharpened to a needle": A. W. Faber, Inc., p. 7.

"a stronger lead-to-wood bond": Berol USA.

"smoothest pencil": Eagle Pencil Company, *1940 Eagle Catalog,* p. 14.

239 29 percent sharper: *New York Times,* August 4, 1950, p. 29.

"To those who have": Seeley, "Manufacturing," p. 686.

240 "the literature on the subject": Mitchell, "Black-Lead Pencils," p. 383T.

"The process of drying": Mitchell, "Black-Lead Pencils," p. 390T.

241 "The introduction": *New York Times,* August 22, 1960, p. 34.

242 "pencil clays": Seeley, "Carbon," p. 331.

"a baked ceramic rod": Seeley, "Carbon," pp. 331–32.

244 spreads the wood apart: Peterson, p. 25.

"pressure point": Venus, "How Venus."

Eagle Pencil: C. L. Walker, p. 90.

hide glue: Peterson, p. 25.

245 breaking up inside: Berol Limited, p. 5.

"Some time ago": Cronquist, p. 653.

247 "The Amateur Scientist": Jearl Walker, pp. 162–64.

248 more detailed analysis: Cowin, p. 453.

characteristic slanted surface: Petroski, "On the Fracture," p. 732.

249 carpenter's pencil: Binns, p. 14.

pencil catalogues show: Johann Faber, p. 11; A. W. Faber, *Price-List,* p. 16.

"For technical": A. W. Faber, *Price-List,* p. 16.

249 analysis predicts: Petroski, "On the Fracture," pp. 731–33.

250 "strength or thickness": Binns, p. 14.

diameter of the lead: see, e.g., Svensen and Street, Fig. 2.2; Hoelscher and Springer, Fig. 3.2; Giesecke, et al., Fig. 60.

Conté himself made: Fleming and Guptill, p. 10. Cf. Mitchell, "Black-Lead Pencils," p. 384T.

253 "First, place the Glasses": reproduced in Turner, p. 387.

"Kitchin-Boys": reproduced in Turner, p. 387.

CHAPTER 17   *Getting the Point, and Keeping It*

254 "I went to Mr. Ross's": Pinck, p. 6.

255 schoolmaster's chief distraction: Thayer, p. 1. Cf. Bell, p. 222.

"The fashioning": *New York Times,* August 21, 1923, p. 9.

"If the point broke": *Harper's,* November 26, 1910, p. 26.

256 "it is better": *Edinburgh Encyclopaedia,* 4th ed., 1830, "Drawing Instruments" article, "black lead pencils" sidenote.

"It is usually held": *Scientific American,* November 26, 1904, p. 380.

knife-guiding device: *Scientific American,* February 4, 1911, p. 122.

small carpenter's plane: *Scientific American,* May 15, 1909, p. 376.

Sherlock Holmes: Doyle, pp. 696, 700.

258 sharpened without a knife: cf. *Scientific American Supplement,* May 27, 1905, p. 24576.

"We have spent": Dixon Crucible Company, [*1891*] *Catalog,* p. 28.

Johann Faber: Johann Faber, pp. 22, 28.

Gem pencil sharpener: *Scientific American,* May 11, 1889, p. 290.

259 "point a red or blue": *Scientific American,* December 20, 1913, pp. 478–79.

"Borrowing neighbor's knife": *Scientific American,* December 20, 1913, pp. 478–79.

260 about one thousand pencils: *Scientific American,* October 29, 1910, pp. 345–46.

Apsco: *Modern Plastics,* June 1958, pp. 111–12; *System,* December 1927, p. 754; *System,* September 1927, p. 320.

"install with screws": *Consumers' Research Magazine,* November 1979, p. 18.

261 José Vila: *New York Times,* November 8, 1980, patents column; *The New Yorker,* December 29, 1980, pp. 29–30.

p. 261 largest and most precise: advertisement, *Mechanical Engineering,* October 1956, p. 31.

262 "To sharpen": Kirby, p. 2.

"If you use": U.S. Bureau of Naval Personnel, p. 148.

263 unsymmetrical cutting: *Engineering,* September 2, 1938, p. 292.

"the long lead": in Fleming and Guptill (New York ed.), pp. 42–43.

"a cheap-looking affair": *Scientific American,* September 13, 1890, p. 160.

Baroque designs: Hambly, *Drawing Instruments,* pp. 65–66.

Sampson Mordan: see Banister.

James Bogardus: *Dictionary of American Biography,* Vol. II.

265 toothpicks and ear spoons: E. S. Johnson.

leads tended to give: *Consumers' Research Bulletin,* December 1944, p. 17.

Eversharp pencil: Frary, pp. 3–8.

266 a company chemist: Back, pp. 571, 579.

267 "a somewhat inferior product": Frary, p. 8.

twelve million Eversharps: Frary, pp. 146, 149.

"producing in the minds": Frary, p. 149.

"whether Canton": *Printers' Ink,* December 13, 1923, pp. 115–16, 119–20.

268 Eversharp ads: quoted from *System,* December 1922, p. 732; *System,* November 1922, inside front cover.

"a seven-inch wood pencil": quoted in C. L. Walker, p. 90.

269 parody: Sykes, pp. 652–53.

Venus Everpointed: see, e.g., *Mechanical Engineering,* November 1922, p. 109.

Charles Wehn: *Sales Management,* November 20, 1951, pp. 74–75.

Scripto: *Business Week,* December 17, 1966, pp. 168, 171–72, 174.

270 one bank: *Business Week,* March 30, 1932, p. 10.

exporters to Argentina: *Foreign Commerce Weekly,* February 22, 1941, p. 332.

Scripto advertised: *Life,* March 11, 1946, p. 64.

Eversharp, remembered: *Business Week,* December 30, 1939, p. 22.

271   "guaranteed not for years": see *New York Times,* September 19, 1941, p. 41.

Consumers' Research Bulletin: January 1944, p. 26, and December 1944, p. 17.

Eversharp repeating pencil: Cliff Lawrence, pp. 49, 51.

272   Federal Trade Commission: *Credit and Financial Management,* August 1953, p. 34; U.S. Federal Trade Commission.

plastics: *Modern Plastics,* December 1972, p. 46.

even finer "fine-line": *Engineering,* November 17, 1961, p. 664.

ultra-thin lead: *Consumer Bulletin,* January 1973, pp. 28–30. Cf. Staedtler Mars GmbH, *Aktuell '83.*

273   Yellow Pencil: *Modern Office Technology,* March 1984, p. 104.

CHAPTER 18   *The Business of Engineering*

274   "A friend who attended": Edward Emerson, pp. 34–35.

275   "I know what": Hosmer, in Hendrick, p. 84.

276   "The father Thoreau": quoted in Hendrick, p. 132.

"and a man engaged": Edward Emerson, pp. 35–36.

Francis Munroe: William Munroe, p. 72.

277   "The engineer is both": Layton, p. 1.

a young physician: Hammer, *Quest,* ch. 1.

278   "The two countries": Hammer, *Quest,* pp. 62–63.

"Why don't you take": Hammer, *Quest,* p. 64.

279   terms of the concession: Hammer, *Quest,* pp. 81–82.

agency for all Ford products: Hammer, *Quest,* p. 109.

"more cheaply": Hammer, *Quest,* p. 179.

"I went into a stationery store": Hammer, *Quest,* pp. 179–80.

280   "had produced nothing": Considine, p. 62.

"a million dollars' worth": Hammer, *Quest,* p. 183.

"record time": Hammer, *Quest,* p. 183.

281   "Now the little old town": Hammer, *Quest,* pp. 186–87.

282   "an engineer who held": Considine, p. 64.

283   "The pencil masters": Considine, p. 65.

p. 283 "toy-shop of the world": *Illustrated London News,* February 22, 1851, p. 148.

"a closed industry": Hammer, *Quest,* pp. 189–90. Cf. Timmins, pp. 633–37; *Illustrated London News,* February 22, 1851, pp. 148–49.

Hammer had imported: *New York Times,* June 9, 1928, p. 21.

demand was so great: Hammer, *Quest,* p. 200.

284 export about 20 percent: Hammer, *Quest,* p. 207.

Statue of Liberty: Hammer, *Hammer,* p. 171.

first-year earnings: Hammer, *Quest,* p. 208; *New York Times,* June 9, 1928, p. 21.

Diamond: Hammer, *Hammer,* p.171.

A. HAMMER: Hammer, *Hammer,* p. 171. Cf. Belyakov, pp. 48–49.

Nikita Khrushchev: Hammer, *Hammer,* p. 171.

"slowness and laxity": Hammer, *Quest,* p. 201.

"Under the spur": Hammer, *Quest,* p. 201.

285 "Professors, authors": quoted in Finder, p. 49.

Sacco and Vanzetti Pencil Factory: Goldman, p. 249.

"imaginary pencils": *New York Times,* November 24, 1938, p. 3, and December 4, 1938, p. 51.

CHAPTER 19 *Competition, Depression, and War*

287 an accurate indicator: *Business Week,* January 2, 1943, p. 60.

increased tariffs: *New York Times,* December 28, 1921, p. 7.

high cost of machinery: U.S. Bureau of Labor, pp. 11, 13.

Argentina: *New York Times,* November 18, 1923, Sect. II, p. 14.

288 England: London *Times,* May 13, 1927, p. 18.

Mr. Kirkwood: London *Times,* December 7, 1933, p. 16.

Board of Trade's: London *Times,* March 14, 1930, p. 11; cf. Great Britain Board of Trade, Cmd. 4278.

Standard Pencil Company: Weaver, p. 514.

General Pencil Company: *Diesel Power and Diesel Transportation,* January 1942, p. 42.

Eberhard Faber: Hartmann, p. 356.

290 Eberhard Faber II: quoted in *Sales Management and Advertising Weekly,* January 26, 1929, p. 219.

Great Depression: U.S. Department of Labor, p. 5 and Fig. I.

"severance of our business": H. B. Elmer to C. H. Watson, letter dated May 11, 1932, in the J. Walter Thompson Company Archives, Manuscript Department, William R. Perkins Library, Duke University.

291 Big Three: *New York Times,* May 23, 1931, p. 31; *Business Week,* July 29, 1931, p. 39.

imported pencils . . . foreign markets: U.S. Tariff Commission, various pages.

292 Japanese pencils: *New York Times,* June 19, 1933, p. 1; Godbole, pp. 42, 47; U.S. Tariff Commission, various pages.

In Argentina: *New York Times,* November 15, 1936, Sect. III, p. 9.

eighteen million per year: U.S. Tariff Commission, p. 17.

cotton rugs and matches: *New York Times,* April 22, 1934, Sect. II, p. 19.

293 thirteen firms: U.S. Tariff Commission, pp. 5–6.

Simplified Practice: U.S. Bureau of Standards. Cf. contemporary bound volumes of Simplified Practice Recommendations.

295 steps toward standardization: *New York Times,* November 4, 1938, p. 37, and August 31, 1939, p. 26; *Oil Paint and Drug Reporter,* November 7, 1938, pp. 3, 64, and September 4, 1939, pp. 5, 54.

"a clearing house": U.S. Department of Labor, p. 19.

Pencil Industry Export Association: *Commerce Reports,* November 15, 1939, p. 1083.

Eagle Pencil Company: see, e.g., *New York Times,* June 21, 1938, p. 42; also, June 23, p. 4; June 24, p. 2; July 12, p. 7; July 16, p. 28; July 17, p. 8; July 29, p. 9; August 9, p. 6. Cf. July 8, 1937, p. 6.

296 about half the workers: U.S. Department of Labor, p. 19.

cut off supplies: U.S. Department of Labor, p. 23.

Pearl Harbor: Metz, p. 1.

misrepresenting their pencils: *New York Times,* December 15, 1942, p. 42.

"Remote from the champagne": *Civil Engineering,* March 1942, p. 31.

p. 297 plastic ferrules: *Modern Plastics,* September 1944, p. 98; *Scientific American,* January 1945, p. 29.

Ticonderogas: see, e.g., *Liberty,* September 1948, inside front cover; September 15, 1945, inside front cover; May 18, 1940, p. 29.

War Production Board: *Business Week,* January 2, 1943, p. 61; *New York Times,* February 21, 1943, pp. 1, 24.

labor force: *Business Week,* January 2, 1943, p. 61.

298 "Few will regret": *Economist,* June 6, 1942, p. 806.

CHAPTER 20   *Acknowledging Technology*

299 The Netherlands: *Foreign Commerce Weekly,* January 11, 1947, p. 25.

an army corporal: *New York Times,* January 6, 1949, p. 47.

300 Atomic Products: *New York Times,* September 21, 1951, p. 40.

"This week pencils": London *Times,* September 26, 1949, p. 5.

"with an eye to": *Sales Management,* September 1, 1945, p. 96.

"featuring its factory": *Printers' Ink,* April 28, 1927, pp. 49–50.

graphologist: *Printers' Ink,* April 28, 1927, pp. 50, 52.

301 "mosey around": *Printers' Ink,* May 2, 1935, p. 21.

Isador Chesler: *Printers' Ink,* May 2, 1935, pp. 23–25. Cf. Callahan.

302 Abraham Berwald: *The New Yorker,* June 27, 1953, pp. 18–19.

303 American consumption: *New York Times,* March 9, 1953, p. 38.

Big Four still dominated: *Business Week,* August 9, 1952, p. 52.

government filed suit: *New York Times,* January 27, 1954, p. 35.

"five-cent pencil": *New York Times,* March 9, 1953, p. 38.

Rising costs: *New York Times,* September 30, 1956, Sect. III, p. 10.

"America's standard": *Business Week,* January 22, 1955, p. 69.

304 sell exclusively: *Sales Management,* March 12, 1932, p. 401.

individual pencil buyer: *New York Times,* September 30, 1956, Sect. III, p. 10.

"people were buying": *Printers' Ink,* March 8, 1957, p. 28; see, e.g., *Life,* March 19, 1956, for ad copy.

Wilkes-Barre: *New York Times,* September 30, 1956, Sect. III, p. 10; *Newsweek,* July 1, 1957, pp. 54–55.

"to distinguish": *Modern Packaging,* October 1956, p. 132.

305 Empire: *Newsweek,* July 1, 1957, p. 54.

Japan: *New York Times,* February 4, 1955, p. 8.

"Wood, graphite and clay": Godbole, p. 6.

$4 million: *Foreign Commerce Weekly,* November 22, 1947, p. 23.

306 Controller of Printing: Joglekar, Nayak, and Verman, p. 75.

indigenous woods: Rehman and Ishaq, p. 1.

deodar: Rehman and Ishaq, pp. 2, 6; Rehman and Kishen, p. 2; Rehman and Kishen, p. 512; Rehman, pp. 1–2; Marathe, Iyenger, and Joglekar, p. 17.

307 electrical resistance: Joglekar, Nayak, and Verman.

strength: Marathe, Iyenger, and Joglekar, pp. 17–19.

blackness: Joglekar and Marathe, pp. 78–79.

wearing quality: Marathe, Iyenger, and Joglekar, pp. 20–22.

308 friction: Marathe, Chand, and Joglekar, p. 132.

"Specification": Indian Standards Institution, pp. 2–3.

American standard: U.S. General Services Administration.

309 "Looking in on Eagle's": Callahan.

310 research papers: e.g., Marathe, Iyenger, and Joglekar; Joglekar, Gopalaswami, and Kumar; Joglekar; Joglekar, Bulsara, and Chari.

311 Blackfeet Indians: *Forbes,* February 16, 1981, pp. 106–10, and July 29, 1985, p. 14.

CHAPTER 21   *The Quest for Perfection*

312 "All goods": quoted in Ecenbarger, p. 16.

Salom Rizk: *New York Times,* January 12, 1954, p. 22 (letter), and March 6, 1952, p. 45.

313 Camp Fire Girls: *New York Times,* July 19, 1947, p. 16.

"He could estimate": Ralph Waldo Emerson, "Thoreau," p. 242.

p. 313   "Emerson spoke": Hosmer, in Hendrick, p. 11.

      "One of the needs": Elbert Hubbard, p. 23.

315   "This is merely a board": *Scribner's,* p. 808.

      "Taking a bundle": *Machinery Market,* 1938, p. 1050.

318   "The Black Lead": reproduced in Whalley, *Writing Implements,* p. 121.

320   "reserves the right": Dixon Crucible Company, *1940–1941 Catalog,* p. 96.

      "directions for use": reproduced in Whalley, *Writing Implements,* p. 121.

323   fountain pen: Cliff Lawrence, pp. 3–19.

      Alonzo Cross: *Nation's Business,* December 1974, p. 56; *Tooling & Production,* April 1978, pp. 94–95.

      ball-point pen was patented: Kane, p. 454.

      "The wood pencil seems": Wharton, p. 156.

324   "vest-pocket high": Callahan.

      "liquid graphite": see, e.g., *New York Times,* January 11, 1955, p. 40, and February 19, 1955, p. 20.

      "wearing down seven": Hemingway, quoted in Winokur, p. 124.

      John Steinbeck: Steinbeck, p. 36.

      poem: Sandburg, p. 199.

      "I watched the girl": Hemingway, pp. 5–6.

325   "They can't move": Steinbeck, p. 61.

      damp day: Steinbeck, p. 118.

      sixty pencils a day: Steinbeck, p. 36.

      "480 #2⅜ round": Steinbeck, p. 131.

      "For years": Steinbeck, pp. 35–36.

326   disowns its title: Murry, p. vi.

      "The pen of my dream": Murry, pp. 43–44.

327   "I am the pencil": Anonymous, quoted by Berol Limited.

329   *"Then felt I"*: Keats, p. 32.

CHAPTER 22   *Retrospect and Prospect*

331   editorialized: *New York Times,* August 22, 1938, p. 12.

      fourteen billion: Ecenbarger, p. 15. Cf. Thomson, p. 31, where the total production of forty countries is put at six billion annually.

332   "We are sure": Schrodt, p. 32.

"To initialize": Schrodt, p. 32.

"Have you considered": Porter, p. 66.

337   "reverse engineering": with thanks to Charles Townshend.

338   carpenter's chalk: Huxley.

"the child who masters": Faraday, p. vii.

"searching for a new pattern": Lindbergh, p. 10.

339   "One man draws the wire": Adam Smith, p. 3.

"perhaps the most important": Babbage, p. 169. Cf. pp. 176–90.

Milton Friedman: Metz, p. 6.

APPENDIX A   *From "How the Pencil Is Made"*

343   "The graphite": Fleming and Guptill, pp. 11–14. Reprinted courtesy of Koh-I-Noor Rapidograph, Inc.

APPENDIX B   *A Collection of Pencils*

347   monthly newsletter: Arthur T. Iberg, editor, 491 Pike Drive East, Highland, Ill. 62249.

Writing Equipment Society: Maureen Greenland, Secretary, 4 Greystones Grange Crescent, Sheffield S11 7JL, England.

collector in North Dakota: *People Weekly,* February 1, 1988, p. 81.

collector in Russia: Belyakov.

Texas man: *Popular Science,* May 1939, p. 126.

King Tut's tomb: *New York Times,* May 2, 1924, Sect. IX, p. 2.

348   "over seven hundred styles": Dixon Crucible Company, *Pencillings,* pp. [10–11].

349   trick pencil: Slocum, p. 114.

350   "deceptive pencil": *Scientific American Supplement,* May 19, 1888, pp. 10322–23.

"Remember Gordon": see, e.g., *Illustrated London News,* October 14, 1899, p. 552, and December 16, 1899, p. 880.

351   "monument to longevity": *ARTnews,* February 1981, p. 89.

"at long last": *ID: Industrial Design,* January–February 1985, p. 77.

kosher butchers: Remington, p. 24.

p. 352   Aristotle's anomaly: see *Journal of the American Medical Association,* December 17, 1982, p. 3095.

"oversize eraser": see, e.g., Eagle Pencil Company, *1940 Catalog,* pp. 16, 23.

felt their ferrules: Steinbeck, p. 47.

John Updike's words: *The New Yorker,* January 23, 1989, p. 34.

353   "Now and then he took": Waugh, p. 65.

"withdrawn from the sheath": Army and Navy Co-op, p. 430, item 951.

354   "the very best pencil": Cornfeld and Edwards.

"It's covered with wood": Stephen Doyle, in *ID: Industrial Design,* November–December 1988, p. 56. With thanks to Dianne Himler.

Bibliography

Acheson, E. G. "Graphite: Its Formation and Manufacture," *Journal of the Franklin Institute,* June 1899: 475–86.

———. *A Pathfinder: Discovery, Invention, and Industry.* New York, 1910.

Adams, Henry. *The Education of Henry Adams: An Autobiography.* Boston, 1918.

Agricola, Georgius. *De Re Metallica.* Translated by Herbert Clark Hoover and Lou Henry Hoover. New York, 1950.

Alibert, J. P. *The Pencil-Lead Mines of Asiatic Siberia. A. W. Faber. A Historical Sketch. 1761–1861.* Cambridge, 1865.

Allen, Andrew J. *Catalogue of Patent Account Books, Fine Cutlery, Stationery, [etc.].* Boston, [1827].

American Society for Testing and Materials. "Standard Practice [D4236-85] for Labeling Art Materials for Chronic Health Hazards," *Annual Book of ASTM Standards,* Vol. 06.01. Philadelphia, 1986.

Andrew, James H. "The Copying of Engineering Drawings and Documents," *Transactions of the Newcomen Society,* 53 (1981–82): 1–15.

Anthony, Gardner C. *Elements of Mechanical Drawing.* Revised and enlarged edition. Boston, 1906.

Aristotle. *Minor Works.* With an English translation by W. S. Hett. Cambridge, Mass., 1936.

Army and Navy Co-operative Society Store. *The Very Best English Goods: A Facsimile of the Original Catalogue of Edwardian Fashions, Furnishings, and Notions Sold at the Army and Navy Co-operative Society Store in 1907.* New York, 1969.

Armytage, W. H. G. *A Social History of Engineering.* London, 1961.

The Art-Journal. "The Crystal Palace Exhibition Illustrated Catalogue, London 1851." New York, 1970.

Asimow, Morris. *Introduction to Design*. Englewood Cliffs, N.J., 1962.

Astle, Thomas. *The Origin and Progress of Writing, [etc.]*. 2nd edition, with additions (1803). New York, 1973.

Atkin, William K., Raniero Corbelletti, and Vincent R. Fiore. *Pencil Techniques in Modern Design*. New York, 1953.

Austen, Jane. *Emma*. Edited with an introduction by David Lodge. London, 1971.

Automatic Pencil Sharpener Company. "From Kindergarten Thru College." [Folder.] Chicago, [1941].

Babbage, Charles. *On the Economy of Machinery and Manufactures*. 4th edition enlarged (1835). New York, 1963.

Back, Robert. "The Manufacture of Leads for the Mechanical Pencil," *American Ceramic Society Bulletin*, 4 (November 1925): 571–79.

Baker, Joseph B. "The Inventor in the Office," *Scientific American*, October 29, 1910: 344–45.

Banister, Judith. "Sampson Mordan and Company," *Antique Dealer and Collectors' Guide*, June 1977: [5 pp.] unpaged.

Basalla, George. *The Evolution of Technology*. Cambridge, 1988.

Bay, J. Christian. "Conrad Gesner (1516–1565), the Father of Bibliography: An Appreciation," *Papers of the Bibliographical Society of America*, 10 (1916): 52–88.

Baynes, Ken, and Francis Pugh. *The Art of the Engineer*. Woodstock, N.Y., 1981.

Bealer, Alex W. *The Tools That Built America*. Barre, Mass., 1976.

Beamish, Richard. *Memoir of the Life of Sir Marc Isambard Brunel*. London, 1862.

Beaver, Patrick. *The Crystal Palace, 1851–1936: A Portrait of Victorian Enterprise*. London, 1970.

Beckett, Derrick. *Stephensons' Britain*. Newton Abbot, Devon., 1984.

Beckmann, Johann. *Beiräge zur Geschichte der Erfindungen*. Five volumes. Leipzig, 1780–1805.

Beckmann, John. *A History of Inventions and Discoveries*. Translated by William Johnston. 3rd edition [four volumes]. London, 1817.

———. *A History of Inventions, Discoveries, and Origins*. Translated by William Johnston. 4th edition [two volumes], revised and enlarged by William Francis and J. W. Griffith. London, 1846.

Bell, E. T. *Men of Mathematics*. New York, 1937.

Belyakov, Vladimir. "The Pencil Is Mightier than the Sword," *Soviet Life*, Issue 348 (September 1985): 48–49.

Berol Limited. "Berol: The Pencil. Its History and Manufacture." Norfolk, n.d.

Berol USA. "The Birth of a Pencil." [Folder.] Danbury, Conn., n.d.

Bigelow, Jacob. *Elements of Technology,* [*etc.*]. 2nd edition, with additions. Boston, 1831.

Binns, William. *An Elementary Treatise on Orthographic Projection,* [*etc.*]. 11th edition. London, 1886.

Birdsall, John. "Writing Instruments: The Market Heats Up," *Western Office Dealer,* February 1983.

Bolton, Theodore. *Early American Portrait Draughtsmen in Crayons.* New York, 1923.

Booker, Peter Jeffrey. *A History of Engineering Drawing.* London, 1963.

Boyer, Jacques. "La Fabrication des Crayons," *La Nature,* 66, part 1 (March 1, 1938): 149–52.

Braudel, Fernand. *The Structures of Everyday Life: The Limits of the Possible.* Translation from the French revised by Siân Reynolds. New York, 1981.

Briggs, Asa. *Iron Bridge to Crystal Palace: Impact and Images of the Industrial Revolution.* London, 1979.

Brondfield, Jerome. "The Marvelous Marking Stick," *Kiwanis Magazine,* February 1979: 28, 29, 48. [Condensed as "*Everything Begins with a Pencil,*" *Reader's Digest,* March 1979: 25–26, 31–33.]

Brown, Martha C. "Henry David Thoreau and the Best Pencils in America," *American History Illustrated,* 15 (May 1980): 30–34.

Brown, Nelson C. *Forest Products: The Harvesting, Processing, and Marketing of Material Other than Lumber,* [*etc.*]. New York, [1950].

Brown, Sam. "Easy Pencil Tricks," *Popular Mechanics,* 49 (June 1928): 993–98.

Bryson, John. *The World of Armand Hammer.* New York, 1985.

Buchanan, R. A. "The Rise of Scientific Engineering in Britain," *British Journal for the History of Science,* 18 (1985): 218–33.

———. "Gentlemen Engineers: The Making of a Profession," *Victorian Studies,* 26 (1983): 407–29.

Buchwald, August. *Bleistifte, Farbstifte, Farbige Kreiden und Pastellstifte, Aquarellfarben, Tusche und Ihre Herstellung nach Bewährten Verfahren.* Vienna, 1904.

The Builder's Dictionary: Or, Gentleman and Architect's Companion. 1734 edition. Washington, D.C., 1981.

Bump, Orlando F. *The Law of Patents, Trade-Marks, Labels and Copy-Rights,* [*etc.*]. 2nd edition. Baltimore, 1884.

California Cedar Products Company. "California Incense Cedar." [Illustrated brochure.] Stockton, Calif., n.d.

Callahan, John F. "Along the Highways and Byways of Finance," *The New York Times,* October 9, 1949: III, 5.

Calle, Paul. *The Pencil.* Westport, Conn., 1974.

Canby, Henry Seidel. *Thoreau.* Boston, 1939.

Caran d'Ache. *50 Ans Caran d'Ache, 1924–1974.* Geneva, [1974].

Carpener, Norman. "Leonardo's Left Hand," *The Lancet,* April 19, 1952: 813–14.

Carter, E. F. *Dictionary of Inventions and Discoveries.* New York, 1966.

Cassell's Household Guide: Being a Complete Encyclopaedia of Domestic and Social Economy, and Forming a Guide to Every Department of Practical Life. London, [ca. 1870].

Cather, Willa. *Alexander's Bridge.* Boston, 1922.

Chambers' Edinburgh Journal. "Visit to the Pencil Country of Cumberland," Vol. VI, No. 145, New Series (October 10, 1864): 225–27.

Chambers's Encyclopaedia. Various editions.

Channing, William Ellery. *Thoreau the Poet-Naturalist.* New edition, enlarged, edited by F. B. Sanborn. Boston, 1902.

Charlton, James, editor. *The Writer's Quotation Book: A Literary Companion.* New York, 1985.

Chaucer, Geoffrey. *The Canterbury Tales.* Verse translation with an introduction and notes by David Wright. Oxford, 1985.

Cicero. *Letters to Atticus.* English translation by E. O. Winstedt. London, 1956.

———. *Letters of Cicero: A Selection in Translation,* by L. P. Wilkinson. New York, 1966.

Clark, Edwin. *The Britannia and Conway Tubular Bridges. With General Inquiries on Beams and on the Properties of Materials Used in Construction.* London, 1850.

Cleveland, Orestes. *Plumbago (Black Lead—Graphite): Its Uses, and How to Use It.* Jersey City, N.J., 1873.

Cliff, Herbert E. "Mechanical Pencils for Business Use," *American Gas Association Monthly,* 17 (July 1935): 270–71.

Cochrane, Charles H. *Modern Industrial Processes.* Philadelphia, 1904.

Coffey, Raymond. "The Pencil: 'Hueing' to Tradition," *Chicago Tribune,* June 30, 1985: V, 3.

Collingwood, W. G., translator. *Elizabethan Keswick: Extracts from the Original Account Books, 1564–1577, of the German Miners, in the Archives of Augsburg.* Kendal, 1912.

Compton's Encyclopedia. 1986 edition.

Considine, Bob. *The Remarkable Life of Dr. Armand Hammer.* New York, 1975.

Constant-Viguier, F. *Manuel de Miniature et de Gouache.* [Bound with Langlois-Longueville.] Paris, 1830.

Cooper, Michael. "William Brockedon, F.R.S.," *Journal of the Writing Equipment Society,* No. 17 (1986): 18–20.

Cornfeld, Betty, and Owen Edwards. *Quintessence: The Quality of Having It.* New York, 1983.

Cowin, S. C. "A Note on Broken Pencil Points," *Journal of Applied Mechanics,* 50 (June 1983): 453–54.

Cronquist, D. "Broken-off Pencil Points," *American Journal of Physics,* 47 (July 1979): 653–55.

Cumberland Pencil Company Limited. "The Pencil Story: A Brief Account of Pencil Making in Cumbria Over the Last 400 Years." [Keswick], n.d.

Daumas, Maurice, editor. *A History of Technology & Invention: Progress Through the Ages.* Translated by Eileen B. Hennessy. New York, 1969.

Day, Walton. *The History of a Lead Pencil.* Jersey City, N.J., 1894.

de Camp, L. Sprague. *The Ancient Engineers.* Garden City, N.Y., 1963.

———. *The Heroic Age of American Invention.* Garden City, N.Y., 1961.

Decker, John. "Pencil Building," *Fine Woodworking,* May–June 1988: 108–9.

Desbecker, John W. "Finding 338 New Uses [for Pencils] Via a Prize Contest," *Printers' Ink,* 156 (July 2, 1931): 86–87.

Deschutes Pioneers' Gazette. "Short Lived Bend Factory Made Juniper Pencil Slats for Export." Vol. 1 (January 1976): 2, 5–6.

Dibner, Bern. *Moving the Obelisks: A Chapter in Engineering History in Which the Vatican Obelisk in Rome in 1586 Was Moved by Muscle Power, and a Study of More Recent Similar Moves.* Cambridge, Mass., 1950.

Dickinson, H. W. "Besoms, Brooms, Brushes and Pencils: The Handicraft Period," *Transactions of the Newcomen Society,* 24 (1943–44, 1944–45): 99–108.

———. "A Brief History of Draughtsmen's Instruments," *Transactions of the Newcomen Society,* 27 (1949–50): 73–84.

Dictionary of American Biography.

Dictionary of National Biography.

Dictionnaire de Biographie Française. Fascicule 103. Paris, 1989.

Diderot, Denis. *A Diderot Pictorial Encyclopedia of Trades and Industry.* Edited by Charles Coulston Gillispie. New York, 1959.

———. *The Encyclopedia: Selections.* Edited and translated by Stephen J. Gendzier, New York, 1967.

Dixon, Joseph, Crucible Company. *Catalog and Price List of Dixon's American Graphite Pencils and Dixon's Felt Erasive Rubbers.* Jersey City, N.J., [1891].

———. *Dixon 1940–1941 Catalog.* Jersey City, N.J., 1940.

———. *Dixon's School Pencils.* Jersey City, N.J., 1903.

———. *Dixon's Standard Graphite Productions.* Jersey City, N.J., [ca. 1916].

———. *Hints of What We Manufacture in Graphite.* Jersey City, N.J., 1893.

———. *Pencillings.* Jersey City, N.J., 1898.

[Dixon Ticonderoga, Inc.] "Manufacture of Pencils and Leads." [Photocopied outline. Versailles, Mo.], n.d.

d'Ocagne, Maurice. "Un Inventeur Oublié, N.-J. Conté," *Revue des Deux Mondes,* Eighth Series, 22 (1934): 912–24.

Doyle, A. Conan. *The Complete Sherlock Holmes.* Garden City, N.Y., 1938.

Drachmann, A. G. *The Mechanical Technology of Greek and Roman Antiquity: A Study of the Literary Sources.* Copenhagen, 1963.

Drake, Stillman. *Cause, Experiment and Science: A Galilean Dialogue Incorporating a New English Translation of Galileo's "Bodies That Stay atop Water, or Move in It."* Chicago, 1981.

Eagle Pencil Company. *Catalog.* London, 1906.

———. *1940 Eagle Catalog.* New York, 1940.

Ecenbarger, William. "What's Portable, Chewable, Doesn't Leak and Is Recommended by Ann Landers?," *Inquirer: The Philadelphia Inquirer Magazine,* June 16, 1985: 14–19. [See also Ecenbarger's "Pencil Technology Gets the Lead Out, But It Can't Erase a Classic," Chicago *Tribune*, November 1, 1985: V, 1, 3.]

The Edinburgh Encyclopaedia. 4th edition, 1830.

The Edinburgh Encyclopaedia. 1st American edition. Philadelphia, 1832.

Eldred, Edward. *Sampson Mordan & Co.* [London], 1986.

Emerson, Edward Waldo. *Henry Thoreau: As Remembered by a Young Friend.* Concord, Mass., 1968.

Emerson, Ralph Waldo. "Thoreau," *The Atlantic Monthly,* August 1862: 239–49.

———. *Journals, 1841–1844.* Edited by Edward Waldo Emerson and Waldo Emerson Forbes. Boston, 1911.

Emmerson, George S. *Engineering Education: A Social History.* Newton Abbot, Devon., 1973.

Empire Pencil Company, "*500,000,000 Epcons.*" [Brochure.] Shelbyville, Tenn., [ca. 1976].

———. "How a Pencil Is Made." [Folder.] Shelbyville, Tenn., [ca. 1986].

Encyclopaedia Americana. Various editions.

Encyclopaedia Britannica. Various editions.

Encyclopaedia Edinensis. 1827 edition.

Encyclopaedia Perthensis. Various editions.

The English Correspondent. "Graphite Mining in Ceylon," *Scientific American,* January 8, 1910: 29, 36–37, 39.

Erskine, Helen Worden. "Joe Dixon and His Writing Stick," *Reader's Digest,* 73 (November 1958): 186–88, 190.

Evans, Oliver. *To His Counsel, Who Are Engaged in Defence of His Patent Rights for the Improvements He Has Invented.* [Ca. 1817.]

Faber, A. W., [Company]. "A. W. Faber, Established 1761." [Typescript, ca. 1969.]

———. *The Manufactories and Business Houses of the Firm of A. W. Faber: An Historical Sketch.* Nuremberg, 1896.

————. *Price-List of Superior Lead and Colored Pencils, Writing and Copying Inks, Slate Manufactures, Rulers, Penholders and Erasive Rubber.* New York, [ca. 1897].

Faber, A. W., Inc. [*Catalog of*] *Drawing Pencils, Drawing Materials,* [*etc.*]. Newark, N.J., [ca. 1962–63].

Faber, E. L. "History of Writing and the Evolution of the Lead Pencil Industry." [Typescript.] August 1921.

Faber, Eberhard. "Words to Grow On," *Guideposts,* August 1988: 40–41.

Faber, Eberhard, Pencil Company. "A Personally Conducted Tour of the World's Most Modern Pencil Plant with Marty the Mongol." [Broadside.] Wilkes-Barre, Pa., [1973].

————. "Since 1849: Quality Products for Graphic Communications." [Folder.] Wilkes-Barre, Pa., [1986].

————. *The Story of the Oldest Pencil Factory in America.* [New York], 1924.

Faber-Castell, A. W., GmbH & Company. *Das Bleistiftschloss: Familie und Unternehmen Faber-Castell in Stein.* Stein, 1986.

————. "Faber Castell." [Illustrated brochure. Stein], n.d.

————. "Faber-Castell: 225 Years of Company History in Short." *Presseinformation,* November 1987.

————. "Origin and History of the Family and Company Name." [Folder.] N.d.

Faber-Castell Corporation. "Date Log: Venus Company History." [Photocopied sheets.] Lewisburg, Tenn., n.d.

————. "Faber-Castell Corporation." [Photocopied stapled sheets.] Lewisburg, Tenn., n.d.

————. "The Story of the Lead Pencil." [Photocopied report. Lewisburg, Tenn.], n.d.

————. "Writing History for Over 225 Years." [Illustrated folder.] [Parsippany, N.J., 1987.]

[Faber, Johann]. *The Pencil Factory of Johann Faber (Late of the Firm of A. W. Faber) at Nuremberg, Bavaria.* Nuremberg, 1893.

Fairbank, Alfred. *The Story of Handwriting: Origins and Development.* New York, 1970.

Faraday, Michael. *The Chemical History of a Candle: A Course of Lectures Delivered Before a Juvenile Audience at the Royal Institution.* New edition, with illustrations. Edited by William Crookes. London, 1886.

Farmer, Lawrence R. "Press Aids Penmanship," *Tooling & Production,* 44 (April 1978): 94–95.

Feldhaus, Franz Maria. "Geschichtliches vom deutschen Graphit," *Zeitschrift für Angewandte Chemie,* 31 (1918): 76.

————. *Geschichte des Technischen Zeichnens.* Wilhelmshaven, 1959.

Ferguson, Eugene S. "Elegant Inventions: The Artistic Component of Technology," *Technology and Culture,* 19 (1978): 450–60.

————. "The Mind's Eye: Nonverbal Thought in Technology," *Science*, 197 (August 26, 1977): 827–36.

————. "La Fondation des Machines Modernes: Des Dessins," *Culture Technique*, No. 14 (June 1985): 182–207.

Feynman, Richard P., as told to Ralph Leighton. *"Surely You're Joking, Mr. Feynman!": Adventures of a Curious Character.* Edited by Edward Hutchin. New York, 1985.

Finch, James Kip. *Engineering Classics.* Edited by Neal Fitz-Simons. Kensington, Md., 1978.

Finder, Joseph. *Red Carpet.* New York, 1983.

Fleming, Clarence C., and Arthur L. Guptill. *The Pencil: Its History, Manufacture and Use.* New York, 1936. [A shorter version of this booklet, with back matter on different Koh-I-Noor products, was published in Bloomsbury, N.J., also in 1936.]

Foley, John. *History of the Invention and Illustrated Process of Making Foley's Diamond Pointed Gold Pens, With Complete Illustrated Catalogue of Fine Gold Pens, Gold,- Silver,- Rubber,- Pearl and Ivory Pen & Pencil Cases, Pen Holders, &c.* New York, 1875.

Fowler, Dayle. "A History of Writing Instruments," *Southern Office Dealer,* May 1985: 12, 14, 17.

Frary, C. A. "What We Have Learned in Marketing Eversharp," *Printers' Ink,* 116 (August 11, 1921): 3–4, 6, 8, 142, 145–46, 149.

Fraser, Chelsea. *The Story of Engineering in America.* New York, 1928.

French, Thomas E., and Charles J. Vierck. *A Manual of Engineering Drawing for Students and Draftsmen.* 9th edition. New York, 1960.

Friedel, Robert. *A Material World: An Exhibition at the National Museum of American History, Smithsonian Institution.* Washington, D.C., 1988.

Friedel, Robert, and Paul Israel, with Bernard S. Finn. *Edison's Electric Light: Biography of an Invention.* New Brunswick, N.J., 1986.

Frost, A. G. "How We Made a Specialty into a Staple," *System, the Magazine of Business,* November 1922: 541–43, 648.

————. "Marketing a New Model in Face of Strong Dealer Opposition," *Printers' Ink,* 128 (July 17, 1924): 3–4, 6, 119–20, 123.

Galilei, Galileo. *Dialogues Concerning Two New Sciences.* Translated by Henry Crew and Alfonso de Salvio. 1914 edition. New York, [1954].

German Imperial Commissioner, editor. *International Exposition, St. Louis 1904: Official Catalogue of the Exhibition of the German Empire.* Berlin, 1904.

Gesner, Konrad. *De Rerum Fossilium Lapidum et Gemmarum Maxime, Figuris et Similitudinibus Liber,* [etc.]. Zurich, 1565.

Getchell, Charles. "Cause for Alarm in Peking," *The New York Times*, February 4, 1977: op-ed page.

Geyer's Stationer. "The Joseph Dixon Crucible Co.—Personnel, Progress and Plant." Vol. 25 (March 19, 1903): 1–11.

Gibb, Alexander. *The Story of Telford: The Rise of Civil Engineering.* London, 1935.

Gibbs-Smith, C. H. *The Great Exhibition of 1851.* London, 1964.

Giedion, Siegfried. *Mechanization Takes Command: A Contribution to Anonymous History.* New York, 1969.

Giesecke, F. E., and A. Mitchell. *Mechanical Drawing.* 4th edition. Austin, Tex., 1928.

Giesecke, F. E., Alva Mitchell, and Henry Cecil Spencer. *Technical Drawing.* 3rd edition. New York, 1949.

Gilfillan, S. C. *The Sociology of Invention.* Cambridge, Mass., 1970.

Gille, Bertrand. *The History of Techniques.* Translated from the French and revised. New York, 1986.

———. *The Renaissance Engineers.* London, 1966.

Gillispie, Charles C. "The Natural History of Industry," *Isis,* 48 (1948): 398–407.

Gimpel, Jean. *The Medieval Machine: The Industrial Revolution of the Middle Ages.* New York, 1976.

Godbole, N. N. *Manufacture of Lead and Slate Pencils, Slates, Plaster of Paris, Chalks, Crayons and Taylors' Chalks (with Special Reference to India).* Jaipur, Rajasthan, 1953.

Goldman, Marshall I. *Détente and Dollars: Doing Business with the Soviets.* New York, 1975.

Gopalaswamy, T. R., and G. D. Joglekar. "Smearing Property of Graphite Powders," *ISI Bulletin,* 11 (1959): 243–46.

Gopalaswamy Iyenger, T. R., B. R. Marathe, and G. D. Joglekar. "Graphite for Pencil Manufacture," *ISI Bulletin,* 10 (1958): 159–62.

Gorringe, Henry H. *Egyptian Obelisks.* New York, 1882.

Great Britain Board of Trade. "Report of the Standing Committee on Pencils and Pencil Strips." Cmd. 2182. London, 1963.

———. "Report of the Standing Committee Respecting Fountain Pens, Stylographic Pens, Propelling Pencils and Gold Pen Nibs." Cmd. 3587. London, 1930.

———. "Report of the Standing Committee Respecting Pencils and Pencil Strips." Cmd. 4278. London, 1933.

Great Britain Forest Products Research Laboratory. *African Pencil Cedar: Studies of the Properties of* Juniperus procera *(Hochst.) with Particular Reference to the Adaptation of the Timber to the Requirements of the Pencil Trade.* London, 1938.

Great Soviet Encyclopedia. Translation of the 3rd edition. New York, 1976.

Greenland, Maureen. "Visit to the Berol Pencil Factory, Tottenham, Wednesday, May 19th, 1982," *Journal of the Writing Equipment Society,* No. 4 (1982): 7.

Guptill, Arthur L. *Sketching and Rendering in Pencil.* New York, 1922.

Haldane, J. W. C. *Life as an Engineer: Its Lights, Shades and Prospects.* London, 1905.

Hall, Donald, editor. *The Oxford Book of American Literary Anecdotes.* New York, 1981.

Hall, William L., and Hu Maxwell. *Uses of Commercial Woods of the United States: I. Cedars, Cypresses, and Sequoias.* U.S. Department of Agriculture, Forest Service Bulletin 95 (1911).

Halse, Albert O. *Architectural Rendering: The Techniques of Contemporary Presentation.* New York, 1960.

Hambly, Maya. *Drawing Instruments: Their History, Purpose and Use for Architectural Drawings.* [Exhibition catalogue.] London, 1982.

———. *Drawing Instruments: 1580–1980.* London, 1988.

Hammer, Armand. *The Quest of the Romanoff Treasure.* New York, 1936.

Hammer, Armand, with Neil Lyndon. *Hammer.* New York, 1987.

Hammond, John Winthrop. *Charles Proteus Steinmetz: A Biography.* New York, 1924.

Harding, Walter. *The Days of Henry Thoreau: A Biography.* New York, 1982.

———. *Thoreau's Library.* Charlottesville, Va., 1957.

———, editor. *A Catalog of the Thoreau Society Archives in the Concord Free Public Library.* Thoreau Society Booklet 29. Geneseo, N.Y., 1978.

Hardtmuth, L. & C. *Retail Price List: L. & C. Hardtmuth's "Koh-I-Noor" Pencils.* New York, [ca. 1919].

Hart, Ivor B. *The World of Leonardo da Vinci: Man of Science, Engineer and Dreamer of Flight.* London, 1961.

Hartmann, Henry. "Blushing on Lacquered Paint Parts Overcome by Gas Fired Dehumidifier," *Heating, Piping and Air Conditioning,* June 1939: 356.

Hauton, Paul S. "Splitting Pennies," *Factory and Industrial Management,* 76 (July 1928): 43–47.

Hayward, Phillips A. *Wood: Lumber and Timbers.* New York, 1930.

Helmhacker, R. "Graphite in Siberia," *Engineering and Mining Journal,* December 25, 1897: 756.

Helphenstine, R. K., Jr. "What Will We Do for Pencils?," *American Forests,* 32 (November 1926): 654.

Hemingway, Ernest. *A Moveable Feast.* New York, 1964.

Hendrick, George, editor. *Remembrances of Concord and the Thoreaus: Letters of Horace Hosmer to Dr. S. A. Jones.* Urbana, Ill., 1977.

Hero of Alexandria. *The Pneumatics.* Translated for and edited by Bennet Woodcroft. London, 1851.

Hill, Donald. *A History of Engineering in Classical and Medieval Times.* La Salle, Ill., 1984.

Hill, Henry. "The Quill Pen." *The Year Book of the London School of Printing & Kindred Trades,* 1924–1925: 73–78.

Hindle, Brooke. *Emulation and Invention.* New York, 1981.

Historische Bürowelt. "L'Histoire d'un Crayon." No. 11 (October 1985): 11–13.

Hofstadter, Douglas R. *Gödel, Escher, Bach: An Eternal Golden Braid.* New York, 1980.

Howard, Seymour. "The Steel Pen and the Modern Line of Beauty," *Technology and Culture,* 26 (October 1985): 785–98.

Hubbard, Elbert. *Joseph Dixon: One of the World-Makers.* East Aurora, N.Y., 1912.

Hubbard, Oliver P. "Two Centuries of the Black Lead Pencil," *New Englander and Yale Review,* 54 (February 1891): 151–59.

Hunt, Robert, editor. *Hunt's Hand-Book to the Official Catalogues: An Explanatory Guide to the Natural Productions and Manufactures of the Great Exhibition of the Industry of All Nations, 1851.* London, [1851].

Huxley, Thomas Henry. *On a Piece of Chalk.* Edited and with an introduction and notes by Loren Eiseley. New York, 1967.

Illustrated London News. "The Manufacture of Steel Pens in Birmingham." February 22, 1851: 148–49.

The Illustrated Magazine of Art. "Pencil-Making at Keswick." Vol. 3 (1854): 252–54.

Indian Standards Institution. *Specification for Black Lead Pencils.* New Delhi, 1959.

International Cyclopaedia. Revised with large additions. New York, 1900.

Israel, Fred L., editor. *1897 Sears, Roebuck Catalogue.* New York, 1968.

Ives, Sidney, general editor. *The Parkman Dexter Howe Library.* Part II. Gainesville, Fla., 1984.

Jacobi, Albert W. "How Lead Pencils Are Made," *American Machinist,* January 26, 1911: 145–46.

[James, George S.] "A History of Writing Instruments," *The Counselor,* July 1978.

Japanese Standards Association. "Pencils and Coloured Pencils," JIS S 6006–1984. Tokyo, 1987.

Jenkins, Rhys. "The Society for the Mines Royal and the German Colony in the Lake District," *Transactions of the Newcomen Society,* 18 (1937–38): 225–34.

Jennings, Humphrey. *Pandaemonium, 1660–1886: The Coming of the Machine as Seen by Contemporary Observers.* Edited by Mary-Lou Jennings and Charles Madge. New York, 1985.

Jewkes, John, David Sawers, and Richard Stillerman. *The Sources of Invention.* London, 1958.

Joglekar, G. D. "A 100 g.-cm. Impact Testing Machine for Testing the Strength of Pencil Leads," *Journal of Scientific and Industrial Research,* 21D (1962): 56.

Joglekar, G. D., T. R. Gopalaswami, and Shakti Kumar. "Abrasion Characteristics of Clays Used in Pencil Manufacture," *Journal of Scientific and Industrial Research,* 21D (1962): 16–19.

Joglekar, G. D., P. R. Nayak, and L. C. Verman. "Electrical Resistance of Black Lead Pencils," *Journal of Scientific and Industrial Research,* 6B (1947): 75–80.

Joglekar, G. D., A. N. Bulsara, and S. S. Chari. "Impact Testing of Pencil Leads," *Indian Journal of Technology,* 1 (1963): 94–97.

Joglekar, G. D., and B. R. Marathe. "Writing Quality of Pencils," *Journal of Scientific and Industrial Research,* 13B (1954): 78–79.

Johnson, E. Borough. "How to Use a Lead Pencil," *The Studio,* 22 (1901): 185–95.

Johnson, E. S. *Illustrated Catalog of Unequaled Gold Pens, Pen Holders, Pencils, Pen and Pencil Cases, Tooth Picks, Tooth & Ear Picks, &c. in Gold, Silver, Pearl, Ivory, Rubber & Celluloid.* New York, [ca. 1895].

Journal of the Writing Equipment Society. Various numbers.

Kane, Joseph Nathan. *Famous First Facts: A Record of First Happenings, Discoveries, and Inventions in American History.* 4th edition, expanded and revised. New York, 1981.

Kautzky, Theodore. *Pencil Broadsides: A Manual of Broad Stroke Technique.* New York, 1940.

Keats, John, and Percy Bysshe Shelley. *Complete Poetical Works.* New York, n.d.

Kemp, E. L. "Thomas Paine and His 'Pontifical Matters,' " *Transactions of the Newcomen Society,* 49 (1977–78): 21–40.

Keuffel & Esser Co. *Catalogue of . . . Drawing Materials, Surveying Instruments, Measuring Tapes.* 38th edition. New York, 1936.

King, Carl H. "Pencil Points," *Industrial Arts and Vocational Education,* 25 (November 1936): 352–53.

Kirby, Richard Shelton. *The Fundamentals of Mechanical Drawing.* New York, 1925.

Kirby, Richard Shelton, and Philip Gustave Laurson. *The Early Years of Modern Civil Engineering.* New Haven, Conn., 1932.

Kirby, Richard Shelton, Sidney Withington, Arthur Burr Darling, and Frederick Gridley Kilgour. *Engineering in History.* New York, 1956.

Kisner, Howard W., and Ken W. Blake. " 'Indelible Lead' Puncture Wounds," *Industrial Medicine,* 10 (1941): 15–17.

Klingender, Francis D. *Art and the Industrial Revolution.* Edited and revised by Arthur Elton. New York, 1968.

Knight's Cyclopaedia of the Industry of All Nations. London, 1851.

Kogan, Herman. *The Great EB: The Story of the Encyclopaedia Britannica.* Chicago, 1958.

Kozlik, Charles J. "Kiln-drying Incense-cedar Squares for Pencil Stock," *Forest Products Journal,* 37 (May 1987): 21–25.

Kranzberg, Melvin, and Carroll W. Pursell, Jr. *Technology in Western Civilization.* New York, 1967.

Laboulaye, C. P. *Dictionnaire des Arts et Manufactures,* [*etc.*]. Paris, 1867.

Lacy, Bill N. "The Pencil Revolution," *Newsweek,* March 19, 1984: 17.

Laliberté, Norman, and Alex Morgan. *Drawing with Pencils: History and Modern Techniques.* New York, 1969.

Landels, J. G. *Engineering in the Ancient World.* Berkeley, Calif., 1978.

Langlois-Longueville, F. P. *Manuel du Lavis à la Sépia, et de l'Aquarelle.* [Bound with Constant-Viguier.] Paris, 1836.

Larousse, Pierre. *Grand Dictionnaire Universel du XIXe Siècle* [*etc.*]. Paris, 1865.

Latham, Jean. *Victoriana.* New York, 1971.

Latour, Bruno. "Visualization and Cognition: Thinking with Eyes and Hands." In Henrika Kuklick and Elizabeth Long, editors, *Knowledge and Society: Studies in the Sociology of Culture Past and Present,* 6 (1986): 1–40.

Lawrence, Cliff. *Fountain Pens: History, Repair and Current Values.* Paducah, Ky., 1977.

Lawrence, D. H. *The Complete Poems.* Collected and edited by Vivian de Sola Pinto and Warren Roberts. New York, 1971.

Layton, Edwin T., Jr. *The Revolt of the Engineers: Social Responsiblity and the American Engineering Profession.* Baltimore, 1986.

Lefebure, Molly. *Cumberland Heritage.* London, 1974.

Die Leistung, 12, No. 95 (1962). [Issue devoted to the J. S. Staedtler Company.]

Leonardo da Vinci. *Il Codice Atlantico.* Edizione in Facsimile Dopo il Restauro dell'originale Conservato nella Biblioteca Ambrosiana di Milano. Florence, 1973–75.

———. *The Drawings of Leonardo da Vinci.* Introduction and notes by A. E. Popham. New York, 1945.

———. *The Literary Works of Leonardo da Vinci.* Compiled and edited by Jean Paul Richter. 2nd edition, enlarged and revised by Jean Paul Richter and Irma A. Richter. London, 1939.

———. *The Notebooks of Leonardo da Vinci.* Arranged, rendered into English, and introduced by Edward MacCurdy. New York, 1939.

Leonhardt, Fritz. *Brücken: Ästhetik und Gestaltung / Bridges: Aesthetics and Design.* Cambridge, Mass., 1984.

Lévi-Strauss, Claude. *The Savage Mind.* Chicago, 1966.

Lewis, Gene D. *Charles Ellet, Jr.: The Engineer as Individualist, 1810–1862.* Urbana, Ill., 1968.

Ley, Willy. *Dawn of Zoology.* Englewood Cliffs, N.J., 1968.

Lindbergh, Anne Morrow. *Gift from the Sea.* New York, 1955.

Lindgren, Waldemar. *Mineral Deposits.* New York, 1928.

Lomazzo, Giovanni Paolo. *A Tracte Containing the Artes of Curious Paintinge Carvinge and Buildinge.* Translated in 1858 by Richard Haydocke. England, 1970.

Lo-Well Pencil Company. ["Lo-Well Pencils." Advertising folder.] New York, [ca. 1925].

Lubar, Steven. "Culture and Technological Design in the 19th-Century Pin Industry: John Howe and the Howe Manufacturing Company," *Technology and Culture,* 28 (April 1987): 253–82.

Lucas, A. *Ancient Egyptian Materials and Industries.* 4th edition, revised and enlarged by J. R. Harris. London, 1962.

Lyra Bleistift-Fabrik GmbH and Company. [*Catalog.*] Nuremberg, 1914.

———. "The Early Days." [Mimeographed notes.] Nuremberg, n.d.

Machinery Market. "Manufacture of Pencils. The Works of the Cumberland Pencil Co., Ltd., of Keswick, Revisited." December 22, 1950: 25–27.

———. "The Manufacture of Pencils and Crayons. Being a Description of a Visit to the Works of the Cumberland Pencil Co., Ltd., Keswick." December 16, 1938: 31–32.

MacLeod, Christine. "Accident or Design? George Ravenscroft's Patent and the Invention of Lead-Crystal Glass," *Technology and Culture,* 28 (October 1987): 776–803.

Maigne, W. *Dictionnaire Classique des Origines Inventions et Découvertes, [etc.].* 3rd edition. Paris, [ca. 1890].

The Manufacturer and Builder. "Lead Pencils." March 1872: 80–81.

Marathe, B. R., Gopalaswamy Iyenger, K. C. Agarwal, and G. D. Joglekar. "Evaluation of Clays Suitable for Pencil Manufacture," *ISI Bulletin,* 10 (1958): 199–203.

Marathe, B. R., Gopalaswamy Iyenger, and G. D. Joglekar. "Tests for Quality Evaluation of Black Lead Pencils," *ISI Bulletin,* 7 (1955): 16–22.

Marathe, B. R., Kanwar Chand, and G. D. Joglekar. "Tests for Quality Evaluation of Black Lead Pencils—Measurement of Friction," *ISI Bulletin,* 8 (1956): 132–34.

Marble, Annie Russell. *Thoreau: His Home, Friends and Books.* New York, 1902.

Marshall, J. D., and M. Davies-Shiel. *The Industrial Archaeology of the Lake Counties.* Newton Abbot, Devon., 1969.

Martin, Thomas. *The Circle of the Mechanical Arts; Containing Practical Treatises on the Various Manual Arts, Trades, and Manufactures.* London, 1813.

Masi, Frank T., editor. *The Typewriter Legend.* Secaucus, N.J., 1985.

Masterson, R. L. "Dip Finishing Pencils and Penholders," *Industrial Finishing,* 4 (September 1928): 59–60, 65.

McCloy, Shelby T. *French Inventions of the Eighteenth Century.* Lexington, Ky., 1952.

McClurg, A. C., & Co. *General Catalogue.* 1908–9.

McDuffie, Bruce. "Rapid Screening of Pencil Paint for Lead by a Combustion-Atomic Absorption Technique," *Analytical Chemistry,* 44 (July 1972): 1551.

McGrath, Dave. "To Fill You In," *Engineering News-Record,* May 13, 1982: 9.

McNaughton, Malcolm. "Graphite," *Stevens Institute Indicator,* 18 (January 1901): 1–15.

McWilliams, Peter A. *The McWilliams II Word Processor Instruction Manual.* West Hollywood, Calif., 1983.

Meder, Joseph. *Mastery of Drawing.* Vol. 1. Translated and revised by Winslow Ames. New York, 1978.

Meltzer, Milton, and Walter Harding. *A Thoreau Profile.* Concord, Mass., 1962.

Metz, Tim. "Is Wooden Writing Soon to Be Replaced by a Plastic Variety?," *Wall Street Journal,* January 5, 1981: 1, 6.

Mitchell, C. Ainsworth. "Black-Lead Pencils and Their Pigments in Writing," *Journal of the Society of Chemical Industry,* 38 (1919): 383T–391T.

———. "Characteristics of Pigments in Early Pencil Writing," *Nature,* 105 (March 4, 1920): 12–14.

———. "Copying-Ink Pencils and the Examination of Their Pigments in Writing," *The Analyst,* 42 (1917): 3–11.

———. "Graphites and Other Pencil Pigments," *The Analyst,* 47 (September 1922): 379–87.

———. "Pencil Markings in the Bodleian Library," *Nature,* 109 (April 22, 1922): 516–17.

Montgomery, Charles F., editor. *Joseph Moxon's Mechanick Exercises: Or, the Doctrine of Handy Works,* [*etc.*]. New York, 1970.

Morgan, Hal. *Symbols of America.* New York, 1986.

Moss, Marcia, editor. *A Catalog of Thoreau's Surveys in the Concord Free Public Library.* Geneseo, N.Y., 1976.

Mumford, Lewis. *The Myth of the Machine: Technics and Human Development.* New York, 1967.

———. *Technics and Civilization.* New York, 1963.

Munroe, William. "Francis Munroe." In Social Circle in Concord. *Memoirs of Members.* Third Series. Cambridge, Mass., 1907.

Munroe, William, Jr. "Memoirs of William Munroe." In Social Circle in Concord. *Memoirs of Members.* Second Series. Cambridge, Mass., 1888.

Murry, J. Middleton. *Pencillings.* New York, 1925.

Nasmyth, James. *James Nasmyth, Engineer: An Autobiography.* Edited by Samuel Smiles. New York, 1883.

Nelms, Henning. *Thinking with a Pencil.* New York, 1985.

New Edinburgh Encyclopaedia. 2nd American edition. New York, 1821.

Newlands, James. *The Carpenter and Joiner's Assistant,* [*etc.*]. London, [ca. 1880].

The New York Times. "Dixon Stands by Jersey City." December 14, 1975: 14.

———. "How Dixon Made Its Mark." January 27, 1974: 74.

———. "Mr. Eberhard Faber's Death. The Man Who Built the First Lead Pencil Factory in America—A Sketch of His Career." March 4, 1879: obituary page.

Nichols, Charles R., Jr. "The Manufacture of Wood-Cased Pencils," *Mechanical Engineering,* November 1946: 956–60.

Noble, David F. *America by Design: Science, Technology, and the Rise of Corporate Capitalism.* New York, 1977.

Norman, Donald A. *The Psychology of Everyday Things.* New York, 1988.

Norton, Thomas H. "The Chemistry of the Lead Pencil," *Chemicals,* 24 (August 31, 1925): 13.

Official Catalogue of the Great Exhibition of the Works of Industry of All Nations, 1851. Corrected edition. London, [1851].

Oliver, John W. *History of American Technology.* New York, 1956.

Oppenheimer, Frank. "The German Drawing Instrument Industry: History and Sociological Background," *Journal of Engineering Drawing,* 20 (November 1956): 29–31.

Ormond, Leonee. *Writing.* London, 1981.

Pacey, Arnold. *The Maze of Ingenuity: Ideas and Idealism in the Development of Technology.* Cambridge, Mass., 1976.

Palatino, Giovambattista. *The Tools of Handwriting: From . . . Un Nuovo Modo d'Imparare* [1540]. Introduced, translated, and printed by A. S. Osley. Wormley, 1972.

The Pencil Collector. Various numbers.

Pentel of America, Limited. "Pentel Brings an End to the Broken Lead Era with New 'Super' Hi-Polymer Lead." [Sales catalogue insert.] Torrance, Ca., 1981.

Peterson, Eldridge. "Mr. Berwald Absorbs Pencils," *Printers' Ink,* 171 (May 2, 1935): 21, 24–26.

Petroski, H. "On the Fracture of Pencil Points," *Journal of Applied Mechanics,* 54 (September 1987): 730–33.

———. "Inventions Spurned: On Bridges and the Impact of Society on Technology," *Impact of Science on Society,* 37 (No. 147, 1987): 251–59.

———. *To Engineer Is Human: The Role of Failure in Successful Design.* New York, 1985.

Pevsner, N. "The Term 'Architect' in the Middle Ages," *Speculum,* 17 (1942) 549–62.

Phillips, E. W. J. "The Occurrence of Compression Wood in African Pencil Cedar," *Empire Forestry Journal,* 16 (1937): 54–57.

Pichirallo, Joe. "Lead Poisoning: Risks for Pencil Chewers?," *Science,* 173 (August 6, 1971): 509–10.

Pigot and Company. *London and Provincial New Commercial Direc-*

tory, for 1827–28; Comprising a Classification of, and Alphabetical Reference to the Merchants, Manufacturers and Traders of the Metropolis, [etc.]. 3rd edition. London, [1827].

———. *Metropolitan New Alphabetical Directory, for 1827; [etc.].* London, [1827].

Pinck, Dan. "Paging Mr. Ross," *Encounter,* 69 (June 1987): 5–11.

Pliny. *Natural History.* With an English translation by H. Rackham. Cambridge, Mass., 1979.

Plot, Rob. "Some Observations Concerning the Substance Commonly Called, Black Lead, " *Philosophical Transactions* (London), 20 (1698): 183.

Porter, Terry. "The Pencil Revolution," *Texas Instruments Engineering Journal,* 2 (January–February 1985): 66.

Pratt, Joseph Hyde. "The Graphite Industry," *Mining World,* July 22, 1905: 64–66.

Pratt, Sir Roger. *The Architecture of Sir Roger Pratt, [etc.].* Edited by R. T. Gunther. Oxford, 1928.

Pye, David. *The Nature and Aesthetics of Design.* London, 1978.

Rance, H. F., editor. *Structure and Physical Properties of Paper.* New York, 1982.

Reed, George H. "The History and Making of the Lead Pencil," *Popular Educator,* 41 (June 1924): 580–82.

Rees, Abraham. *The Cyclopaedia; or, Universal Dictionary of Arts, Sciences, and Literature.* Philadelphia, n.d.

Rehman, M. A., and S. M. Ishaq. "Indian Woods for Pencil Making," *Indian Forest Research Leaflet,* No. 66 (1945).

Rehman, M. A., and Jai Kishen. "Chemical Staining of Deodar Pencil Slats," *Indian Forester,* 79 (September 1953): 512–13.

———. "Deodar as Pencil Wood," *Indian Forest Bulletin,* No. 149, [ca. 1953].

———. "Treatment of Indian Timbers for Pencils and Hand Tools for Pencil Making," *Indian Forest Leaflet,* No. 126 (1952).

Remington, Frank L. "The Formidable Lead Pencil," *Think,* November 1957: 24–26. [Condensed as "The Versatile Lead Pencil," *Science Digest,* 43 (April 1958): 38–41.]

Rennie, John. *The Autobiography of Sir John Rennie, F.R.S., [etc.].* London, 1875.

Rexel Limited. "Making Pens and Pencils: A Story of Tradition." [Illustrated folder.] Aylesbury, Bucks., n.d.

Richards, Gregory B. "Bright Outlook for Writing Instruments," *Office Products Dealer,* June 1983: 40, 42, 44, 48.

Riddle, W. "Lead Pencils," *The Builder,* August 3, 1861: 537–38. [See also, *The Builder,* July 27, 1861: 517.]

Rix, Bill. "Pencil Technology." A paper prepared for a course taught by Professor Walter G. Vincenti, Stanford University, ca. 1978.

Robinson, Tho. *An Essay Towards a Natural History of Westmorland and Cumberland, [etc.].* London, 1709.

Rocheleau, W. F. *Great American Industries. Third Book: Manufactures.* Chicago, 1908.

Roe, G. E. "The Pencil," *Journal of the Writing Equipment Society,* No. 5 (1982): 12.

Rolt, L. T. C. "The History of the History of Engineering," *Transactions of the Newcomen Society,* 42 (1969–70): 149–58.

Root, Marcus Aurelius. *The Camera and the Pencil; or the Heliographic Art.* [1864 ed.] Pawlet, Vt., 1971.

Rosenberg, Harold. *Saul Steinberg.* New York, 1978.

Rosenberg, N., and W. G. Vincenti. *The Britannia Bridge: The Generation and Diffusion of Knowledge.* Cambridge, Mass., 1978.

Ross, Stanley. "Drafting Pencil—A Teaching Aid," *Industrial Arts and Vocational Education,* February 1957: 52–53.

Russell and Erwin Manufacturing Company. *Illustrated Catalogue of American Hardware.* 1865 edition. [Washington, D.C., 1980.]

Russo, Edward, and Seymour Dobuler. "The Manufacture of Pencils," *New York University Quadrangle,* 13 (May 1943): 14–15.

Sackett, H. S. "Substitute Woods for Pencil Manufacture," *American Lumberman,* January 27, 1912: 46.

Sandburg, Carl. *The Complete Poems.* Revised and expanded edition. New York, 1970.

Scherer, J. S. "More than 55% Replies," *Printers' Ink,* 188 (August 11, 1939): 15–16.

Schodek, Daniel L. *Landmarks in American Civil Engineering.* Cambridge, Mass., 1987.

Schrodt, Philip. "The Generic Word Processor," *Byte,* April 1982: 32, 34, 36.

Schwanhausser, Eduard. *Die Nürnberger Bleistiftindustrie von Ihren Ersten Anfängen bis zur Gegenwart.* Greifswald, 1893.

Scribner's Monthly. "How Lead Pencils Are Made." April 1878: 801–10.

Sears, Roebuck and Company. *Catalogue.* Various original and reprinted editions.

Seeley, Sherwood B. "Carbon (Natural Graphite)." In *Encyclopedia of Chemical Technology,* 4 (2nd edition, 1964): 304–55.

———. "Manufacturing pencils," *Mechanical Engineering,* November 1947: 686.

Silliman, Professor. "Abstract of Experiments on the Fusion of Plumbago, Anthracite, and the Diamond," *Edinburgh Philosophical Journal,* 9 (1823): 179–83.

Singer, Charles, et al., editors. *A History of Technology.* Oxford, 1954–78.

Slocum, Jerry, and Jack Botermans. *Puzzles Old and New: How to Make and Solve Them.* Seattle, 1986.

Smiles, Samuel. *Lives of the Engineers.* Popular edition. London, 1904.

————. *Selections from Lives of the Engineers: With an Account of Their Principal Works.* Edited with an introduction by Thomas Parke Hughes. Cambridge, Mass., 1966.

Smith, Adam. *An Inquiry into the Nature and Causes of the Wealth of Nations.* Chicago, 1952.

Smith, Cyril Stanley. "Metallurgical Footnotes to the History of Art," *Proceedings of the American Philosophical Society,* 116 (1972): 97–135.

Smithwick, R. Fitzgerald. "How Our Pencils Are Made in Cumberland," *Art-Journal,* 18, n.s. (1866): 349–51.

Social Circle in Concord, [Mass.]. *Memoirs of Members.* Second series: From 1795 to 1840. Cambridge, Mass., 1888.

————. Third series: From 1840 to 1895. Cambridge, Mass., 1907.

————. Fourth series: From 1895 to 1909. Cambridge, Mass., 1909.

Speter, Max. "Wer Hat Zuerst Kautschuk als Radiergummi Verwendet?," *Gummi-Zeitung,* 43 (1929): 2270–71.

Staedtler, J. S., [Company]. *275 Jahre Staedtler-stifte.* Nuremberg, 1937.

Staedtler Mars. *Design Group Catalog.* Montville, N.J., [1982].

Staedtler Mars GmbH & Co. *The History of Staedtler.* Nuremberg, [1986].

————. Various catalogues and reports.

Stafford, Janice. "An Avalanche of Pens, Pencils and Markers!," *Western Office Dealer,* March 1984: 18–22.

Steel, Kurt. "Prophet of the Independent Man," *The Progressive,* September 24, 1945: 9.

Steinbeck, John. *Journal of a Novel: The* East of Eden *Letters.* New York, 1969.

Stephan, Theodore M. "Lead-Pencil Manufacture in Germany," *U.S. Department of State Consular Reports. Commerce, Manufactures, Etc.,* 51 (1896): 191–92.

Stern, Philip van Doren, editor. *The Annotated* Walden. New York, [1970].

Stowell, Robert F. *A Thoreau Gazetteer,* edited by William L. Howarth. Princeton, N.J., 1970.

Stuart, D. G. "Listo Works Back from the User to Build Premium Market," *Sales Management,* November 20, 1951: 74–78.

Sutton, F. Colin. "Your Pencil Unmasked," *Chemistry and Industry,* 42 (July 20, 1923): 710–11.

Svensen, Carl Lars, and William Ezra Street. *Engineering Graphics.* Princeton, N.J., 1962.

Sykes, M'Cready. "The Obverse Side," *Commerce and Finance,* 14 (April 8, 1925): 652–53.

Talbot, William Henry Fox. *The Pencil of Nature.* New York, 1969.

Tallis's History and Description of the Crystal Palace, and the Exhibition of the World's Industry in 1851. [Three volumes.] London, [ca. 1852].

Taylor, Archer. *English Riddles from Oral Tradition.* Berkeley, Calif., 1951.

Thayer, V. T. *The Passing of the Recitation.* Boston, 1928.

Thomson, Ruth. *Making Pencils.* London, 1987.

Thoreau, Henry David. *The Correspondence.* Edited by Walter Harding and Carl Bode. New York, 1958.

———. *Journal.* Vols. 1 and 2. John C. Broderick, general editor. Princeton, N.J., 1981, 1984.

———. *A Week on the Concord and Merrimack Rivers. Walden; or, Life in the Woods. The Maine Woods. Cape Cod.* [In one volume.] New York, 1985.

Thoreau Society Bulletin. "A Lead Pencil Diploma . . ." No. 74 (Winter 1961): 7–8.

Thoreau's Pencils: An Unpublished Letter from Ralph Waldo Emerson to Caroline Sturgis, 19 May 1844. Cambridge, Mass., 1944.

Tichi, Cecelia. *Shifting Gears: Technology, Literature, Culture in Modernist America.* Chapel Hill, N.C., 1987.

Timmins, Samuel, editor. *Birmingham and the Midland Hardware District.* 1866 edition. New York, 1968.

Timoshenko, Stephen P. *History of Strength of Materials: With a Brief Account of the History of Theory of Elasticity and Theory of Structures.* 1953 edition. New York, 1983.

Todhunter, I., and K. Pearson. *A History of the Theory of Elasticity and of the Strength of Materials from Galilei to Lord Kelvin.* 1886 edition. New York, 1960.

Townes, Jane. "Please, Some Respect for the Pencil," *Specialty Advertising Business,* March 1983: 61–63.

Turnbull, H. W., editor. *The Correspondence of Isaac Newton. Vol. 1: 1661–1675.* Cambridge, 1959.

Turner, Gerard L'E. "Scientific Toys," *The British Journal for the History of Science,* 20 (1987): 377–98.

Turner, Roland, and Steven L. Goulden, editors. *Great Engineers and Pioneers in Technology. Vol 1: From Antiquity through the Industrial Revolution.* New York, 1981.

Ullman, David G., Larry A. Stauffer, and Thomas G. Dietterich. "Toward Expert CAD," *Computers in Mechanical Engineering,* November–December 1987: 56–70.

U.S. Bureau of Naval Personnel. *Draftsman 3.* Washington, D.C., 1955.

U.S. Centennial Commission. *International Exhibition, 1876: Official Catalogue.* Philadelphia, 1876.

U.S. Court of Customs. "United States v. A. W. Faber, Inc. (No. 3105)," *Appeals Reports,* 16 [ca. 1929]: 467–71.

U.S. Department of Agriculture. "Seeking New Pencil Woods." Forest Service report, [ca. 1909].

U.S. Department of Commerce. "Simplified Practice Recommendation R151-34 for Wood Cased Lead Pencils." Typescript attached to memorandum, from Bureau of Standards Division of

Simplified Practice to Manufacturers et al., dated September 28, 1934.

———. Bureau of the Census. "Current Industrial Reports: Pens, Pencils, and Marking Devices (1986)." [1987.]

U.S. Department of Labor. "Economic Factors Bearing on the Establishment of Minimum Wages in the Pens and Pencils Manufacturing Industry." Report . . . prepared for Industry Committee No. 52. November 1942.

U.S. Federal Trade Commission. "Amended Trade Practice Rules for the Fountain Pen and Mechanical Pencil Industry." Promulgated January 28, 1955.

U.S. General Services Administration. *Federal Specification SS-P-166d: Pencils, Lead.* 1961.

U.S. International Trade Commission. *Summary of Trade and Tariff Information: Pens, Pencils, Leads, Crayons, and Chalk.* 1983.

———. *Supplement to Summary of Trade and Tariff Information: Writing Instruments.* 1981.

U.S. Tariff Commission. "Wood-Cased Lead Pencils." Report to the President under the Provisions of Section 3(e) of the National Industrial Recovery Act. With Appendix: Limitations of Imports. No. 91 (Second Series). 1935.

U.S. Tobacco Review. "From Forests to Pencils." [1977.]

Urbanski, Al. "Eberhard Faber," *Sales and Marketing Management,* November 1986: 44–47.

Ure, Andrew. *A Dictionary of Arts, Manufactures, and Mines; Containing a Clear Exposition of Their Principles and Practice.* New York, 1853.

Usher, Abbott Payson. *A History of Mechanical Inventions.* New York, 1929.

V. & E. Manufacturing Company. *Note on Drawing Instruments.* Pasadena, Calif., 1950.

van der Zee, John. *The Gate: The True Story of the Design and Construction of the Golden Gate Bridge.* New York, 1986.

Vanuxem, Lardner. "Experiments on Anthracite, Plumbago, &c.," *Annals of Philosophy,* 11 (1826): 104–11.

Veblen, Thorstein. *The Engineers and the Price System.* 1921 edition. New York, 1963.

———. *The Instinct of Workmanship: And the State of the Industrial Arts.* New York, 1918.

Venus Pen & Pencil Corporation. "How Venus—the World's Finest Drawing Pencil—Is Made." [Illustrated folder.] N.d.

———. "List of Questions Most Frequently Asked, With Answers." [Undated typescript.]

———. "The Story of the Lead Pencil." [Undated typescript.]

———. "Venus 100 Years." [Report. New York, 1961.]

Vincenti, Walter G. *What Engineers Know and How They Know It: Historical Studies in the Nature and Sources of Engineering Knowledge.* Baltimore, 1990.

Viollet-le-Duc, Eugène Emmanuel. *Discourses on Architecture.* Translated, with an introductory essay, by Henry van Brunt. Boston, 1875.

————. *The Story of a House.* Translated by George M. Towle. Boston, 1874.

Vitruvius. *De Architectura (The Ten Books on Architecture).* Translated by Morris Hicky Morgan. 1914 edition. New York, 1960.

Vivian, C. H. "How Lead Pencils Are Made," *Compressed Air Magazine,* 48 (January 1943): 6925–31.

Vogel, Robert M. "Draughting the Steam Engine," *Railroad History,* 152 (Spring 1985): 16–28.

Voice, Eric H. "The History of the Manufacture of Pencils," *Transactions of the Newcomen Society,* 27 (1949–50 and 1950–51): 131–41.

Vossberg, Carl A. "Photoelectric Gage Sorts Pencil Crayons," *Electronics,* July 1954: 150–52.

Wahl Company. "Making Pens and Pencils," *Factory and Industrial Management,* October 1929: 834–35.

Walker, C. Lester. "Your Pencil Could Tell a Sharp Story," *Nation's Business,* March 1948: 54, 56, 58, 90–91.

Walker, Derek, [editor]. *The Great Engineers: The Art of British Engineers 1837–1987.* New York, 1987.

Walker, Dick. "Elastomer = Eraser," *Rubber World,* 152 (April 1965): 83–84.

Walker, Jearl. "The Amateur Scientist," *Scientific American,* February 1979: 158, 160, 162–66. (See also November 1979: 202–4.)

Walls, Nina de Angeli. *Trade Catalogs in the Hagley Museum and Library.* Wilmington, Del., 1987.

Watrous, James. *The Craft of Old-Master Drawings.* Madison, Wisc., 1957.

Watson, J. G. *The Civils: The Story of the Institution of Civil Engineers.* London, 1988.

Waugh, Evelyn. *Brideshead Revisited: The Sacred and Profane Memories of Captain Charles Ryder.* Boston, 1946.

Weaver, Gordon. "Electric Oven Reduces Cost of Baking Pencil Leads," *Electrical World,* 78 (September 10, 1921): 514.

Whalley, Joyce Irene. *English Handwriting, 1540–1853: An Illustrated Survey Based on Material in the National Art Gallery, Victoria and Albert Museum.* London, 1969.

————. *Writing Implements and Accessories: From the Roman Stylus to the Typewriter.* Detroit, 1975.

Wharton, Don. "Things You Never Knew About Pencils," *Saturday Evening Post,* December 5, 1953: 40–41, 156, 158–59.

White, Francis Sellon. *A History of Inventions and Discoveries: Alphabetically Arranged.* London, 1827.

White, Lynn, Jr. *Medieval Religion and Technology: Collected Essays.* Berkeley, Calif., 1978.

————. *Medieval Technology and Social Change.* New York, 1966.

Whittock, N., et al. *The Complete Book of Trades, or the Parents' Guide and Youths' Instructor, [etc.].* London, 1837.

Wicks, Hamilton S. "The Utilization of Graphite," *Scientific American,* 40 (January 18, 1879): 1, 34.

Wilson, Richard Guy, Dianne H. Pilgrim, and Dickran Tashjian. *The Machine Age in America, 1918–1941.* New York, 1986.

Winokur, Jon. *Writers on Writing.* 2nd edition. Philadelphia, 1987.

Wolfe, John A. *Mineral Resources: A World View.* New York, 1984.

Wright, Paul Kenneth, and David Alan Bourne. *Manufacturing Intelligence.* Reading, Mass., 1988.

The Year-Book of Facts in Science and Art, [etc.]. London, various years, but especially the 1840s.

Zilsel, Edgar. "The Sociological Roots of Science," *American Journal of Sociology,* 47 (January 1942): 544–62.

Illustrations

with Sources

Acknowledgments

Bibliographies have always seemed to me to be implicit acknowledgments, but I would like to make explicit my debt to some particular sources. While no definitive history of the pencil seems to have been written, the pioneering efforts of a few scholars gave me an initial orientation that proved to be invaluable. John Beckmann's chapter on black lead, Clarence Fleming's story of the pencil, Molly Lefebure's chapter on wadd, Joseph Meder's chapter on graphite, and Eric Voice's article on the history of the pencil are among the contents of my bibliography that I would single out.

At one point my intention was to make separate bibliographies of items pertaining to the pencil and of those pertaining to engineering generally, but I decided against that dichotomy for several reasons. While various entries might have fallen easily into one category or the other, the designation of some of the most important works would have been arbitrary. That this is true reinforces for me the very idea of this book—namely, that to write about the pencil is to write about engineering, and vice versa. Furthermore, it seems to me to be potentially misleading to compartmentalize in any way a book like George Hendrick's *Remembrances of Concord and the Thoreaus,* in which I found considerable information about nineteenth-century pencil making, or Cicero's letters, in which I found a paradigm for engineering.

Some of the entries in my bibliography were first brought to my attention directly or indirectly by individuals whose minds clicked when I mentioned that I wanted to write about the pencil. Armand Hammer kindly sent me a photocopy of his out-of-print book *The Quest of the Romanoff Treasure* and gave me permission to quote extensively from it. Daniel Jones of the National Endowment for the Humanities called my attention to Michael Faraday's *Chemical History of a Candle*, which reinforced my commitment to write about the pencil.

William Ecenbarger, Malcomb M. Ferguson, and Walter Harding, secretary of the Thoreau Society, gave me some early help. Anne McGrath, curator of the Thoreau Lyceum, and Mrs. William Henry Moss, archivist in the Concord Free Public Library, not only directed me to important sources of information about pencil making in Concord but also allowed me to inspect Thoreau pencils in their respective collections. Carolyn Newton, of the Encyclopaedia Britannica Corporation, provided me with a definitive chronology of "pencil" entries in the various editions of the *Britannica*. Jerry Slocum sent me an example of a pencil puzzle and some of its lore. Harold K. Steen, executive director of the Forest History Society, gave me some articles on pencil woods that I might not otherwise have found, as well as a copy of the single print pertaining to pencil making in the society's collection of a half million Forest Service photographs. Eugene Ferguson and Walter Vincenti, whom I wrote to about the pencil only after my manuscript was essentially complete, sent me relevant papers that added to my bibliography. Florence Letouzey-Dumont kindly provided some galley proof for *Dictionnaire de Biographie Française*.

Maureen Greenland, secretary of the Writing Equipment Society, sent me some relevant items from her files, including the address of Arthur Iberg, editor of *The Pencil Collector,* through whom I was introduced to the American Pencil Collectors Society as a unique source of information and artifacts. Chris Hardy of Monadnock Media put me on to the article in *Scribner's Monthly* that provided much information about pencil making in America in the 1870s. Robert Post, editor of *Technology and Culture,* brought my interests to the attention of David Shayt, who provided much useful information about the Smithsonian Institution's pencil displays and who in turn contacted Kay Youngflesh, also of the National Museum of American History, who sent me still further references to the pencil. James Bitler, resident naturalist on Little St. Simons Island, gave me an oral history of its relationship to the Berolzheimer family. Peter Kohn, project manager, showed me around the Dixon Mills renovation and allowed me to inspect some old slides and artifacts relating to crucible and pencil making in Jersey City. And John Striker, a collector, gave me information about early pencil sharpeners.

Without libraries and librarians, conventional scholarly documentation for this book would be slim indeed. My bibliography began, and will no doubt continue through proofreading, in the libraries of Duke University. Above all, Eric Smith, librarian of the Vesić Engineering Library at Duke, has helped me with this project, as with others, in more ways than I can trust myself to remember. In the reference department of Duke's Perkins Library, Bessie Carrington helped me early on to get oriented in the confusing world of encyclopedias, and Joe Rees helped me clarify some references at the end. Stuart Basefsky helped locate some inscrutable government documents, and Linda Withrow kept track of innumerable interlibrary loan requests. Linda McCurdy and Ellen Gartrell, of the manuscript department, searched for and located pencil-related materials in the J. Walter Thompson Company archives, and Sam Hammond helped me in the Rare Book Room. Albert Nelius understood my continuing need for a carrel in Perkins Library, where the final draft of this book was written.

Diane K. Portnoy, reference librarian in the Hagley Museum and Library, responded to my inquiry about trade catalogues with items that proved to be sources of unique information. Mary S. Smith, reference librarian in the Harvard College Library, was very helpful in my search for the encyclopedia that Thoreau is said to have consulted in Cambridge. Sidney Ives, rare books and manuscripts librarian at the University of Florida, provided me with information about the Thoreau pencils and broadside in the university's Parkman Dexter Howe Library. The staff of the National Museum of American History Branch of the Smithsonian Institution Libraries was very cooperative in allowing me to locate and photocopy some of their uncatalogued trade catalogues. And the library staff at the National Bureau of Standards located a unique document in their files. While the National Humanities Center does not have a collection of its own, its library service is a scholar's dream, and its library staff of Jean Houston, Alan Tuttle, and Rebecca Vargha kept a steady stream of essential books and articles coming from the Triangle University libraries in Durham, Chapel Hill, and Raleigh—and from around the country. The institution of interlibrary loan, generally unheralded and anonymous, is in the end what has enabled this book's bibliography to reach beyond the collections of any single one of the libraries I have visited.

Much of the story of the pencil is not contained in books or journal articles, and therefore is not generally to be found in libraries. Pencil manufacturers want to distinguish and sell their products, however, and so they do produce a lot of printed material in the form of labels, boxes, folders, brochures, catalogues, and other promotional media that might be classified as advertising and ephemera. While it is the rare such text that will carry the name of its author or the documentation for its assertions, these materials seldom if ever contain an outright lie. Even if manufacturers have a bias toward their own product and against that of their competition, the contents of their literature are not without value as a source of information about the history and manufacture of the pencil. Many such items in my bibliography seem to be available only from the companies that are listed as their authors and publishers. Among those that have responded to my requests for information with illustrations, literature, and pencils in various stages of manufacture are: Berol USA, Berol Limited (and especially John Storrs), Blackfoot Indian Writing Company, Caran d'Ache, Cumberland Pencil Company (and David Sharrock), Dixon Ticonderoga (and Mayellen Ahneman and Bill Spratt), Eberhard

Faber Corporation (and Thelma Marshall), Empire Pencil Corporation (and Harold Hassenfeld), A. W. Faber-Castell (and Peter Schafhauser), Faber-Castell Corporation (and Robert Gooch, chief plant engineer in Shelbyville, Tennessee, as well as Barbara Moss of the company's public relations firm, Grant Marketing), Koh-I-Noor Rapidograph (and John Wollman), Lyra-Bleistift-Fabrik (and W. H. Kring), Mallard Pen and Pencil Company, J. R. Moon Pencil Company, National Pen & Pencil Company, Pentel of America (and Mark Welfley), Rexel Limited, J. S. Staedtler (and Raymond Urmston, Jr.), Staedtler Mars (and engineers V. Schüren and A. Rauchenberger), and the pencil sharpener manufacturer Wilhelm Dahle. While it has been difficult to know how to acknowledge in my bibliography some of the more unusual items provided by these firms, I have listed what I consider to be the more important and useful things, giving the company as author. The Pencil Makers Association gave me a great number of articles from retail-store trade journals, newspapers, and popular magazines, and I have included in my bibliography what I consider to be the most appropriate of these. The Writing Instrument Manufacturers Association provided material that was helpful in placing the pencil in the broader context of modern writing implements, but a large number of anonymous articles on pens and pencils provided by the various associations do not appear in my bibliography. The German trade association Industrieverband Schreib- und Zeichengeräte was also helpful, as was the Federation of European Pencil Manufacturers' Associations.

There is one final source of artifacts, words, slang, references, quotes, anecdotes, and personal experiences related to the pencil that my bibliography does not reflect. And since I do not trust my memory to reconstruct when and from whom I may have first heard this or that item, I will not attempt to name names here. But many, many of the members of the staff and the fellows of the Class of 1987–88 of the National Humanities Center can no doubt read something in this book that sounds familiar. While some of them will find my explicit thanks in the notes, I really owe thanks to virtually everyone at the Center, on whose grounds an unharvested red cedar still grows, for thinking of me whenever they heard or read anything that was remotely connected to the pencil, or to engineering.

Among others to whom I feel a special debt for their encouragement and support of my work, both here and elsewhere, as well as for inspiration from their own work, are: Freeman Dyson, William Gass, James Gordon, Alec Nisbett, Clifford Truesdell, and, especially for his enthusiasm for this book when it was only an idea, Leon Kass. So many of my colleagues at Duke University have helped me over the years that I do not trust myself to make a list that would be complete, but I can single out Seymour Mauskopf and Alex Roland, who have served as examples to me of scholarship in the history of science and technology.

I am grateful to Duke University, the National Endowment for the Humanities, and the National Humanities Center, which made my sabbatical leave not only possible but also real. Although I worked in a study but ten miles from Durham, my colleagues at Duke, especially those in my own Department of Civil and Environmental Engineering, kindly spared me the distractions of committee assignments and meetings. My graduate students worked with a mature independence, and they understood when I took an extra-long time to return their manuscripts. Finally, the department secretaries protected me from telephone calls, but were always there when I called upon them.

The production of a book does not begin with its writing; nor does it end with the manuscript. His family seldom escapes an author's obsession with his project, and so they should not be without recognition in its product. William Petroski, my brother and fellow engineer, provided an ongoing and wonderful variety of pencil facts, artifacts, ephemera, and catalogues, along with his insightful interpretations of and speculations about them. Marianne Petroski, my sister, sent me architect's pencils with the longest points, and my mother and many of my relatives hunted through their desks for pencils and more pencils. Karen Petroski, my daughter, compiled for me an early list of articles from *The New York Times* and other sources that provided information on the pencil and pencil making that I have found nowhere else. Stephen Petroski, my son, helped me keep my project in perspective by providing pencil jokes and tricks. And Catherine Petroski, my wife, passed on literary references to the pencil and was, as usual, the first reader of my earliest draft. Both she and Karen were also meticulous proofreaders, and Catherine helped with the index.

For all these advantages, my bibliography, let alone the book it is bound in, might have suffered the fate of an engineer's pencil sketch had no publisher been willing to associate himself with it. When Ashbel Green showed an interest in my idea, I was greatly encouraged, and I am grateful to him for his perceptive and critical reading of my work and for giving me my head in producing this history of the pencil and of engineering, however idiosyncratic and quirky it might be. Finally, I am grateful to Staci Capobianco and Virginia Tan at Alfred A. Knopf, who saw this book through production and design; but, needless to say, I remain responsible for any of its shortcomings.

H. P.

Index

Italicized page numbers refer to illustrations.